Petroleum and Petrochemical Engineering

Edited by **Andy Margo**

SYRAWOOD
PUBLISHING HOUSE
New York

Published by Syrawood Publishing House,
750 Third Avenue, 9th Floor,
New York, NY 10017, USA
www.syrawoodpublishinghouse.com

Petroleum and Petrochemical Engineering
Edited by Andy Margo

International Standard Book Number: 978-1-68286-032-8 (Hardback)

The publisher's policy is to use permanent paper from mills that operate a sustainable forestry policy. Furthermore, the publisher ensures that the text paper and cover boards used have met acceptable environmental accreditation standards.

Trademark Notice: Registered trademark of products or corporate names are used only for explanation and identification without intent to infringe.

Printed in the United States of America.

Contents

Preface

This book has been an outcome of determined endeavour from a group of educationists in the field. The primary objective was to involve a broad spectrum of professionals from diverse cultural background involved in the field for developing new researches. The book not only targets students but also scholars pursuing higher research for further enhancement of the theoretical and practical applications of the subject.

Petroleum and petrochemical engineering is an emerging field aimed at production of fuels, natural gases and petrochemicals. There has been a tremendous surge in the last few decades for exploration of new hydrocarbon deposits as well as improving the refining and distillation processes for maximum recovery of crude deposits from the reservoirs. It is a multidisciplinary field that includes concepts and technological aspects of geological, mechanical, civil and chemical engineering. This book provides an in-depth explanation of the various processes involved in petroleum and petrochemical engineering such as drilling, processing and technical analysis of petrochemicals. Some of the significant topics included in this book are design of petroleum plants and reservoirs, catalysis and synthesis of petrochemicals, reaction engineering, etc. Students, researchers, experts and engineers associated with petroleum engineering will benefit alike from this book.

It was an honour to edit such a profound book and also a challenging task to compile and examine all the relevant data for accuracy and originality. I wish to acknowledge the efforts of the contributors for submitting such brilliant and diverse chapters in the field and for endlessly working for the completion of the book. Last, but not the least; I thank my family for being a constant source of support in all my research endeavours.

Editor

Transformation of ethylene to higher hydrocarbons on silica-supported Ir catalysts: the nature of carbonaceous deposits

Hongwei Yang · Shik Chi Edman Tsang

Abstract The first stage of ethylene decomposition followed by second stage of temperature-programmed surface reduction (H_2-TPSR) to produce higher hydrocarbons at different temperatures over silica-supported iridium catalysts has been investigated. The catalysts for the two stepwise reactions are characterized by X-ray diffraction, Raman and Fourier transformed infrared spectroscopies, temperature-programmed reduction, and mass spectroscopy. These studies reveal that ethylene decomposition at low temperatures (≤ 673 K) in the first stage produces mainly C1 hydrocarbon moieties on the Ir surface via dissociative adsorption, the sequential hydrogenation in the second stage will give arise to CH_4. The surface polymerization of C_1 to higher hydrocarbon species and metathesis reactions under these temperatures are also clearly evident. When ethylene is decomposed at 773–973 K, stable graphitic carbon deposits with poor propensity for hydrogenation are obtained. Interestingly, water formed from surface dehydroxylation on silica can produce a significant quantity of CO/H_2 with these carbons during the H_2-TPSR at elevated temperature.

Keywords Ethylene homologation · Iridium catalysts · Hydrocarbon species · Propylene metathesis · Mass spectroscopy

H. Yang · S. C. E. Tsang (✉)
Department of Chemistry, Wolfson Catalysis Centre, University of Oxford, Oxford OX1 3QR, UK
e-mail: edman.tsang@chem.ox.ac.uk

Introduction

The dependence on oil over last century is expected to be gradually offset in this century by an increasing dependence on natural gas (the main constituent of which is methane). Much attention has therefore been paid to methane conversion to more commercially useful chemicals [1]. However, direct CH_4 conversion into condensable chemicals such as oxygenates (methanol, formaldehyde) or higher hydrocarbons pose key technical challenges [2, 3]. It was demonstrated that higher hydrocarbons can be formed from methane by a two-stage procedure, namely catalytic deposition of carbonaceous species, followed by Fischer–Tropsch like hydrogenation. They were operated at different reaction temperatures and conditions in order to overcome thermodynamic limitations [4, 5]. Similar processes have also been reported using other light alkanes [6]. It is interesting to extend this fundamental research to ethylene molecule, which is one of the key petrochemicals. It is expected that the use of ethylene, an unsaturated hydrocarbon can facilitate the first stage of carbonaceous deposition process hence enabling their surface coupling to higher hydrocarbons in the second stage. It is noted from previous research that a significant quantity of propylene was selectively formed when ethylene was in contact with cobalt catalyst at elevated temperature [7]. Similar results could also be obtained over Ru catalyst under defined reaction conditions, suggesting the possibility of obtaining selective higher hydrocarbon products [8]. The homologation of ethylene on metal surfaces, more often, was reported in the presence of H_2 [9]. The use of two-step sequence is expected to be more favorable to maximize the hydrocarbon chain length than those of co-feeding C_2H_4 and H_2 [10], which should be systematically explored.

It is well known that the metal–carbon bond strength is crucial to determine the type of carbonaceous deposit produced during the low-temperature two-step CH_4 conversion [1]. Conceivably, the stepwise conversion of C_2H_4 may also inherit this feature of methane homologation. Compared to other group VIII noble metals, Ir catalysts are scarcely studied. However, due to its characteristic properties, the C_2H_4 activation and the sequential hydrogenation on Ir surface are expected to give different activities and selectivities.

In this paper, a systematic study of the two-step C_2H_4 conversion sequence was therefore performed over iridium-supported catalysts which were then characterized by different techniques. Preliminary results reveal that the type of carbonaceous deposits from ethylene decomposition and the nature of hydrocarbons synthesized during the temperature-programmed reduction (H_2-TPR) on Ir surfaces are strongly dependent on the temperature of C_2H_4 decomposition used. As a result, a mechanism for the stepwise C_2H_4 conversion is hereby proposed.

Experimental

Catalyst preparation

The 10 % Ir/SiO_2 catalyst (10 % stands for the nominal loading weight of Ir in catalyst) was prepared by an impregnation method. Typically, 107.54 mg $IrCl_3$ (of 99.99 % purity, supplied from Johnson Matthey) was dissolved in 30 mL deionized water, and then 0.50 g silica gel (99+ % purity, supplied from Aldrich Chemical Company Inc.) was added as the carrier. After stirring at room temperature for 2 h, the water in the solution was evaporated at 353 K. The light yellow sample thus obtained was further dried at 353 K overnight. The product was then calcined in air at 873 K for 3 h, and sequentially treated under a H_2 flowing stream at 573 K for 3 h, yielding the 10 % Ir/SiO_2 catalyst.

Decomposition of C_2H_4 on the Ir-supported catalyst was performed at ambient pressure as follows: 30 mg 10 % Ir/SiO_2 was placed in a vertical quartz tube. A helium flow (10 mL/min) was first used to flush the tube at 313 K for 15 min, removing all the remaining air inside. Pure C_2H_4 flow (10 mL/min) was then introduced. The temperature was subsequently increased at a rate of 10 K/min to a designated value and kept for 1 h to obtain a 10 % Ir/SiO_2-x sample, where x stands for the temperature of C_2H_4 decomposition. Afterwards, the temperature was decreased to 313 K under the He flow. The solid product was hydrogenated by switching the gas flow to 20 mL/min of 5 % H_2–95 % Ar. The temperature was increased gradually to 1,273 K at a rate of 10 K/min.

Catalyst characterization

XRD patterns were acquired on a PANalytical PW3719 diffractometer, using Cu K_α radiation ($\lambda = 1.5418$ Å) from a generator operating at 40 kV and 30 mA. All Raman spectra were recorded on a Jobin–Yvon spectrometer (Labram 1B) equipped with a microscope using 20 mW He–Ne laser (632.8 nm) in the range of 200 cm^{-1} up to 3,600 cm^{-1}, by adding four sets of spectra together. The spectral resolution was set at 4.0 cm^{-1}. FTIR spectra were collected on a Nicolet 6700 FTIR spectrometer with a liquid nitrogen-cooled MCT detector. FTIR spectra were obtained by averaging 128 scans at a resolution of 4 cm^{-1}. H_2–TPR measurement was performed immediately after the decomposition of C_2H_4 on the Ir catalysts on a ThermoQuest CE INSTRUMENTS TPDRO1100 equipment, using a 5 % H_2–95 % Ar gas mixture in the temperature range of 313–1,273 K. The gaseous products leaving the tubular reactor were also analyzed on line by a HIDEN ANALYTICAL HPR20 mass spectrometer, where the signals of H_2, H_2O, CO, CO_2, and the hydrocarbon species C_1 to C_6 were detected. The MS signal changes for CO_2 and C_6 species were not obvious during the whole process, and therefore are not shown in the figures.

Results and discussion

Characterization of the Ir catalysts

The XRD patterns of different 10 % Ir/SiO_2 samples are shown in Fig. 1. Diffraction lines due to the metallic Ir were observed at 40.7°, 47.2°, 69.3°, 83.3°, and 88.0°, respectively [11], while silica gave diffraction around 23.5° [12]. There was no obvious change in the intensity or shape of the diffraction peaks before and after the C_2H_4 decomposition. This indicates that there were strong interactions between Ir nanoparticles (NPs) and the silica support, and the characteristic structure of Ir NPs was not altered by the high-temperature treatment under C_2H_4 flow over the whole range of the ethylene decomposition temperatures. Furthermore, there was no observable shift in the diffraction peak positions to suggest any lattice expansion of Ir nanocrystal due to carbon atoms inclusion in the crystal interstice as those observed in the case of fcc Pd (C) [13]. Thus, the carbonaceous deposition during the ethylene decomposition appears to take place only on the catalyst surface instead of lattice insertion.

Raman measurement was then performed to explore the type of carbonaceous species produced after the C_2H_4 decomposition. As seen in Fig. 2, there is no obvious peak found on the surfaces of 10 % Ir/SiO_2-x ($x = 473, 573$, and 673 K) samples, suggesting the absence of less reactive

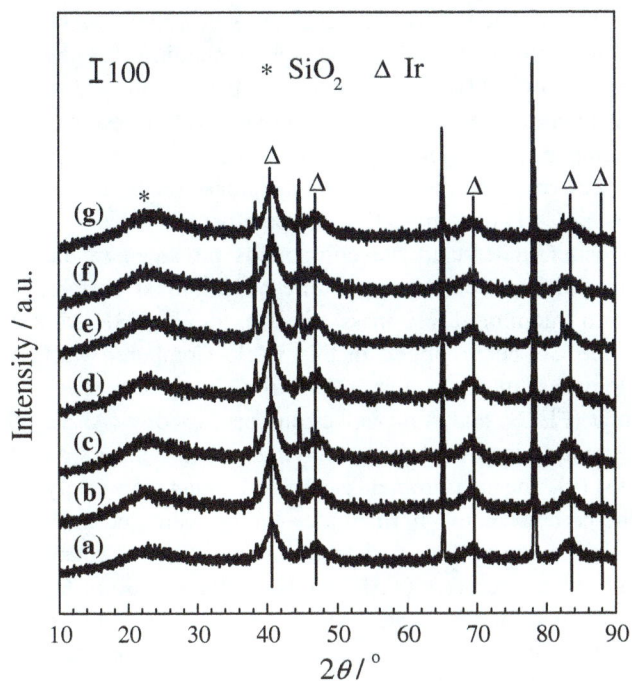

Fig. 1 XRD patterns of different 10 % Ir/SiO$_2$ samples: *a* 10 % Ir/SiO$_2$; *b* 10 % Ir/SiO$_2$-473; *c* 10 % Ir/SiO$_2$-573; *d* 10 % Ir/SiO$_2$-673; *e* 10 % Ir/SiO$_2$-773; *f* 10 % Ir/SiO$_2$-873; *g* 10 % Ir/SiO$_2$-973. 2θ of 45°, 65° and 78° are from sample holder

carbons (C$_\beta$) and graphitic carbon deposits (C$_\gamma$) previously identified during methane decomposition at these temperatures [5]. Further increasing the temperatures of C$_2$H$_4$ decomposition to 773–973 K resulted in the appearance of the characteristic D and G bands around 1,323 and 1,589 cm^{-1}, respectively of a typical carbonaceous material [14]. Moreover, the I_D/I_G value decreased on increasing the temperature of C$_2$H$_4$ decomposition, implying the gradual formation of stable graphitic carbons from reactive carbons.

FTIR spectra were collected in order to identify the existence of hydrocarbon species on the catalysts. However, no absorption band was attributable to any of the vibration modes of CH$_x$ for all the samples in Fig. 3. This means that all the above CH$_x$ fragments may have been reacted or decomposed quickly at high temperature with their surface concentrations below the detection limit. Similar results were also reported by Solymosi and co-workers [15], where in situ measurements or sudden cooling of the sample in a continuous CH$_4$ + H$_2$ flow to 200 K did not produce detectable IR peaks. Nevertheless, it is noted that the intensity of band around 3,379 cm^{-1}, which is characteristic of –OH groups on the silica surface [16] was decreased with the increasing C$_2$H$_4$ decomposition temperature, especially at temperatures above 673 K. Similar trend was also observed for the band at around 1,631 cm^{-1}, which can be attributed to the scissoring mode

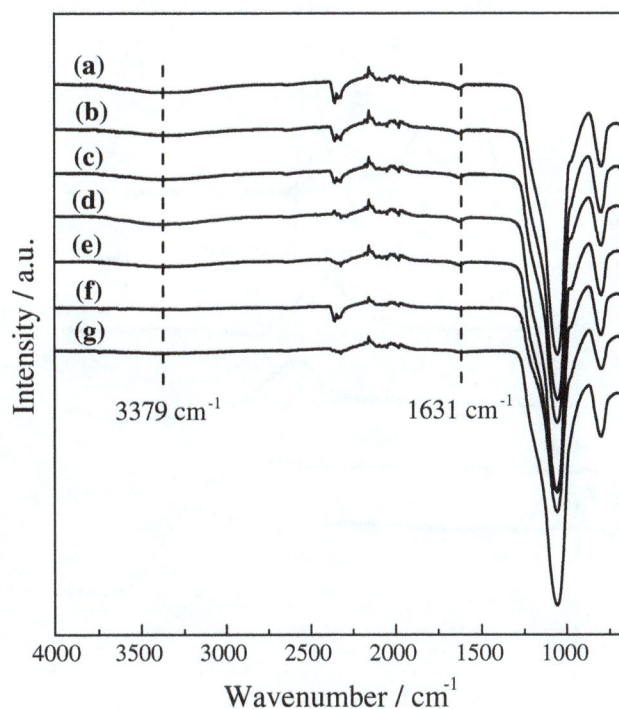

Fig. 2 Raman spectra of different 10 % Ir/SiO$_2$ samples: *a* 10 % Ir/SiO$_2$-473; *b* 10 % Ir/SiO$_2$-573; *c* 10 % Ir/SiO$_2$-673; *d* 10 % Ir/SiO$_2$-773; *e* 10 % Ir/SiO$_2$-873; *f* 10 % Ir/SiO$_2$-973

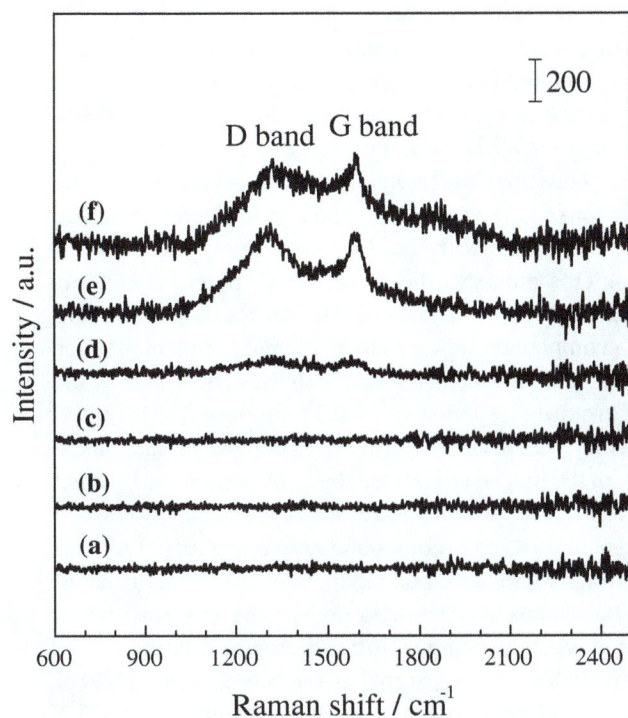

Fig. 3 FTIR spectra of different 10 % Ir/SiO$_2$ samples: *a* 10 % Ir/SiO$_2$; *b* 10 % Ir/SiO$_2$-473; *c* 10 % Ir/SiO$_2$-573; *d* 10 % Ir/SiO$_2$-673; *e* 10 % Ir/SiO$_2$-773; *f* 10 % Ir/SiO$_2$-873; *g* 10 % Ir/SiO$_2$-973

of H_2O molecules [17]. All these results suggest the loss of H_2O and $-OH$ groups from the catalyst surface at high temperatures.

It is reported that temperature-programmed reduction (TPR) can separate various types of carbon residing on the metal surface [6]. Thus, H_2–TPR spectra related to the hydrogenation of surface carbons over various catalysts were collected in this work and displayed in Fig. 4. First, negative peaks at the programming temperatures below 373 K were observed for all samples, which can be attributed to the release of hydrogen from Ir NPs similar to those cases of other group VIII noble metals [13]. In addition, there were some reduction peaks detected at the temperature range of 373–923 K. In the case of 10 % Ir/SiO_2-473, two peaks centered at 547 and 674 K, respectively, corresponded to the hydrogenation of at least two different carbonaceous species on the catalyst surface. When increasing the temperature of C_2H_4 decomposition up to 573 K, reduction peaks appeared to shift to higher temperatures. Furthermore, the intensity of the reduction peak increased drastically, suggesting an increase in the amount of reactive carbonaceous species on the Ir surfaces during the higher temperature decomposition. Further increasing the decomposition temperature to 673 K, the reduction peak on 10 % Ir/SiO_2-673 shifted toward further higher temperature but the intensity this time decreased suggesting that the reactive carbonaceous species were

somehow converted to less reactive forms that could not be easily hydrogenated. This trend was maintained as the temperature of C_2H_4 decomposition increased from 773 to 973 K. Interestingly, new but distinctive negative peaks at high temperatures, indicative of hydrogen gas evolution could be observed and their intensities increased at increasing C_2H_4 decomposition temperature.

To better understand the compounds produced in the hydrogenation process, temperature-programmed surface reduction monitored by mass spectroscopy, TPSR–MS technique has been applied in this work. Compared with other physical methods such as electron energy loss spectroscopy (EELS) and Auger electron spectroscopy (AES), this technique is powerful for studying carbon deposits on surface [18]. The measurement can be performed more easily and the carbonaceous intermediates can be identified indirectly. As shown in Fig. 5, detection of positive hydrogen peak at low temperature (<373 K) due to the release of H_2 was confirmed for all samples, which are consistent with the observation of the negative peaks in the H_2–TPR profiles. Over 10 % Ir/SiO_2-473, two broad CH_4 peaks were detected accompanying the inverted H_2 consumption profile at the same temperatures. This suggests that the catalyst surface must be covered with reactive carbonaceous species during ethylene decomposition at the first stage for the gaseous CH_4 production under hydrogen at the second stage. In addition, small amounts of higher hydrocarbons of C_2–C_4 species were clearly produced, suggesting a degree of surface polymerization of adsorbed hydrocarbon species akin to those of Fischer–Tropsch catalysis. The quantities of gaseous higher hydrocarbons produced were higher from the same sample at the decomposition temperature of 573 K, implying a higher level of surface coverage of the reactive carbonaceous species. Interestingly, we also noted that the production of gaseous ethylene and butylene molecules were always associated with the decrease in propylene concentration. This indicates the favorable propylene metathesis reaction ($2C_3H_6 \rightarrow C_2H_4 + C_4H_8$) on the Ir surface. CH_4 and C_2 compounds appeared to be the predominant products than C_{3+} alkanes/olefins over our Ir catalyst which is different from the reported Ni and Pt catalysts [10]. Further increasing the decomposition temperature at and above 673 K in the first step rendered the consumption of H_2 much reduced together with the decrease in CH_4 signal. In addition, the signals for C_2–C_5 compounds were attenuated to zero. This strongly indicates the fundamental change in the nature of carbonaceous species deposited on the Ir surface at the higher ethylene decomposition temperatures. In the case of 10 % Ir/SiO_2-773, CH_4 signal arose only at around 850 K. Also, a broad peak assigned to H_2 appeared above 923 K during the temperature-programmed heating. For 10 % Ir/SiO_2-873, the signal for H_2 at high temperature further increased with a clear decrease in H_2O concentration. At the

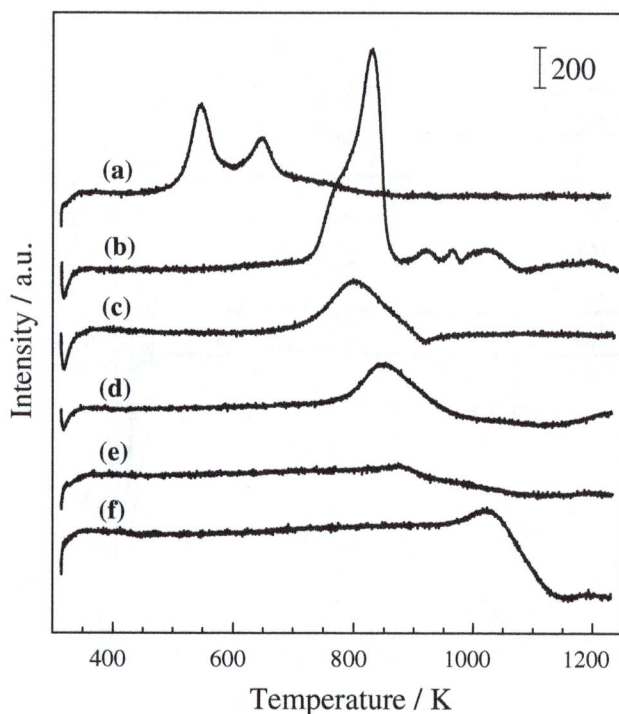

Fig. 4 TPR profiles of different 10 % Ir/SiO_2 samples: a 10 % Ir/SiO_2-473; b 10 % Ir/SiO_2-573; c 10 % Ir/SiO_2-673; d 10 % Ir/SiO_2-773; e 10 % Ir/SiO_2-873; f 10 % Ir/SiO_2-973

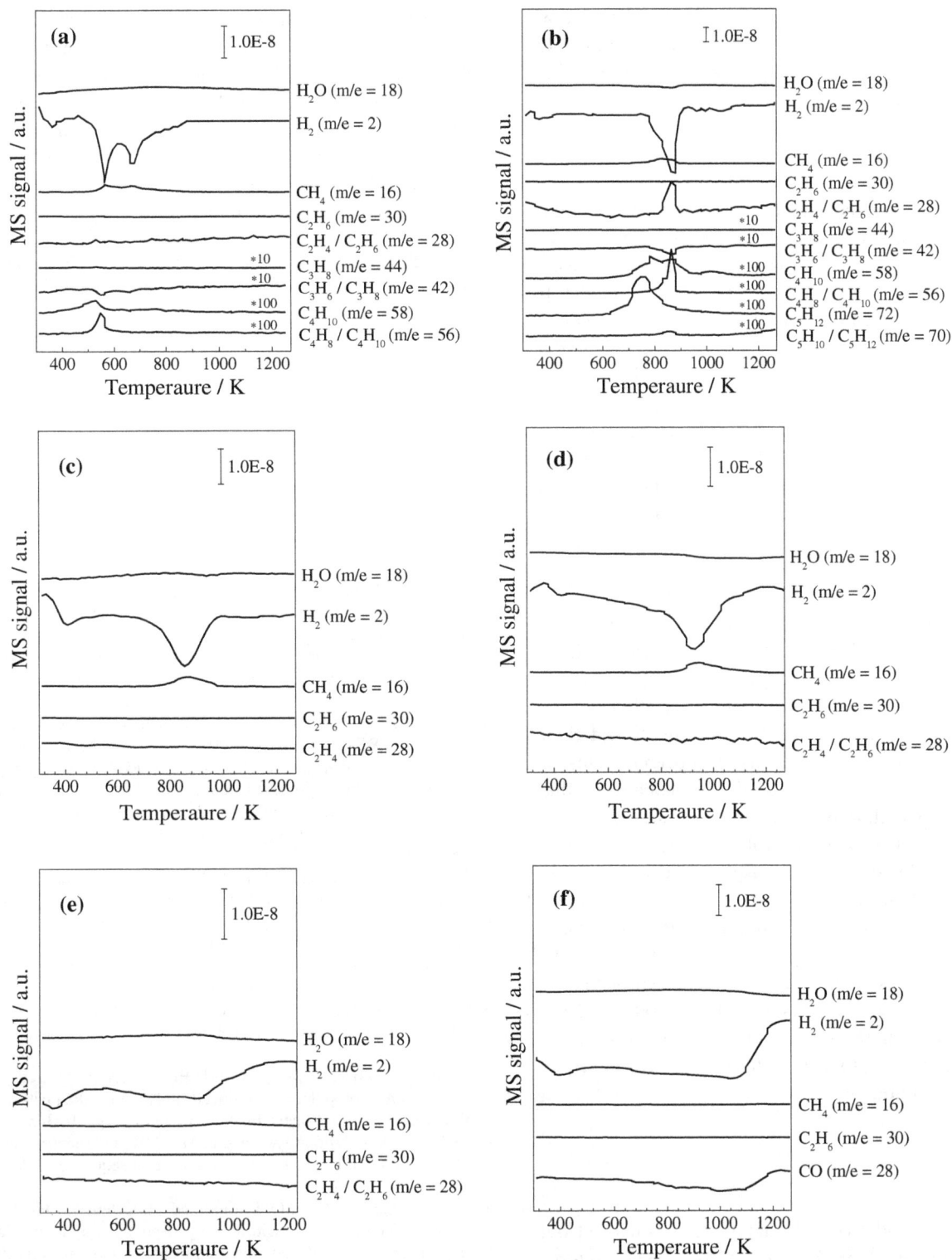

Fig. 5 MS signals for different 10 % Ir/SiO$_2$ samples: **a** 10 % Ir/SiO$_2$-473; **b** 10 % Ir/SiO$_2$-573; **c** 10 % Ir/SiO$_2$-673; **d** 10 % Ir/SiO$_2$-773; **e** 10 % Ir/SiO$_2$-873; **f** 10 % Ir/SiO$_2$-973

decomposition temperature of 973 K, a similar observation of increase in H_2 signal together with the reduction of H_2O signal was noted, but in this case, production of gaseous CO molecules was detected.

Mechanistic issues

With reference to the above studies and previous research, this short paper demonstrates that when C_2H_4 decomposition takes place at 473 K, its dissociative adsorption leads to the formation of hydrocarbon CH_x species on the Ir surfaces. These species can then be readily hydrogenated into gaseous CH_4 in the second step. As the decomposition temperature reaches 573 K, CH_x groups on the catalyst surface can assemble to ethylidyne- or vinylidene-like surface species similar to those surface reactions reported by Bond et al. and Goodman et al. [6, 18]. Their presence can be clearly reflected by the TPSR–MS results, where both gaseous CH_4 and C_2 compounds were detected upon hydrogenation (Fig. 5). In addition, surface C_1 polymerization and metathesis reactions clearly took place at ≤573 K on the Ir surface, resulting in the formation of a range of higher hydrocarbons. But, the gaseous hydrocarbon distribution formed appears to be quite different from that reported on Ni and Pt catalysts [10]. Stepwise C_2H_4 adsorption can produce surface ethylidyne, ethynyl, and carbon and hydrogen atoms on the metallic Ir and their surface interactions (i.e. metathesis) at different rates as compared to other metal surfaces [6]. We thus believe that both the iridium–carbon bond strength and the silica support must play some important roles during this process. More detailed studies on these aspects are currently in progress. Nevertheless, these hydrocarbon species are not stable at higher temperatures and can be transformed or decomposed into more stable graphitic carbon deposits. Interestingly, despite their reluctance for hydrogenation these carbons can react readily with H_2O produced from the dehydroxylation reaction on the catalyst surface via steam gasification [19] to form H_2/CO in the second step at elevated temperature.

Conclusion

A two-step non-oxidative C_2H_4 conversion is carried out over silica-supported iridium catalysts. XRD patterns reveal that there is no obvious lattice expansion of the Ir nanocrystal after the C_2H_4 decomposition over a whole temperature range, indicative of the purely surface deposition of carbonaceous species on the supported catalyst with no inclusion of sub-layer carbon atoms. Raman, FTIR, TPR, and TPSR–MS studies reveal that methyl and higher hydrocarbons like species (CH_x and C_nH_y) are formed on

the Ir surface at relative low ethylene decomposition temperatures (≤673 K). These lead to the formation of methane and higher hydrocarbons via surface hydrogenation, polymerization and metathesis reactions in the second hydrogenation stage. In contrast, much less reactive graphitic carbons are formed at higher ethylene decomposition temperatures (>673 K), which can give rise to H_2/CO via steam gasification in the second step.

Acknowledgments This work was supported by the EPSRC and the authors wish to thank China Scholarships Council for a studentship to H. Yang.

References

1. Bradford MCJ (2000) Two-step methane conversion to higher hydrocarbons: comment on the relevance of metal-carbon bond strength. J Catal 189:238
2. Ross JRH, van Keulen ANJ, Hegarty MES, Seshan K (1996) The catalytic conversion of natural gas to useful products. Catal Today 30:196
3. Soulivong D, Copéret C, Thivolle-Cazat J, Basset JM, Maunders BM, Pardy RBA, Sunley GJ (2004) Cross-metathesis of propane and methane: a catalytic reaction of CC bond cleavage of a higher alkane by methane. Angew Chem Int Ed 43:5366
4. Belgued M, Pareja P, Amariglio A, Amariglico H (1991) Conversion of methane into higher hydrocarbons on platinum. Nature 352:789
5. Koerts T, van Santen RA (1991) A low temperature reaction sequence for methane conversion. J Am Chem Soc 113:1281
6. Bond GC (1997) The role of carbon deposits in metal-catalysed reactions of hydrocarbons. Appl Catal A Gen 149:3
7. Suzuki T (2007) Selective ethene homologation reaction on silica supported cobalt catalyst. React Kinet Catal Lett 90:61
8. Suzuki T (2004) Homologation of ethylene without metathesis on silica supported ruthenium catalyst. React Kinet Catal Lett 81:327
9. Suzuki T, Hirai T, Hayashi S (1991) Enhancement of the ethylene conversion to propylene on reduced molybdena silica catalyst in the presence of hydrogen. Int J Hydrogen Energy 16:345
10. Lefort L, Amariglio A, Amariglio H (1994) Oligomerization of ethylene on platinum by a two-step reaction sequence. Catal Lett 29:125
11. Silvennoinen RJ, Jylhä OJT, Lindblad M, Österholm H, Krause AOI (2007) Supported iridium catalysts prepared by atomic layer deposition: effect of reduction and calcination on activity in toluene hydrogenation. Catal Lett 114:135
12. Yang HW, Tang DL, Lu XN, Yuan YZ (2009) Superior performance of gold supported on titanium-containing hexagonal mesoporous molecular sieves for gas-phase epoxidation of propylene with use of H_2 and O_2. J Phys Chem C 113:8186
13. Chan CWA, Xie Y, Cailuo N, Yu KMK, Cookson J, Bishop P, Tsang SC (2011) Palladium with interstitial carbon atoms as a catalyst for ultraselective hydrogenation in the liquid phase. Chem Commun 47:7971

14. Demir-Cakan R, Baccile N, Antonietti M, Titirici MM (2009) Carboxylate-rich carbonaceous materials via one-step hydrothermal carbonization of glucose in the presence of acrylic acid. Chem Mater 21:484

15. Solymosi F, Erdöhelyi A, Cserényi J (1992) A comparative study on the activation and reactions of CH_4 on supported metals. Catal Lett 16:399

16. Mrityunjoy K, Vijayakumar PS, Prasad BLV, Gupta SS (2010) Synthesis and characterization of poly-L-lysine-grafted silica nanoparticles synthesized via NCA polymerization and click chemistry. Langmuir 26:5772

17. Yu XP, Chu W, Wang N (2011) Hydrogen production by ethanol steam reforming on NiCuMgAl catalysts derived from hydrotalcite-like precursors. Catal Lett 141:1228

18. Choudhary TV, Goodman DW (2002) Methane activation on ruthenium: the nature of the surface intermediates. Top Catal 20:35

19. Choudhary TV, Goodman DW (2000) CO-free production of hydrogen via stepwise steam reforming of methane. J Catal 192:316

2

Study the efficiency of some compounds as lubricating oil additives

Rasha S. Kamal · Nehal S. Ahmed ·
Amal M. Nasser

Abstract In the present work, some Mannich bases have been prepared by using different polyethylenepolyamines. Phosphosulphurized Mannich bases have been also prepared by reaction of prepared Mannich bases with P_2S_5. Structure of the prepared compounds was confirmed by infrared spectroscopy and the molecular weights determination. The efficiency of the prepared additives as antioxidants and detergents/dispersants was investigated. It was found that the efficiency increases with increasing the NH groups in the amine used. The best prepared additive as antioxidant and detergent/dispersant was obtained by using triethylenetetramine.

Keywords Mannich base lube oil additives ·
Antioxidants · Detergents/dispersants

Introduction

Lubrication is the act of applying lubricants and lubrication substances, which are capable of reducing friction between moving mechanical parts. The main function of a lubricant is forming a film that separates two surfaces that roll or slide between one another, to reduce friction and eliminate wear. Lubricant additives are compounds or mixtures when incorporated in base lubricating fluid, supplement its natural characteristics and improve their field service performance in existing applications. The functions of lubrication additives are to reduce the oxidative or thermal degradation of oil, reduce wear, minimize rust and corrosion, lessen the

R. S. Kamal · N. S. Ahmed (✉) · A. M. Nasser
Department of Petroleum Applications, Egyptian Petroleum
Research Institute, Nasr City, Cairo, Egypt
e-mail: mynehal@yahoo.com

deposition of harmful deposits on lubricated parts and prevent destructive metal-to-metal contact [4]. Detergent and dispersant additives are used primarily in internal combustion engines to keep metal surfaces clean by preventing deposition of oxidation products. Dispersants are typically the highest treat additives in an engine oil formulation. They are similar to detergents in that they have a polar head group with an oil-soluble hydrocarbon tail. While detergents are used to clean engine surfaces and neutralize acidic byproducts, their effectiveness is limited when it comes to dispersing oil-insoluble products resulting from the byproducts of combustion. Detergents are an integral part of any engine oil formulation. They are rarely the sole additive in a lubricant package and are often blended together with other types of additives [1]. Typically, they are blended with antiwear (AW) and extreme-pressure (EP) additives for use in engine oils, hydraulic fluids and metalworking fluids. Several studies have examined the effects of combining detergents with AW/EP additives such as zinc dialkyl dithiophosphates (ZDDPs), one of the most important chemically-active additives used in engines oils today. The purpose of detergents in crankcase oils is suspend/disperse oil-insoluble combustion products, such as sludge or soot and oxidation products, to neutralize combustion products (inorganic acids), to neutralize organic acids products of oil degradation processes and to control rust, corrosion, and deposit-forming resinous species [2]. One of the most important modes of lubricant degradation is oxidation. This oxidation is the primary cause of increase in viscosity, pour point, sludge, and enhanced engine corrosion and conventional liquid lubricants show poor oxidative and thermal stability at higher temperature ranges, which result in the formation of a vast quantity of volatile and solid products in the lubrication system [8]. The oxidation reactions that occur in a lubricant at elevated temperatures in the presence of atmospheric oxygen may lead to declined lubricant performance such as significant increase in kinematic viscosity observed in severe service conditions or during extended drain intervals [7]. In general, all types of base oil require addition of antioxidants depending on the amount of unsaturation and "natural inhibition" present. The refined mineral base oils contain "natural inhibitors" in the form of sulphur and nitrogen compounds sufficient for many applications [10]. Oxidation stability plays only a minor role in the process by which combustion chamber deposits are formed.

In the present work, some lube oil additives were prepared by synthesis of Mannich bases; which prepared by using alkyl phenol, formaldehyde and amines, then phosphosulphurization of the prepared Mannich bases by using phosphorous pentsulphide (P_2S_5). The efficiency of the prepared compounds as antioxidant and detergent/dispersant additives for lube oil was investigated [9].

Experimental

Preparation of Mannich Bases

In a 5-necked round bottom flask, fitted with an efficient mechanical stirrer, a ground joint thermometer, a condenser and a dropping funnel, the appropriate amounts of alkyl phenol in methanol and aqueous solution of polyethylenepolyamines were added. A formaldehyde solution (37 %) was then added dropwise during 1 h at 25–30 °C under nitrogen gas. After complete addition of the formaldehyde, the temperature of the reactants was elevated up to reflux temperature for 3 h. The mixture was then cooled to room temperature and the upper methanol/water layer was separated from the yellow viscous lower layer. The product (lower layer) was dissolved in 500 ml benzene and washed four times with 300 ml portions of distilled water to eliminate the excess of amine and formaldehyde.

Abbreviation of prepared Mannich bases

Mannich Bases	Abbreviation	Polyethylenepolyamines used
I	A_1	Ethylenediamine
II	A_2	Diethylenetriamine
III	A_3	Triethylenetetramine

Phosphosulphurization of prepared Mannich Bases

A 4-necked round bottom flask, fitted with a mechanical stirrer, efficient condenser, thermometer and inlet for passing nitrogen gas was used. The reactor was charged with four moles of each of prepared Mannich bases and one mole of P_2S_5. The reaction mixture was maintained at about 185–190 °C with continuous stirring for 4 h. The whole reaction has been carried out under nitrogen gas atmosphere. Three products of phosphosulphurized Mannich bases have been prepared using three Mannich bases to give B_1, B_2 and B_3, respectively.

Elemental analysis

The composition of each product has been determined quantitatively in weight percentages using the following standard methods of analysis:

(a) Carbon: Dumas method.
(b) Hydrogen: Dumas method.
(c) Nitrogen: Kjeldahl method.
(d) Phosphorus: IP 148/58 method.
(e) Sulphur: X-Ray Sulphur Meter Model RX-500S TANAKA Scientific instrument.
(f) Oxygen: By difference.

IR spectroscopic analysis

IR spectra of the synthesized compounds were determined by using FTIR spectrometer Model Type Mattson Infinity Series Top 961.

Determination of the Molecular Weights

The molecular weights of the prepared compounds were determined by using Gel Permeation Chromatography (GPC), Water 600 E.

Evaluation of the prepared compounds as lube oil additives

As antioxidants

The lube oil samples as well as its blends with 2 % by weight of each of the prepared additives were subjected to severe oxidation condition in the presence of copper and iron strips at 165.5 °C for 72 h. The Indiana test method of oxidation was used [5]. The oxidation stability of the lube oil blends was determined by taking samples at intervals of 24 h and up to 72 h of oxidation and the samples were tested for:

1. Variation of viscosity ratio V/V_o

The variation of viscosity ratio (V/V_o) was determined using IP 48/86 method, where V = Kinematic viscosity at 40 °C of sample after oxidation, V_o = Kinematic viscosity at 40 °C of sample before oxidation.

2. Change in total acid number (ΔT.A.N.)

The change was calculated according to IP 177/83 method, where ΔT.A.N. = total acid number of sample after oxidation—total acid number of sample before oxidation.

3. Optical density using IR technique

The IR spectra of oxidized oils have been determined in the range of the carbonyl group absorbance $(1{,}500–1{,}900 \text{ cm}^{-1})$ (Fig. 1). The spectra have been superimposed upon that of the unoxidized oil. The absorbance (A) has been calculated according to:

$$A = Log I_o/I$$

where I = % transmittance of the oil after oxidation, I_o = % transmittance of the oil before oxidation.

As dispersants

By using spot method [3] and [6]. Drops were taken from the Indiana oxidation apparatus after each 24 h intervals of oxidation and up to 72 h to make spots on special filter paper (Durieux 122) and the dispersancy of the sample were measured as follows:

$$\% \text{ Dispersancy} = \frac{\text{Diameter of blank spot}}{\text{Diameter of the total spot}} \times 100$$

The efficiency of dispersants has been classified as follows:

Up to 30 % no dispersancy
30–50 % medium dispersancy
50–60 % good dispersancy
60–70 % very good dispersancy
Above 70 % excellent dispersancy.

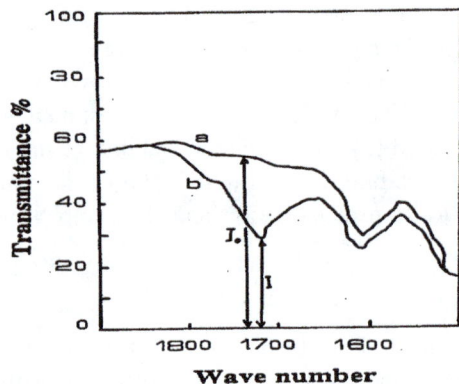

Fig. 1 IR Spectra of SAE in the Region 1,500–1,900 cm^{-1}. *a* Oil before oxidation, *b* oil after oxidation

Results and discussions

Preparation of Mannich Bases

Three Mannich bases have been prepared by using dodecylphenol, formaldehyde and three types of polyethylenepolyamines, namely, ethylenediamine, diethylenetriamine and triethylenetetramine. The molar ratio of dodecylphenol, formaldehyde and amine is always 1:1:1.

The determined mean molecular weights of all products were found to be very near from that calculated theoretically for the chemical structures of prospected Mannich bases which could be illustrated as follows:

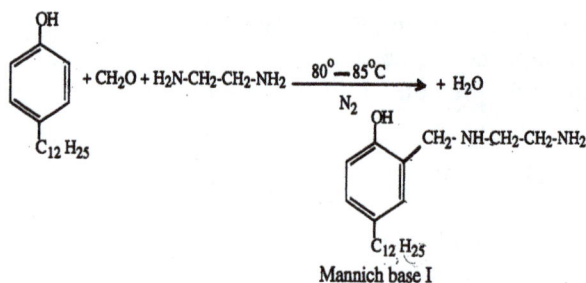

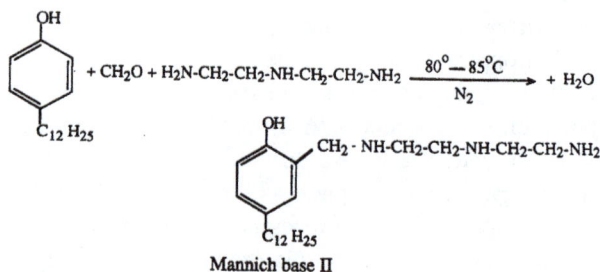

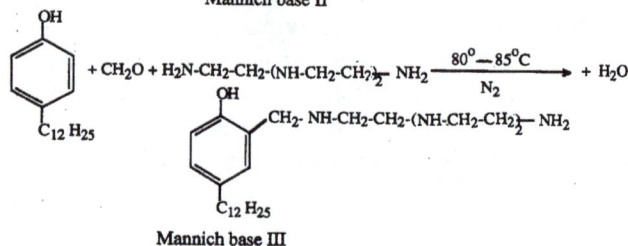

Elemental analysis of the prepared products was found also to be in accordance with that of prospected Mannich bases. Calculated mean numbers of different elements of prepared products, are given in Tables (1, 3), support the preparation of Mannich bases of chemical structures as given here in above. Results show an increase in molecular weight of Mannich bases with increasing the molecular weight of used ethylenepolyamines. The weight percentage of nitrogen- as well as- the mean number of nitrogen atoms/mole has also been found to be in accord with the above mentioned chemical structures of prepared Mannich bases.

Infra red spectrum of prepared Mannich bases is given in Fig. (2), which illustrate the following:

The N–H and O–H regions are overlapping (O–H at 3,650–3,200 cm^{-1} and the N–H at 3,500–3,300 cm^{-1}). C–N band is in the range from 1,350 to 1,000 cm^{-1}, C–O band is in the range from 1,250 to 1,000 cm^{-1}, C–H alkanes in the range from 3,000 to 2,850 cm^{-1}, C–H aromatic in the range from 3,000 to 3,150 cm^{-1}, CH$_3$ group appears at 1,450 and 1,375 cm^{-1}, CH$_2$ group appears at 1,465 and 1,375 cm^{-1}, C=C aromatic appears at 1,475 and 1,600 cm^{-1}. Para disubstituted ring appears as one strong band from 800 to 850 cm^{-1}.

Phosphosulphurization of the prepared Mannich bases

Prepared Mannich bases were reacted with P$_2$S$_5$ to prepare phosphosulphurized Mannich bases. The determined mean molecular weights of all products were found to be very near from that calculated theoretically for the chemical structures of prospected phosphosulphurized Mannich bases which can be illustrated as follows:

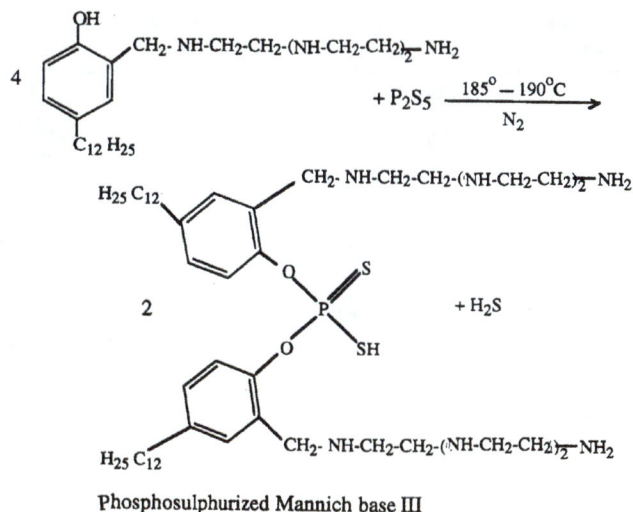

Phosphosulphurized Mannich base III

Elemental analysis of prepared products were found also to be in accordance with that of prospected phosphosulphurized Mannich bases. Calculated mean numbers of different elements of prepared products are given in Tables 2 and 3),

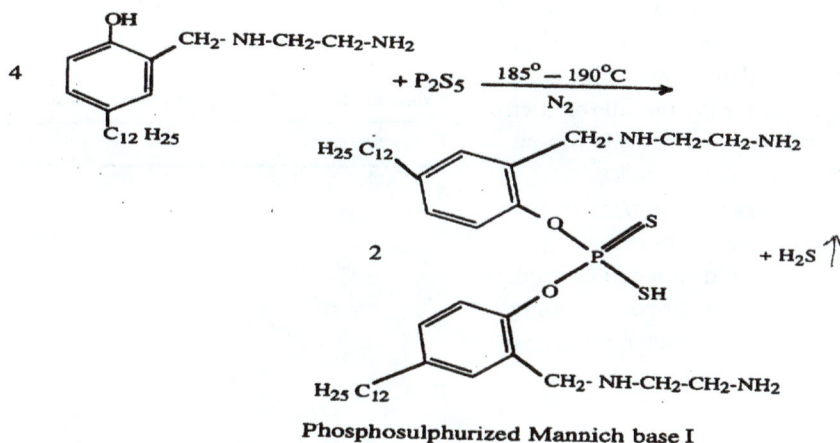

Phosphosulphurized Mannich base I

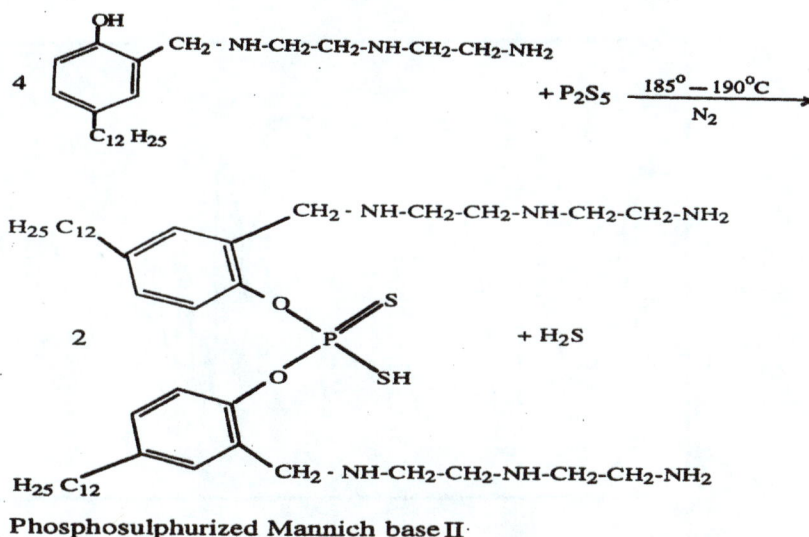

Phosphosulphurized Mannich base II

Table 1 Elemental analysis of the prepared Mannich bases compounds

Characteristics	A1		A2		A3	
Elemental analysis, wt%	Calc.	Found	Calc.	Found	Calc.	Found
Carbon	75.44	76.08	73.21	73.85	71.26	71.72
Hydrogen	11.38	11.53	11.41	11.54	11.64	11.72
Nitrogen	8.38	8.07	11.14	10.77	13.30	12.87
Oxygen	4.79	4.61	4.24	4.10	3.81	3.68
Mean number of C atoms/mol	21	22	23	24	25	26
Mean number of H_2 atoms/mol	38	40	43	45	49	51
Mean number of N_2 atoms/mol	2	2	3	3	4	4
Mean number of O_2 atoms/mol	1	1	1	1	1	1

Table 2 Elemental analysis of the prepared phosphosulphurized Mannich bases compounds

Characteristics	B1		B2		B3	
Elemental analysis, wt %	Calc.	Found	Calc.	Found	Calc.	Found
Carbon	66.14	65.03	65.09	65.43	64.10	64.42
Hydrogen	9.84	9.68	10.02	10.09	10.36	10.42
Nitrogen	7.35	7.23	9.91	9.74	11.97	11.79
Phosphorus	4.07	4.00	3.66	3.60	3.31	3.26
Sulphur	8.4	8.26	7.55	7.42	6.84	6.74
Oxygen	4.20	4.31	3.77	3.71	3.42	3.37
Mean number of C atoms/mol	42	42	46	47	50	51
Mean number of H_2 atoms/mol	75	75	85	87	97	99
Mean number of N_2 atoms/mol	4	4	6	6	8	8
Mean number of P atoms/mol	1	1	1	1	1	1
Mean number of S atoms/mol	2	2	2	2	2	2
Mean number of O_2 atoms/mol	2	2	2	2	2	2

support the preparation of phosphosulphurized Mannich bases of chemical structures as given here in above. Results show an increase in molecular weight of phosphosulphurized Mannich bases with increasing the molecular weight of used polyethylenepolyamine. The weight percentage of nitrogen as well as the mean number of nitrogen atoms/mole were found to be in accord with the above mentioned chemical structures of prepared phosphosulphurized Mannich bases. Also the weight percentage of phosphorus and sulphur as well as the mean number of phosphorus atoms/mole and sulphur atoms/mole were found to be in accord with the above mentioned chemical structures of prepared phosphosulphurized Mannich bases. In all products, the ratio between mean number of phosphorus and sulphur is 1:2.

Table 3 Mean molecular weights of the prepared compounds

The prepared compounds	Mean M.wt
A1	347
A2	390
A3	435
B1	775
B2	862
B3	950

Fig. 2 Infrared spectrum of the prepared Mannich base (I)

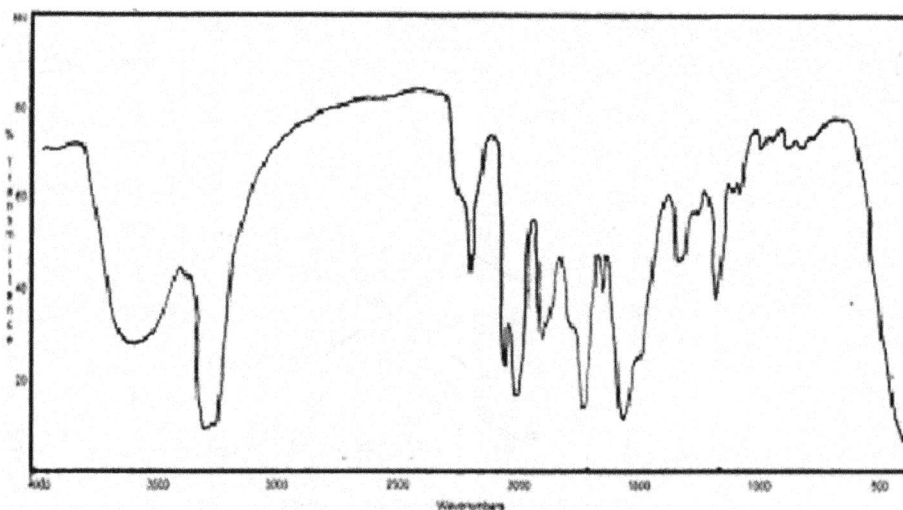

Evaluation of the prepared compounds as lube oil additives

As antioxidants

Effect of Mannich bases with different ethylenepolyamines Mannich bases were prepared by using three different ethylenepolyamines. The prepared Mannich bases were added to a sample of "SAE-30" lubricating oil free from any additives and the blends obtained were subjected to severe oxidation conditions using the Indiana test methods at 165.5 °C with continuous and constant rate of stirring. Samples were taken at intervals of 24 h and up to 72 h of oxidation and tested for their oxidation stability expressed as increase in viscosity ratio (V/V_o), change in total acid number ($\Delta T.A.N.$) and optical density log (I/I_o) compared with lube oil sample free from additives. Results are given in Figs. 3, 4, 5 which indicate the following:

All the prepared compounds impart better oxidation resistance properties to the lube oil compared with the undoped oil. Their efficiencies as lube oil antioxidants are very close to each other, since variations in the total acid number ($\Delta T.A.N.$), viscosity ratio (V/V_o) and optical density log (I/I_o) is very small. This may be attributed to the

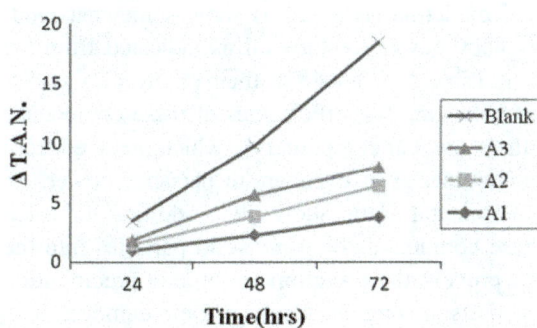

Fig. 3 Variation of $\Delta T.A.N.$ with oxidation time of lube oil without and with Mannich bases additives (A_1, A_2 and A_3)

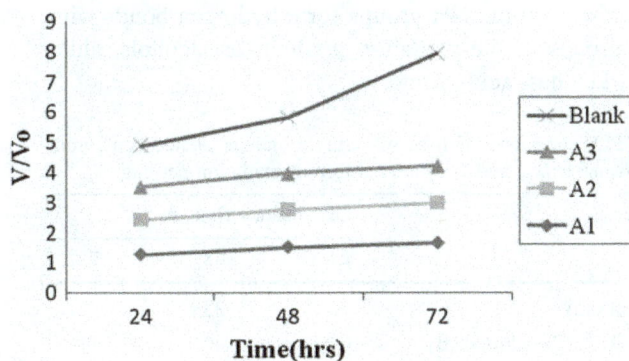

Fig. 4 Variation of V/V_o with oxidation time of lube oil without and with Mannich bases additives (A_1, A_2 and A_3)

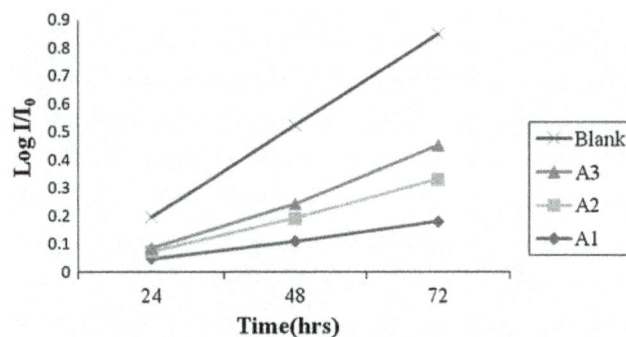

Fig. 5 Variation of log I/I_o with oxidation time of lube oil without and with Mannich bases additives (A_1, A_2 and A_3)

presence of phenolic and amino groups in their structures. The mechanism of phenolic aromatic amine compounds as antioxidants were explained by phenolic and certain aromatic amine inhibitors function by donation of labile hydrogen from the groups (OH or NH) to stabilize the chain radicals; i.e., these inhibitors destroy the peroxide radicals and thus, the oxidation chain is broken. The marked efficiency of hindered phenols is attributed to their radicals being insufficiently reactive to abstract hydrogen from the hydrocarbon and initiate new oxidation chains. The presence of amine part in the structure of prepared Mannich base compounds neutralize some of the acidic products of oil oxidation. The change in total acid number ($\Delta T.A.N.$), viscosity ratio (V/V_o) and optical density log (I/I_o) decrease with increasing the NH groups in the molecular of amine used. Thus, using triethylenetetramine gave better results than diethylenetriamine and ethylene diamine. In all cases, results indicate that there is a quite wide difference between the lube oil samples mixed with prepared products and that of using lube oil sample without additives.

Effect of phosphosulphurized Mannich bases reaction products Mannich bases have been reacted with P_2S_5 to produce some products as lube oil additives containing sulphur and phosphorus elements. The prepared additives were added to the same lube oil sample "SAE-30" which free from additives. The lube oil samples mixed with prepared products were subjected to sever oxidation conditions using the Indiana test method of oxidation previously discussed and the results obtained are compared with those of undoped lube oil sample. Results are given in Figs. (6, 7, 8) which indicate that increasing the number of NH group enhance to some extent the oxidation properties of lube oil sample ($\Delta T.A.N.$), log (I/I_o) and (V/V_o). Results also show that, using triethylenetetramine gave better results than diethylenetriamine and ethylenediamine. However, somewhat better results were obtained in case of phosphosulphurized Mannich bases due to the presence of

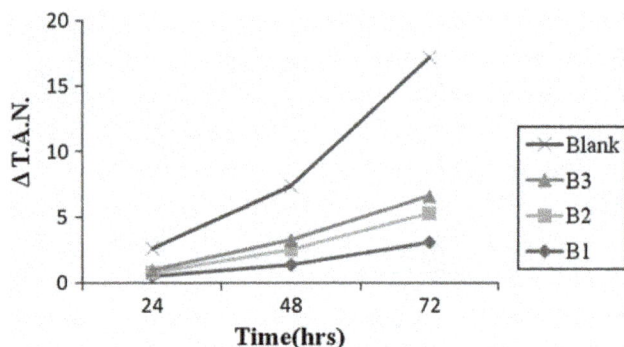

Fig. 6 Variation of ΔT.A.N. with oxidation time of lube oil without and with phosphosulphurized Mannich bases additives (B_1, B_2 and B_3)

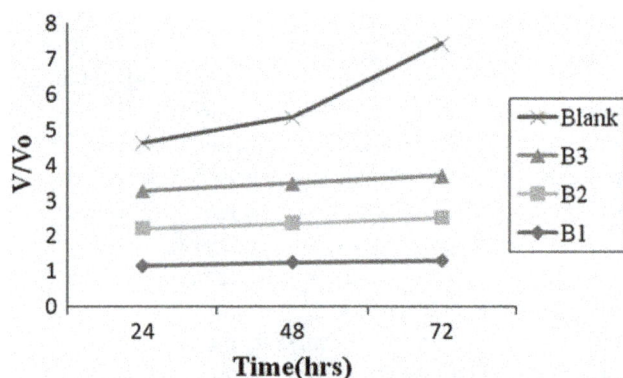

Fig. 7 Variation of V/V_o with oxidation time of lube oil without and with phosphosulphurized Mannich bases additives (B_1, B_2 and B_3)

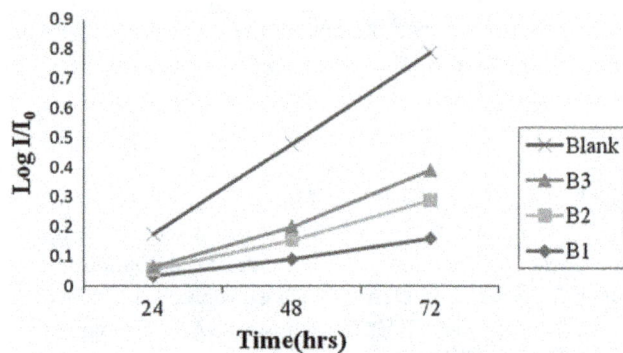

Fig. 8 Variation of log I/I_o with oxidation time of lube oil without and with phosphosulphurized Mannich bases additives (B_1, B_2 and B_3)

phosphorus and sulphur. It is well known that sulphur compounds occurring naturally in mineral oils inhibit oxidation by reduction of peroxides. These compounds decompose hydroperoxide far in excess of amounts that could be accounted for by oxidation of the original inhibitor because the oxidized inhibitor also retards oxidation. It is suggested also that sulphur and phosphorus compounds

Table 4 Dispersancy of lube oil sample and its blends containing additives (A1, A2 and A3) after different oxidation periods

Sample	Dispersancy time, h		
	24	48	72
Lube oil only	35	33	32
Lube oil + 2 % additive A_1	60	65	69
Lube oil + 2 % additive A_2	63	69	74
Lube oil + 2 % additive A_3	65	76	80

react with metals to form sulphide and phosphide films that prevent the contact between metal surface and lube oil, hence prevent the catalytic action of metal on lube oil oxidation process.

As detergents/dispersants

The different prepared Mannich base additives were added to lube oil samples in a concentration of 2 wt%, using the spot test method. Results are given in Table 4 for Mannich base additives with different polyethylenepolyamines A_1, A_2 and A_3, show clearly that the prepared additives have moderate to very good dispersancy power (60–80 %) for the sludge and solid particles formed during lube oil oxidation compared with lube oil only. Compounds A_1- A_3 at the beginning of oxidation (after 24 h) show somewhat moderate dispersancy power for lube oil as indicated from the data given in Table 4. After 48 h, their efficiencies as dispersant become clear. The efficiencies of these compounds are nearly the same, but compound A_3 which prepared from triethylenetetramine, gives excellent dispersancy power.

The experimental data show that addition of such Mannich base compounds disperse solid particles into the oil and thus prevent their agglomeration and precipitation on metallic parts of engines causing their damage. It is clear from data which given in Table 4 that increasing the NH groups in the structures of the prepared compounds, increase their capacity in dispersing sludge and solid particles into the lube oil samples used. This may be explained by the fact that the NH groups form hydrogen bonds with polar groups of the oxidation products as alcohols, aldehydes, ketones, acids... etc.

Table 5 Dispersancy of lube oil sample and its blends containing additives (B1, B 2 and B 3) after different oxidation periods

Sample	Dispersancy time, h		
	24	48	72
Lube oil only	35	33	32
Lube oil + 2 % additive B_1	64	66	71
Lube oil + 2 % additive B_2	67	71	75
Lube oil + 2 % additive B_3	76	78	82

Results given in Table 5 for the phosphosulphurized Mannich base additives with different polyethylenepolyamines B_1, B_2 and B_3 indicate the following:

After 24 h of oxidation, all additives give good dispersancy power, this may be attributed to the fact that the solid particles produced through this period of oxidation are very little and thus easily dispersed in the lube oil.

After 48 h of oxidation, there is very little difference between their efficiencies. The difference is very clear when comparing their efficiencies with that of the lube oil sample without additives.

After 72 h, as the quantity of oxidation products becomes greater, the difference between the efficiency of these compounds becomes clear. Data given in Table 5 indicate also that increasing the number of the effective NH groups in the structure of the prepared additives increases their capacity in dispersing sludge and solid particles into the lube oil samples used. Results show also that additive B_3 prepared from triethylenetetramine, gives the best efficiency as lube oil dispersant compared with the other additives.

Conclusions

The conclusions of the study are:

1. The Mannich base obtained by using triethylenetetramine gave good results than the others.
2. Phosphosulphurized Mannich base prepared from triethylenetetramine gave also good results than others.
3. Phosphosulphurized Mannich bases impart good properties of Mannich bases as lube oil antioxidants.

4. The prepared additives were evaluated as lube oil antioxidants as well as lube oil detergents/dispersants and found to be good lube oil additives.

References

1. Abdel-Azim A-AA, Nassar AM, Ahmed NS, Kamal RS (2009) Multifunctional additives viscosity index improvers, pour point depressants and dispersants for lube oil. Petrol Sci Technol 27: 20–32
2. Nassar AM, Ahmed NS, Abdel Aziz KI, El-Kafrawy AF, Abdel-Azim A-AA (2006) Synthesis and evaluation of detergent/dispersant additives from polyisobutylene succinimdes. Int J Polym Mater 55:703–713
3. Gatis VA, Bergstrom RF, Wendt LA (1955) Society of automative engineers (SAE), Na 572
4. Hus SM (2004) Molecular basis of lubrication. Tribol Int 37:553
5. Lamp GG, Loance CM, Gaynor JW (1941) Indiana stirring oxidation test for lubricating oil. Ind Eng Chem Anal Ed 13:317–321
6. Najman M, Kasrai M, Bancroft GM, Davidson R (2006) Combination of ashless antiwear additives with metallic detergents. Tribol Int 39:342
7. Ahmed NS, Nassar AM, Abdel-Azim A-AA (2008) Synthesis and evaluation of some detergent/dispersant additives for lube oil. Int J Polym Mater 57:114–124
8. Pirro DM, Wessol AA (2001) Lubrication fundamentals, vol 3. Marcel Dekker Inc., New York, p 37
9. Rundnick LR (2003) Lubricant additives: chemistry and applications. Marcel Dekker, New York, p 1293
10. Santos JCO, Santos VJF, Souza AG, Sobrinho EV, Fernandes VJ Jr, Silva AJN (2004) Thermoanalytical and rheological characterization of automotive mineral lubricants after thermal degradation. Fuel 83:2393–2399

3

Characterization and activity study of the Rh-substituted pyrochlores for CO_2 (dry) reforming of CH_4

Devendra Pakhare · Hongyi Wu · Savinay Narendra · Victor Abdelsayed · Daniel Haynes · Dushyant Shekhawat · David Berry · James Spivey

Abstract Isomorphic substitution of Rh at varying levels on the B site of lanthanum zirconate pyrochlore ($La_2Zr_2O_7$; designated LZ) resulted in the formation of thermally stable catalysts suitable for fuel reforming reactions operating at 900 °C. Three specific catalysts are reported here: (a) unsubstituted lanthanum zirconate (LZ), (b) LZ with 2 wt% substituted Rh (L2RhZ), and (c) LZ with 5 wt% substituted Rh (L5RhZ). These catalysts were characterized by XRD, XPS, and H_2-TPR. XRD of the fresh, calcined catalysts showed the formation of the pyrochlore phase ($La_2Zr_2O_7$) in all three materials. In L5RhZ, the relatively high level of Rh substitution led to the formation of $LaRhO_3$ perovskite phase which was not observed in the L2RhZ and LZ pyrochlores. TPR results show that the L5RhZ consumed 1.57 mg H_2/g_{cat}, which is much greater than the 0.508 H_2/g_{cat} and 0.155 mg H_2/g_{cat} for L2RhZ and LZ, respectively, suggesting that the reducibility of the pyrochlore structure increases with increasing Rh-substitution. DRM was studied on these three catalysts at three different temperatures of 550, 575, and 600 °C. The results showed that CH_4 and CO_2 conversion was significantly greater for L5RhZ compared to L2RhZ and no activity was observed for LZ, suggesting that the surface Rh sites are required for the DRM reaction. Temperature programmed surface reaction showed that L5RhZ had light-off temperature 80 °C lower than L2RhZ. The spent catalysts after runs at each temperature were characterized by temperature programmed oxidation (TPO) followed by temperature programmed reduction and XRD. The TPO results showed that the amount of carbon formed over L5RhZ is almost half of that formed on L2RhZ.

Keywords Dry reforming · Lanthanum zirconate · Pyrochlores · Lattice oxygen · Isomorphic substitution · Perovskite · Reverse water gas shift

D. Pakhare · J. Spivey (✉)
Department of Chemical Engineering, Louisiana State University, Baton Rouge, LA 70803, USA
e-mail: jjspivey@lsu.edu

H. Wu
Department of Chemistry, Southern University, Baton Rouge, LA 70816, USA

S. Narendra
Department of Mechanical Engineering, Indian Institute of Technology, Kharagpur 721302, India

V. Abdelsayed · D. Haynes · D. Shekhawat · D. Berry
National Energy Technology Laboratory, U. S. Department of Energy, Morgantown, WV 26507, USA

Introduction

Pyrochlores are a class of ternary metal oxides based on the fluorite structure with a cubic unit cell with a general formula of $A_2B_2O_7$. An important property of these materials is that catalytically active noble metals can be substituted isomorphically on the B site to form a crystalline catalyst. These materials consist of vacancies at the A and O sites, which facilitate oxygen ion migration within the structure [1]. The A site is usually a large cation (typically rare earth elements) and the B site cation has a smaller radius (usually transition metal) [2]. For the pyrochlore structure to be stable it is necessary that the ionic radius ratio of A and B site cations be between 1.46 and 1.78 [1]. The ratio of the ionic radii for $La_2Zr_2O_7$ is 1.61 [3]. If the ratio of the ionic radii is greater than 1.78, a perovskite phase can be formed. Below a ratio of 1.46 a fluorite structure is formed [4]. Catalytically active metals like Ru, Rh, Pt can be substituted into the B site of the pyrochlore structure because they meet this ionic radius constraint and have the required oxidation state [5]. The resulting materials possess the thermal stability inherent in the pyrochlore structure, which also constrains the active metal within the pyrochlore structure even at high temperatures.

Steam reforming, autothermal reforming, and partial oxidation of methane are used to reform methane to synthesis gas [5–7]. CO_2 reforming of CH_4 is a highly endothermic reaction and has been widely studied on a number of catalysts [8–11]. For fuel reforming, one study shows that the activity decreases in the order Ru, Rh > Ir > Ni, Pt, Pd > Co > Fe, Cu [12], with noble metals also showing higher activity and greater resistance to deactivation by carbon deposition [13]. Carbide catalysts have also been used for studying this reaction [6, 14]. Economic evaluations have suggested a cost advantage for DRM as a route to the production of synthesis gas [15].

There are two major problems associated with dry reforming, (a) deactivation due to carbon formation on the catalyst, and (b) thermal degradation of the catalyst and/or support at the high temperatures required for this reaction, typically above 700 °C. Studies using non-noble metals like Fe, Ni, have consistently shown rapid deactivation by carbon deposition [16–18], although this can be minimized in some cases by maintaining high metal dispersion [17]. Because temperatures well above 700 °C are required to reach high syngas yields, traditional supported metals are not stable, suggesting the need to develop an inherently stable material that is catalytically active. Resistance to carbon formation is also related to the oxygen conductivity of substituted pyrochlores [19]. Oxygen mobility within the pyrochlore structure is a strong function of the La content in the structure. It is shown by Diaz-Guillen et al. [20] that the activation energy for oxygen ion conductivity decreases from 1.13 eV for $Gd_2Zr_2O_7$ to 0.81 eV for $GdLaZr_2O_7$ with an increase in the La substitution in the pyrochlore structure.

Although pyrochlores have the thermal stability and potential for active metals to be substituted into the structure, we are aware of only one report of DRM on any pyrochlore. Ashcroft et al. [21] studied pyrochlores based on Eu, Ru, Gd, but found that they decomposed into the various oxides at DRM conditions.

The present study focuses on the characterization and activity of Rh-substituted pyrochlores, which have been studied for reactions such as fuel reforming [19, 22]. Specifically, lanthanum zirconates (LZ) into which 2 and 5wt% Rh have been substituted at the B site of the pyrochlore structure are characterized by ICP, XRD, XPS, TPO and H_2-TPR. $La_2Zr_2O_7$ pyrochlores with Rh substitution on the Zr site (known as the B site, based on the general formula for pyrochlores as $A_2B_2O_7$) were tested for

their activity at 550, 575 and 600 °C to study the kinetics of the reaction. Post-run temperature programmed oxidation (TPO) is used to determine the coke formation.

Experimental section

Catalyst synthesis

The LZ, L2RhZ and L5RhZ pyrochlores were synthesised by modified Pechini method [19]. The synthesis and ICP procedure had been reported earlier [5, 19].

Catalyst characterization

The equipment and experimental procedure details for X-ray diffraction (XRD), X-ray photoelectron spectroscopy (XPS), H_2 temperature programmed reduction (TPR), temperature programmed surface reaction (TPSR) are reported in our earlier work [5]. For this work XPS spectra were obtained for the C 1 s, O 1 s, La 3d, Zr 3d, and Rh 3d. In each case, the binding energy (BE) and the area of the corresponding peaks were measured.

Activity study

The composition of reactant gases used for the reaction over the catalyst was 10 mol% of CO_2/He and 10 mol% of CH_4/He. We studied the activity of the catalysts for DRM at different temperatures 550, 575, and 600 °C. The mass spectrometer (MS) connected to the reactor gave the mole fractions of the reactants in the blank condition. DRM was performed with an equimolar reactant feed of 20 mL/min of each of the reactant gases to give a total space velocity of 48,000 mL/g_{cat}/h. For each run, 50 mg of the catalyst was loaded in the U-tube reactor. Before each reaction run, the catalysts were heated to the reaction temperature in flowing He; no reduction was conducted before subjecting the catalysts to DRM. The reactants CO_2 and CH_4 were introduced after this into the reactor at desired flow rates. The mole fractions of the reactants and products from the mass spectrometer helped us to compare the results with the blank conditions.

Temperature programmed oxidation (TPO)

After DRM, a TPO was conducted for studying the carbon formed during the reaction. For conducting the TPO, the catalyst was cooled to room temperature (ca. 35 °C) in flowing He at 20 mL/min. Then it was oxidized in flowing 5 % O_2/He at 30 mL/min from room temperature to 950 °C and the ramp rate was 5 °C/min. The conditions were maintained isothermal at 950 °C for 30 min. The CO

($m/z = 28$) and CO_2 ($m/z = 44$) emitted during the TPO were tracked using the mass spectrometer hooked up to the reactor outlet.

Results and discussions

X-ray diffraction study of fresh catalyst

Figure 1 shows the XRD patterns for the freshly calcined pyrochlores. The star marked peaks represent the $La_2Zr_2O_7$ (ICSD no: 50-0837) pyrochlore phase and the diffraction angle for these peaks is similar to the diffraction patterns observed in the literature [19, 23].

Haynes et al. [19] used L2RhZ pyrochlores for the partial oxidation of n-tetradecane, and their XRD pattern for pyrochlores match those in Fig. 1 for LZ and L2RhZ in this study. However, for L5RhZ there was an extra peak observed at ~32°. Gallego et al. [17] and Arauj et al. [11]. studied perovskites for DRM and observed main diffraction peak for $LaCoO_3$ and $LaRuO_3$, respectively, at ~32°. The similar peak observed at 32° in the XRD pattern for L5RhZ pyrochlore (Fig. 1) suggests that 5 wt% Rh substitution resulted in the formation of a separate $LaRhO_3$ perovskite phase.

To study effect of Rh substitution in further detail, XRD with a slow sweep rate was conducted on the fresh catalysts. Figure 2 shows the XRD pattern for a slow sweep rate; it is observed that there is a small peak for $LaRhO_3$ for the L2RhZ pyrochlore, but a prominent one for L5RhZ. The amount of Rh in the 5 % and the 2 % sample appears to be in excess of the maximum substitution limit of the pyrochlore structure, and thus resulted in the formation of a separate $LaRhO_3$ perovskite phase. Isomorphic substitution

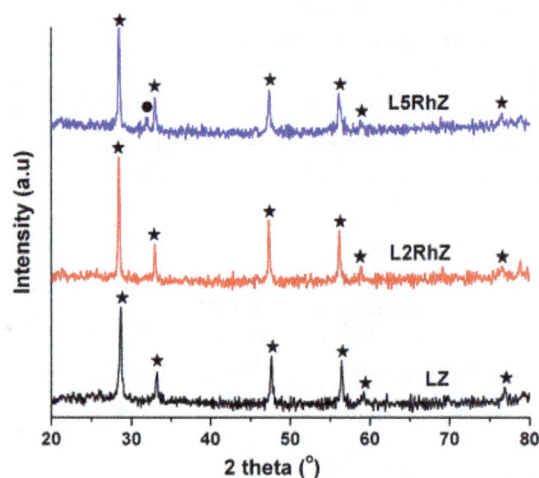

Fig. 2 Slow scan XRD pattern for freshly calcined LZ, L2RhZ, and L5RhZ pyrochlores

of Rh on the B site has caused a small shift in the diffraction peak to a smaller angle for L2RhZ and L5RhZ compared to LZ. Lower diffraction angle corresponds to an increase in the lattice parameter of LRhZ catalysts due to Rh substitution.

X-ray photoelectron spectroscopy of the fresh catalysts

XPS spectra for the Rh 3d core level obtained from L2RhZ and L5RhZ pyrochlores are shown in Fig. 3a, b, respectively.

In the deconvolution process, the relative intensity and separation of the spin–orbit for Rh $3d_{5/2}$–$3d_{3/2}$ doublet were fixed at ratio of 3:2 and 4.8 eV, respectively [24]. According to the literature, the BE of Rh^0 valence state is 307.1–307.6 eV and BE of Rh^{3+} valence state is in a wide range from 308.8 to 311.3 eV depending on the surrounding environment [24–27]. For L2RhZ (see Fig. 3a), Rh^{3+} was the only detected species as deduced from the binding energy of the Rh $3d_{5/2}$ photoelectron peak at 308.9 eV. Compared to L2RhZ, the valence state of Rh for L5RhZ is more complicated. For L5RhZ, the Rh^{3+} is the dominant valance state with peak at 309.0 eV (see Fig. 3b); a smaller Rh $3d_{5/2}$ peak is observed at a lower binding energy of 308.3 eV which indicates the presence of another valence state of Rh. This peak at 308.3 eV is attributed to partially oxidized $Rh^{\delta+}$ species [25]. The relative distributions of Rh^{3+} and $Rh^{\delta+}$ in L5RhZ are 82.8 and 17.2 %, respectively (see Table 1).

It should be noted that the FWHM (full width half maxima) of Rh peaks was significantly broader (2.0–2.7 in our measurement) when compared to those of Rh standard (about 0.7 for pure bulk Rh_2O_3). The broadening of Rh peaks may suggest a high dispersion of Rh in catalyst with little local aggregation [25, 28].

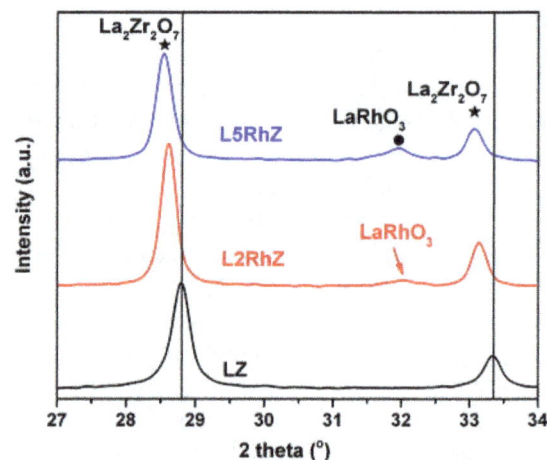

Fig. 1 XRD pattern for freshly calcined LZ, L2RhZ, and L5RhZ pyrochlores (star symbols $La_2Zr_2O_7$, and circles $LaRhO_3$)

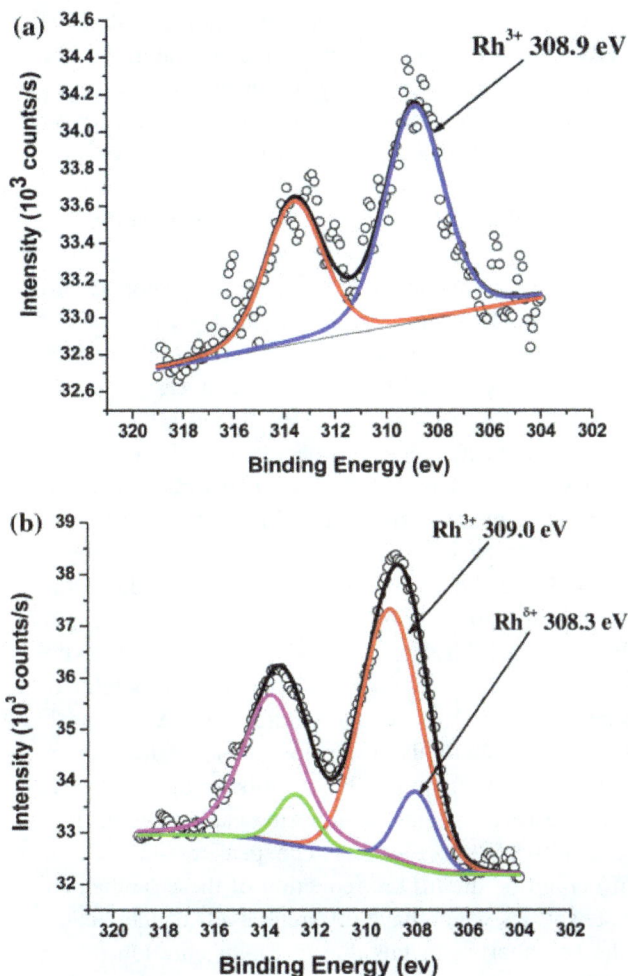

Fig. 3 Photoelectron Rh 3d spectra for **a** L2RhZ, **b** L5RhZ pyrochlores

Table 1 XPS determined relative atomic ratio of surface Rh species in 2 and 5wt% Rh catalyst

Catalyst	$Rh^{\delta+}/Rh_{total}$ (%)	Rh^{3+}/Rh_{total} (%)
2 wt% Rh	0	100
5 wt% Rh	17.2	82.8

Table 2 Rh concentration (wt%) obtained by different methods

Catalyst	Method for obtaining Rh concentration (wt%)	
	ICP-OES (bulk)	XPS (surface)
L2RhZ	1.7	0.78
L5RhZ	4.4	3.15

depths from the surface, a higher surface concentration would mean that the surface Rh on L5RhZ is well dispersed. If there was any aggregation of Rh on the surface then the detected amount of Rh on L5RhZ surface would be less or close to that of L2RhZ. But a higher surface concentration indicates that the greater Rh loading in L5RhZ did not cause any local surface aggregation of Rh, which generally would decrease the detected surface concentration of the metal.

XPS analysis shows that for both L2RhZ and L5RhZ pyrochlores, the $3d_{5/2}$ peaks of La and Zr elements were 833.4–833.6 and 182.1–182.3 eV, respectively. These peaks positions indicated Zr and La in the pyrochlore structure and were in the Zr^{4+} (ZrO_2) and La^{3+} (La_2O_3) oxidation states, respectively [29, 30].

Temperature programmed reduction (TPR)

Figure 4 shows the hydrogen TPR profiles for the three catalysts. The LZ pyrochlore reduction profile shows reduction peaks at 490 and 580 °C, corresponding to an H_2 consumption of 0.155 mg H_2/g_{cat}. This corresponds to 0.6 % reduction of the lanthanum zirconate. The TPR profile of L2RhZ shows three distinct peaks at 380, 455, and 570 °C. All three catalysts show two peaks above 450 °C, which can be attributed to reduction of the LZ itself.

Comparison of the L2RhZ and LZ TPR results suggests that the 380 °C peak is due to reduction of Rh that is interacting strongly with the pyrochlore. The small additional peak 280 °C for L2RhZ may be due to reduction of Rh that is less strongly interacting with the pyrochlore [19]. For L5RhZ the 410 °C peak is close to that at 380 °C for the L2RhZ, and its larger area is consistent with the larger amount of reducible Rh in this catalyst. The similarity of the peak temperatures for this peak and that of the L2RhZ (410 versus 380 °C) indicates that the strength of the

Quantitative analysis indicates the atomic percent of Rh on the surface of L2RhZ and L5RhZ pyrochlores is 0.78 and 3.15 %, respectively. This surface Rh concentration for individual pyrochlores is smaller than the theoretical levels obtained from inductively couple plasma-optical emission spectroscopy (ICP-OES) results (see Table 2). Since XPS is a surface sensitive technique which only can detect elements several nanometers under the surface, this result suggests that the remaining Rh lies within the lattice, as expected.

Both pyrochlores have surface Rh in similar oxidation states, but the absence of $Rh^{\delta+}$ signal for L2RhZ may be due to its lower Rh loadings, which limits the accuracy in deconvolution of the Rh peak due to high signal-to-noise level. The ratio of surface Rh concentration (by wt) for L5RhZ and L2RhZ is about (3.15:0.78 = 4), which is greater than their bulk ratio i.e., 4.4:1.7 = 2.6, as observed in the ICP-OES analysis. This indicates that the surface of L5RhZ is enriched with Rh compared to the surface of L2RhZ. As XPS detects elements within a few nanometers

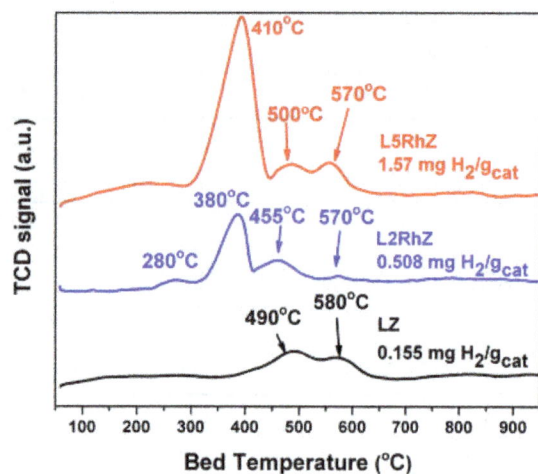

Fig. 4 Temperature programmed reduction of freshly calcined LZ, L2RhZ, L5RhZ pyrochlores

interaction of this reducible Rh is similar on both catalysts. The L2RhZ consumed 0.508 mg H_2/g_{cat}, which includes H_2 consumed for the reduction of the Rh species at 380 °C and a small portion of the lanthanum zirconate at 455 °C and 570 °C. The H_2 consumption for L5RhZ is 1.57 mg H_2/g_{cat}. It is difficult to quantify exactly the percentage of reduction of Rh in the pyrochlore structure due to overlapping reduction peaks of Rh and lanthanum zirconates. However, the H_2 consumption by L2RhZ and L5RhZ is smaller than their respective theoretical consumption assuming complete Rh reduction (and no reduction of the pyrochlore) i.e., 0.66 mg H_2/g_{cat} for L2RhZ and 1.7 mg H_2/g_{cat} for L5RhZ. This means that a significant portion of the Rh is substituted in the bulk of the pyrochlore and is not available during the reduction reaction, as expected.

TPR profiles obtained in this study are similar to the ones observed by Haynes et al. [19]. The peak observed for LZ in our study shows two types of reducing species, one at 490 °C and the other at 580 °C. Whereas, the one observed by Haynes et al. [19], for LZ has a single peak at 527 °C. This difference in the reduction peaks could be due to the difference in the hydrogen concentration in the two TPR procedures. The concentration of the gas used for TPR in the work by Haynes et al. [19] was 5 % H_2/Ar and the one used in this work was higher 10 % H_2/Ar keeping the same ramp rate and flow rate. The higher partial pressure of the reducing gas in the present work resulted in a faster reduction reaction, allowing a distinction to be made between reduction peaks that were not visible in the Haynes et al. [19] study. The single broad peak observed by Haynes et al., thus appeared as a double peak in the present work.

The deconvoluted TPR profiles of LZ, L2RhZ, and L5RhZ pyrochlroes are shown in Fig. 5a–c, respectively (these figures differ in y-axis scale). Figure 5a shows that

the LZ reduction peaks involve the reduction of four species. XPS results show that La and Zr are present in +3 and +4 oxidation states, respectively, in the pyrochlore structure. Peaks at 490 and 580 °C could be due to the reduction of La^{+3} and Zr^{+4} species [31]. The lower temperature peaks at 396 and 430 °C (not visible in Fig. 4 due to the y-axis scale) could be due to the reduction of partially coordinated lanthanum or zirconium cations at the surface. Hoang et al. [31] conducted TPR of the ZrO_2 support and lanthana promoted zirconia structure (La_2O_3–ZrO_2). They observed a reduction peak at 574 °C for ZrO_2 and at 554 °C for La_2O_3–ZrO_2 [31]. The high temperature peaks at 490 and 580 °C can be attributed to the reduction of La_2O_3–ZrO_2 and ZrO_2 phase, respectively [31]. In Fig. 5b, L2RhZ has a small peak at 280 °C appears to be due to the reduction of a weakly interacting Rh species [19]. The intensity of this peak is very low and thus this reducing species could not be accurately determined and deconvoluted during the analysis of the XPS peaks. Peaks at 375 and 394 °C can be assigned to the reduction of Rh with varying degrees of interaction with oxygen and neighboring atoms in the bulk of the pyrochlores. The 394 °C peak could also be due to the reduction of some lanthanum zirconate species reducing at 396 °C as seen in Fig. 5a. As mentioned earlier, the lanthanum zirconate is reduced at 430 and 490 °C (see Fig. 5a). The peak at 455 °C (in Fig. 5b) could be due to the reduction of these lanthanum zirconate species which have different levels of interaction with the neighboring metals due to Rh substitution in the pyrochlore structure compared to the unsubstituted LZ pyrochlore.

Deconvoluted TPR peaks for L5RhZ (Fig. 5c) show that there are peaks at 352, 396, and 416 °C. The 352 °C peak is likely to be due to the reduction of $Rh^{\delta+}$ species as observed in XPS results. The 396 and 416 °C peaks can be assigned to the reduction of bulk Rh with varying interaction with the pyrochlore structure. These species reducing at 396 and 416 °C have similar oxidation states and thus could not be distinctly determined during XPS analysis. Some portion of the 396 °C peak could be due to the reduction of the lanthanum zirconates as seen in Fig. 5a. The high temperature peaks at 500 and 570 °C are primarily due to the reduction of the same lanthanum and zirconium species as seen in Fig. 5a.

There are some apparent differences in the oxidized species observed in the TPR results and the XPS results. This is because the fresh catalysts, after calcination at 1,000 °C for 8 h as the final step in the synthesis process, were pretreated prior to the TPR in flowing oxygen up to 950 °C, whereas the catalysts used for XPS were not pretreated, although they were also calcined at 1,000 °C for 8 h. It is clear from the above deconvoluted TPR peaks that no peak can be assigned solely to the reduction of a particular metal

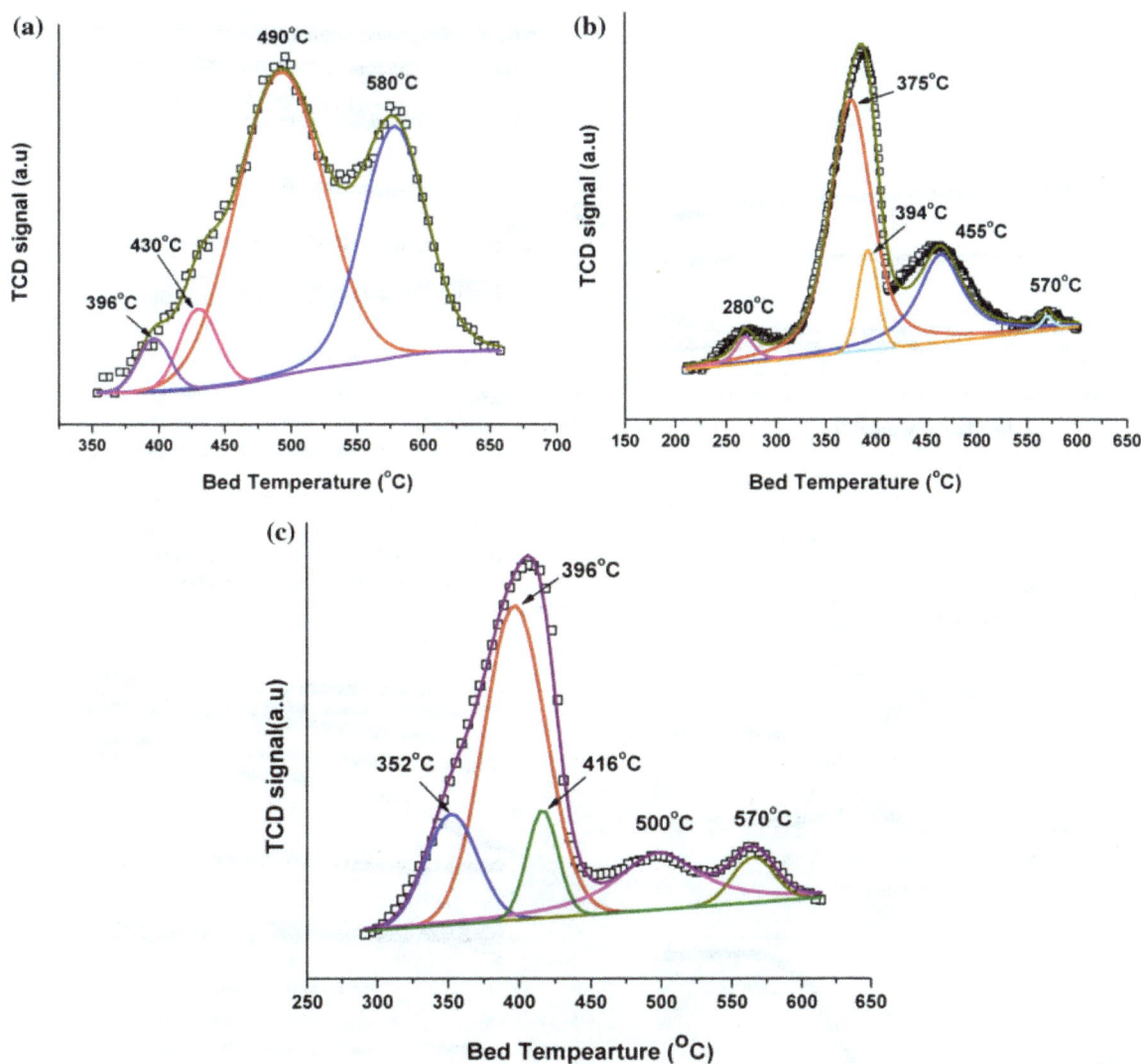

Fig. 5 Deconvoluted TPR profiles of freshly calcined **a** LZ, **b** L2RhZ, and **c** L5RhZ

species. Lanthanum zirconate reduction peaks overlap with the reduction peaks for Rh in the pyrochlore. Thus, quantification of the percentage of Rh reduction in these pyrochlores cannot be performed using TPR.

Temperature programmed surface reaction (TPSR)

TPSR was performed on LZ, L2RhZ, and L5RhZ and the product composition is plotted in Fig. 6a–c, respectively, as a function of temperature. For L2RhZ (Fig. 6b), there is no CO or H_2 formation observed until $\sim$490 °C, but over L5RhZ (Fig. 6c) product formation begins at $\sim$410 °C. Small but measurable water formation is observed over both the catalysts after light-off, which suggests that the RWGS takes place over both the catalysts up to $\sim$700 °C. Assuming the rate determining step is the breaking of the C–H bond in CH_4 to form CH_x surface species [32], and that this step occurs over Rh sites [33], the L5RhZ, which

has more surface Rh sites (shown by XPS and TPR), enhances CH_4 activation and accelerates the reforming reaction rate compared to L2RhZ. This faster reaction rate over L5RhZ results in a lower light-off temperature i.e., 410 °C, compared to 490 °C for L2RhZ.

Temperature effects on activity

Effect on CH_4 conversion (X_{CH_4})

Figure 7 shows the conversion of CH_4 (X_{CH_4}) for all three catalysts at each temperature as a function of time. The values on the right hand y-axis show the thermodynamic equilibrium values at that particular temperature. X_{CH_4} over LZ pyrochlores was between 0.5 and 0.8 %, and was constant at all temperatures. This lower conversion is due to absence of any catalytically active Rh site on the surface of the LZ pyrochlore.

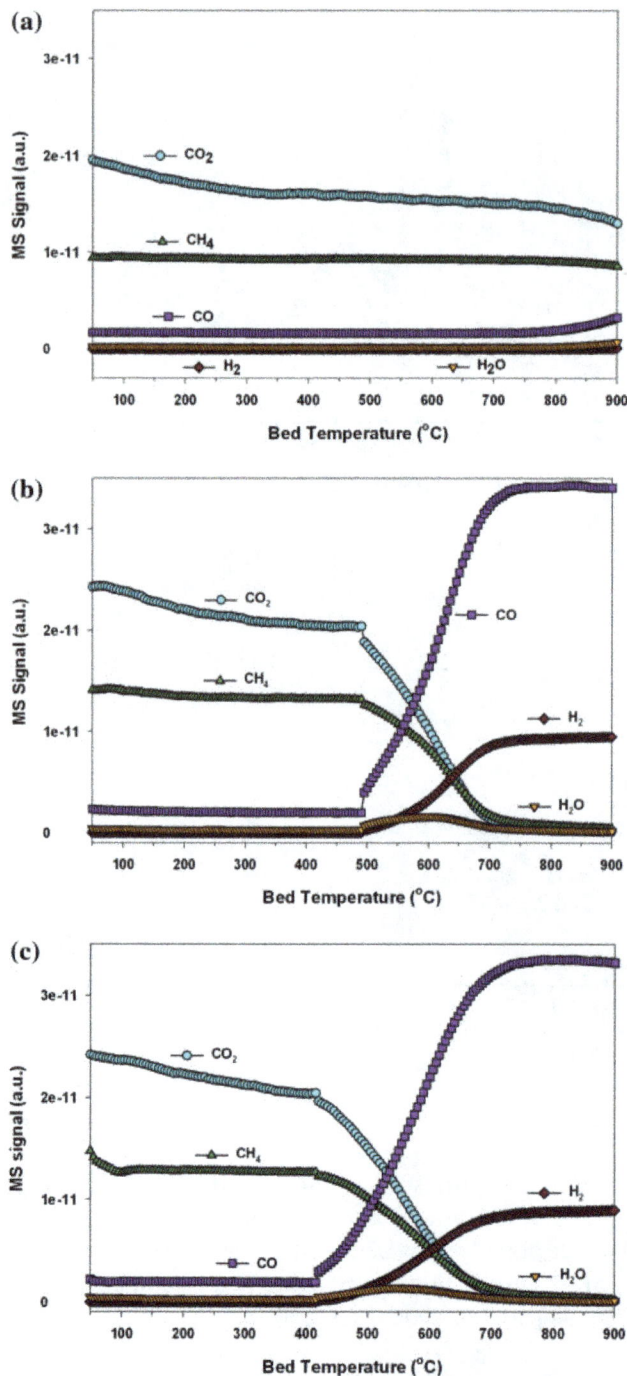

Fig. 6 TPSR plots for **a** LZ, **b** L2RhZ, and **c** L5RhZ at in the temperature range 50–900 °C at 1 atm and GHSV = 48,000 mL/g$_{cat}$/h

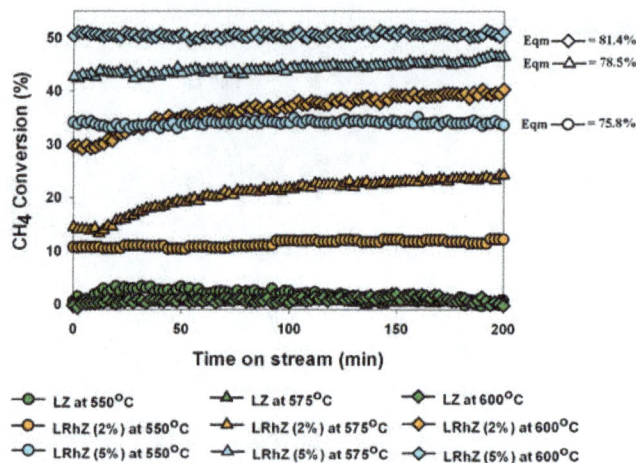

Fig. 7 CH$_4$ conversion for LZ, L2RhZ and L5RhZ pyrochlores at 550, 575, and 600 °C at 1 atm and GHSV = 48,000 mL/g$_{cat}$/h. The values on the *right* hand y-axis show the thermodynamic equilibrium values at that particular temperature as obtained from equilibrium calculations

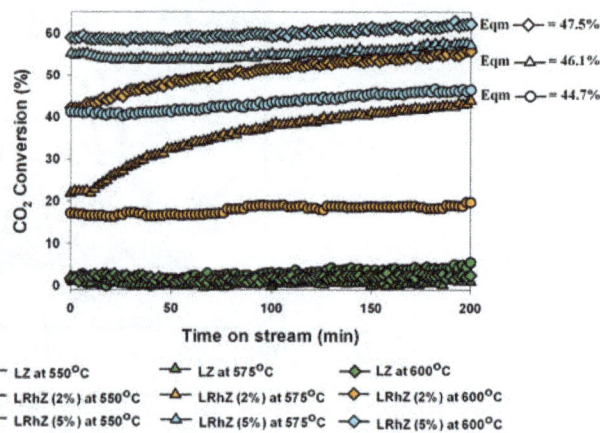

Fig. 8 CO$_2$ conversion for LZ, L2RhZ and L5RhZ pyrochlores at 550, 575, and 600 °C at 1 atm and GHSV = 48,000 mL/g$_{cat}$/h. The values on the *right* hand y-axis show the thermodynamic equilibrium values at that particular temperature as obtained from equilibrium calculations

The catalysts do not reach equilibrium at these temperatures. For the two Rh-containing catalysts, X$_{CH_4}$ increases with time. This increase could be attributed to the in situ reduction of the catalyst by CH$_4$. The catalysts used in this study were not reduced as a part of pre-treatment before conducting the reaction. As a result, the catalysts are likely initially reduced in situ by CH$_4$ then reduced by CO

and H$_2$ as they are formed [21]. However, this in situ reduction by CO and H$_2$ could be slower than that in H$_2$-TPR due to lower concentration of H$_2$ during reaction. Ashcroft et al. [21] proposed this in situ reduction in their study of DRM over Eu$_2$Ir$_2$O$_7$ pyrochlores[21]. In-situ reduction of the catalyst would increase the number of available active metal sites with time thus increasing the conversion of CH$_4$ with time-on-stream.

Iglesia and co-workers [32, 34], demonstrated that the rate of methane consumption on Rh/Al$_2$O$_3$ is first order in CH$_4$ concentration, and is independent of CO$_2$ concentration i.e., r$_{CH4}$ = kP$_{CH4}$. They also demonstrated that the active site for DRM is Rh site and the lack of any significant activity for the LZ catalyst shows that Rh sites are required to catalyze this reaction. The TPR and XPS results show that L5RhZ has more active Rh on the surface compared to L2RhZ and LZ. Thus, it would be expected that L5RhZ will have higher X_{CH_4} than L2RhZ and LZ, as shown in Fig. 7. This is consistent with the results of Verykios et al. [35], who showed that during DRM, breaking of CH_4 to CH_x ($x = 1$–3) on the Rh sites is the slow step in the reaction mechanism and determines the overall kinetics of the reaction over Rh/Al_2O_3. Thus, higher metal loading would kinetically favor the activation of methane and DRM.

Effect on CO_2 conversion (X_{CO_2})

The conversion of CO_2 (X_{CO_2}) as a function of time for these catalysts is shown in Fig. 8. The average X_{CO_2} for LZ was insignificant and independent of temperature. X_{CO_2} for L5RhZ is substantially greater than that for L2RhZ at all temperatures.

For L2RhZ, the experimental X_{CO_2} value at 550 °C is constant with time at ~ 18 %, which is substantially lower than the equilibrium value of 44.7 % (see Fig. 8). When the temperature is further increased to 575 and 600 °C, the experimental X_{CO_2} for L2RhZ increases with time, and reaches a value close to equilibrium at 575 °C and greater than equilibrium value of 47.5 % at 600 °C after 200 min on stream. For L5RhZ, the experimental X_{CO_2} increases slightly with time at all temperatures, and is consistently greater than the equilibrium values at all temperatures except 550 °C, where it is ~ 41 % versus equilibrium of 44.7 %. Equilibrium values of X_{CH_4} and X_{CO_2} were computed by considering $C_{(s)}$ as one product. Thermodynamically, $C_{(s)}$ formation is significant at these temperatures and the conversion of CO_2 is limited. It appears that carbon formation is kinetically limited on these catalysts, (as will be seen in the later H_2/CO ratio results) compared to DRM and the reverse water gas shift (RWGS), allowing X_{CO_2} to be greater than the thermodynamic equilibrium values calculated when $C_{(s)}$ is included in the calculation.

Previous studies show that CO_2 is activated by the support to form carbonate species and not by the active metal during DRM over conventional supported catalysts such as Rh/Al_2O_3 and Rh/La_2O_3 [33, 35–37]. Verykios and co-workers [35] while comparing Ni/La_2O_3 and Ni/Al_2O_3 catalysts for DRM showed that in the presence of La_2O_3; activation of CO_2 occurs via formation of $La_2O_2CO_3$. They proposed that the basic nature of La_2O_3 assists in the activation of CO_2 in the presence of surface CH_x species on the metal or the metal support interface. It can be postulated that, in our case, the lanthanum zirconate assists in the activation of CO_2 to form adsorbed carbonate species. The adsorbed carbonate species are then reduced to form

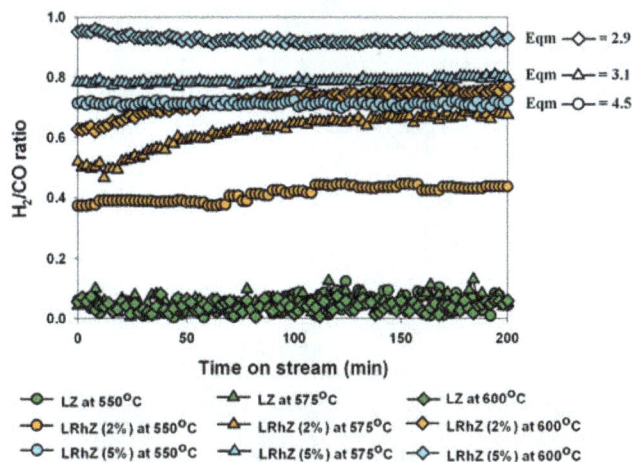

Fig. 9 H_2/CO ratio for LZ, L2RhZ and L5RhZ pyrochlores at 550, 575, and 600 °C at 1 atm and GHSV = 48,000 mL/g_{cat}/h. The values on the *right* hand y-axis show the thermodynamic equilibrium values at that particular temperature as obtained from equilibrium calculations

CO by the adsorbed CH_x species formed on the Rh sites [35, 36, 38]. In case of LZ pyrochlore, there is no activation of CH_4 molecule since there are no Rh sites, thus reduction of CO_2 is limited at all temperatures, consistent with the results in Fig. 8.

If CH_x species enhance the reduction of CO_2 to CO, it would be expected that X_{CO_2} for L5RhZ will be greater than L2RhZ. The experimental results are in agreement with this hypothesis; X_{CO_2} is greater for L5RhZ than for L2RhZ (Fig. 8).

Effect on H_2/CO ratio

The H_2/CO ratio plots for the three catalysts at 550, 575, and 600 °C are shown in Fig. 9. The H_2/CO ratio for LZ pyrochlore was found to be close to 0.05 at all bed temperatures which is negligible and was constant throughout the time on stream. The H_2/CO ratio for L5RhZ was considerably greater than for L2RhZ at all reaction temperatures (see Fig. 9).

The product stream consisted mainly of H_2 and CO, with a consistently lower H_2/CO ratio than equilibrium at all temperatures. As mentioned earlier, thermodynamic calculations show that equilibrium amounts of $C_{(s)}$ are significant at these conditions. High levels of carbon formation would limit the formation of CO and thus increases equilibrium H_2/CO ratios. However, it can be postulated that the rate of carbon formation on these catalysts is kinetically limited, favoring DRM and RWGS. DRM results in the H_2/CO ratio of unity but due to the simultaneous occurrence of RWGS, the H_2/CO ratio drops below equilibrium. Thus, comparing the H_2/CO ratio for L2RhZ and L5RhZ; the H_2/CO ratio of L5RhZ is

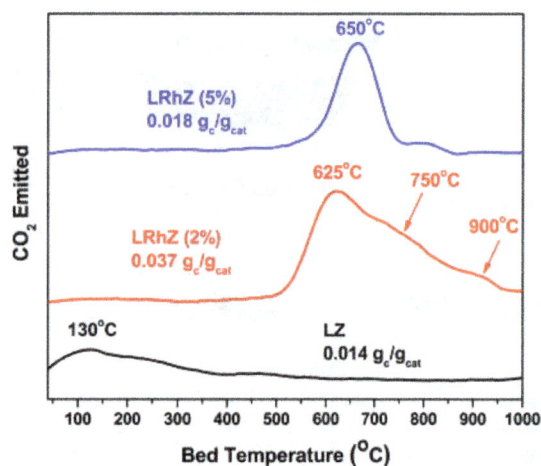

Fig. 10 TPO profile for the catalysts spent at 550 °C, GHSV = 48,000 mL/g$_{cat}$/h at 1 atm

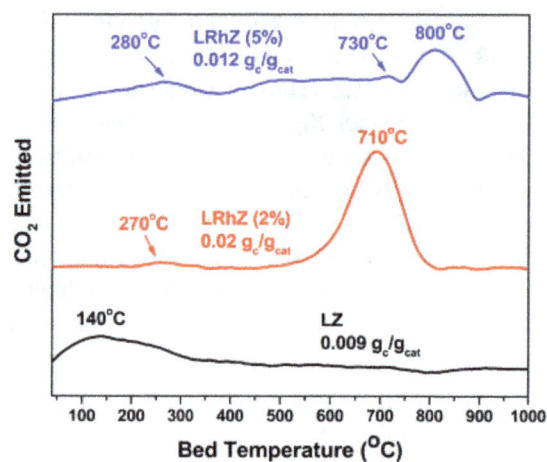

Fig. 11 TPO profile for the catalysts spent at 575 °C, GHSV = 48,000 mL/g$_{cat}$/h at 1 atm

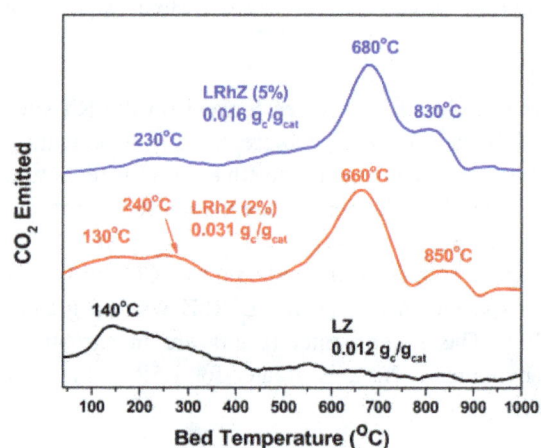

Fig. 12 TPO profile for the catalysts spent at 600 °C, GHSV = 48,000 mL/g$_{cat}$/h at 1 atm

Table 3 Summary of the carbon formed over the catalysts during the time on stream

Bed temperature (°C)	Carbon formed (g$_{carbon}$/g$_{catalyst}$)		
	LZ	L2RhZ	L5RhZ
550	0.014	0.037	0.018
575	0.012	0.031	0.016
600	0.009	0.02	0.012

CO_2 signal during TPO of spent catalysts tested at 550, 575, and 600 °C is plotted in Figs. 10, 11, and 12, respectively. The amount of carbon formed during the reaction was quantified and is summarized in Table 3. The total amount of carbon formed over L5RhZ is roughly half of the amount of carbon formed over L2RhZ in each experiment.

A general mechanism can be postulated based on activation of CH_4 molecule on metallic sites to produce adsorbed CH_x species ($x = 1$–3) [33]. These CH_x species can be further reduced to surface carbon on the metal sites, which can react with CO_2 (DRM), or form surface carbon, leading to deactivation [36]. The TPO results (Figs. 10, 11, 12) show that although there was carbon deposited on all three catalysts, there was no observable decrease in activity with time on stream, up to 200 min, likely due to the slow axial growth of the deactivated portion of the catalyst bed.

Catalyst spent at 550 °C TPO profile for the catalyst spent at 550 °C (Fig. 10) shows that for LZ, a single broad peak was observed at 130 °C. This peak at ∼ 100–130 °C is presumably due to the oxidation of the carbon with a relatively high H/C ratio [39]. For L2RhZ, single broad peak was observed at 625 °C. This broad peak overlaps other small peaks at higher temperatures of 750 and

consistently greater and closer to unity than L2RhZ. This suggests greater rate of DRM than RWGS over L5RhZ compared to L2RhZ. Thus, increasing Rh substitution helps in limiting simultaneous reactions like RWGS and favors DRM.

The H$_2$/CO ratio for L2RhZ increases with time, particularly at 575 and 600 °C suggesting increase in the rate of DRM compared to RWGS. A similar increasing trend was also observed for the conversion of CH_4 (see Fig. 7), suggesting that the rate of DRM reaction increases with that of the activation of CH_4.

Characterization of the spent catalyst

TPO (carbon burn-off) of the spent catalysts

Immediately after performing DRM over these three catalysts, the spent catalysts were subjected to in situ TPO. The

900 °C. This suggests that there are at least two more forms of carbon formed over L2RhZ during the reaction. The peak at 625 °C is attributed to the oxidation of dehydrogenated form of carbon deposited on or near the metal site [22]. The higher temperatures that are overlapped by the 625 °C peak could be due to the oxidation of less reactive carbon species or the graphitic form of carbon which could be present away from the Rh site [22, 40]. TPO profile for L5RhZ shows a single identifiable peak at 650 °C attributed to dehydrogenated form of carbon which is very similar to the 625 °C peak observed over L2RhZ. A shoulder was observed for L5RhZ at 830 °C. This shoulder is most likely due to the oxidation of the less reactive graphitic carbon which is similar to that observed for L2RhZ at higher temperatures. Peaks at 750 and 900 °C (for L2RhZ) are not observed for L5RhZ; this could be due to the greater surface coverage of Rh on L5RhZ compared to L2RZ.

Catalyst spent at 575 °C Figure 11 shows the TPO profile for the catalyst spent at 575 °C for the three catalysts. The unsubstituted LZ pyrochlore shows a single broad oxidation peak at 140 °C. This peak is most likely due to the oxidation of reactive hydrogenated polymeric carbon with greater H/C ratio compared to the high temperature carbon. This peak at 140 °C is similar to the one observed over LZ spent at 550 °C (see Fig. 10). There were four types of oxidation peaks observed for L2RhZ, indicating formation of a corresponding number of different species of surface carbon. The peak at 130 °C could be due to the reactive form of carbon with a high H/C ratio as seen on the LZ pyrochlore. The 240 °C peak could be attributed to carbon with H/C ratio lower than that corresponding to the peak at 140 °C. It could also be the same carbon as 140 °C but situated away from the surface metal [39]. Thus, the peak at 660 °C may be attributed to hydrogenated carbon deposited on the metal atom while the 850 °C peak could be due to the oxidation of highly unsaturated carbon deposited over the non-active sites (i.e., lanthanum zirconate in our case) or further away from the Rh site. Peaks observed for L5RhZ are at 230, 680 and 830 °C which are similar to the ones observed for L2RhZ. The species of the carbon oxidized at these temperatures would be qualitatively the same as those oxidized over L2RhZ.

Catalyst spent at 600 °C TPO profile for LZ pyrochlores spent at 600 °C (shown in Fig. 12) shows a peak at 140 °C which is similar to the peaks observed over LZ spent at 550 °C (see Fig. 10) and 575 °C (see Fig. 11). L2RhZ TPO profile shows a very small peak at 270 °C followed by a large peak at 710 °C. The peak at 270 °C is due to the oxidation of hydrogenated carbon which is qualitatively similar to the 240 °C peak for L2RhZ in Fig. 11. The peak

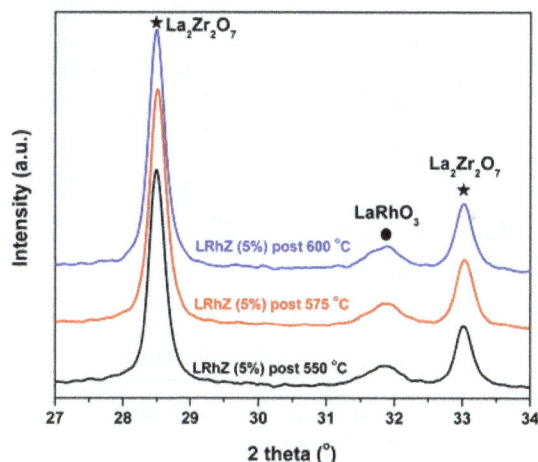

Fig. 13 Plot of the XRD pattern for the L5RhZ pyrochlores spent for DRM at 500, 575, and 600 °C with GHSV = 48,000 mL/g$_{cat}$/h at 1 atm followed by TPO showing in particular the formation of the LaRhO$_3$ perovskite phase

at 710 °C could be due to the dehydrogenation of the carbon formed further away from the active metal site which is oxidized at high temperatures. The graphitic nature of the carbon oxidized at 710 °C could be similar to the one observed in the superimposed peaks at 750, and 900 °C in Fig. 10 for L2RhZ. The TPO profile for L5RhZ has a small hump at 280 °C which is similar to the 270 °C peak for L2RhZ in the same plot. There is a very small peak observed at 730 °C which could be assigned to the same species as seen over the L2RhZ at 710 °C. A peak at 800 °C is observed for L5RhZ which was not seen over L2RhZ, this could be due to the deposition of dehydrogenated or graphitized carbon which is not in the proximity of the metallic site [40].

XRD of the spent catalyst from DRM and TPO

The diffraction pattern of the spent LZ and L2RhZ pyrochlores resembled the pattern for the fresh catalyst (thus not shown here). There was no apparent shift in the peaks for La$_2$Zr$_2$O$_7$ or formation of any perovskite phase observed for LZ and L2RhZ. This shows that LZ and L2RhZ pyrochlores maintained their structure (La$_2$Zr$_2$O$_7$) after catalyzing the reaction under reducing reforming condition at these temperatures.

However, for L5RhZ spent catalysts, there was a peak observed at about 32° which could be assigned to the formation of a separate perovskite (LaRhO$_3$) phase. The magnified image of the XRD pattern for 27°–34° is shown in Fig. 13; this plot shows clearly the formation of the perovskite peak at these reaction temperatures. This peak (LaRhO$_3$) was also observed in the fresh L5RhZ catalysts but it was not as prominent compared to the other La$_2$Zr$_2$O$_7$ peaks. After subjecting the catalysts to the reducing

(a)

(b)

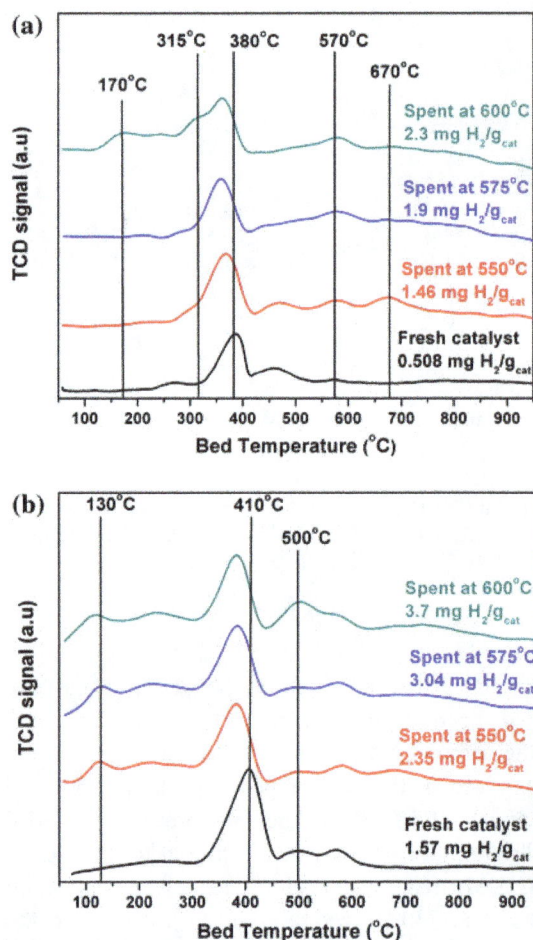

Fig. 14 Temperature programmed reduction by H_2 of the catalysts spent for DRM at different temperatures and TPO **a** L2RhZ, and **b** L5RhZ. Reduction conducted from 50 to 950 °C ramping at 5 °C/min

reaction conditions followed by oxidation, this particular peak for $LaRhO_3$ at 32° becomes apparent.

TPR by H_2 of the spent catalysts from DRM and TPO

To study the changes in the reducibility of LZ and of Rh in the L2RhZ and L5RhZ pyrochlores, TPR was conducted on each spent catalyst after (a) DRM at different temperatures and (b) TPO. The TPR plots obtained from these spent catalysts are compared to the plots of the freshly calcined catalysts and the plots for L2RhZ and L5RhZ in Fig. 14a, b, respectively.

A qualitative and quantitative change was observed in the reduction profiles of L2RhZ pyrochlore (see Fig. 14a). The fresh catalyst had a reduction peak at 380 °C which is shifted consistently to a lower temperature in the reduction profiles of the spent catalysts, suggesting a slight increase in the reducibility Rh in the pyrochlore structure. When the reaction temperature was 600 °C; there was a low temperature peak observed at 170 °C and a shoulder at 315 °C,

which was not seen in the other L2RhZ profiles. These peaks could be attributed to Rh that is less strongly bound to the pyrochlore structure which developed after DRM/TPO conditions. The quantitative increase in the H_2 consumption after DRM/TPO could be attributed partly to the reduction of the lanthanum zirconate at 570 °C and reduction of Rh at lower temperatures.

A comparison of the reduction profiles for fresh L5RhZ and those spent after DRM/TPO is shown in Fig. 14b. For the spent L5RhZ pyrochlores; a low temperature reduction peak was observed at 130 °C which was absent in the profile of fresh L5RhZ. In the TPR study by Haynes et al. [19] a reduction peak at 136 °C for Rh/Al_2O_3 was attributed to the reduction of supported Rh with weaker interaction with the support. This suggests that Rh that was substituted in the pyrochlore structure during calcination; apparently comes out of the structure to the surface of the pyrochlore as a result of the DRM/TPO reactions. This surface Rh is similar to the Rh observed on the supported Rh/Al_2O_3 catalysts in terms of the reducibility [41]. There was a continuous increase observed in the quantity of the H_2 consumed during reduction for L5RhZ (see Fig. 14b) as for L2RhZ (Fig. 14a). This increase in the H_2 consumption is partly due to the increase in the reduction of the lanthanum zirconate at 500 °C and partly due to the reduction of Rh metal that interacted less strongly with the pyrochlore structure.

The temperature of reduction of spent LZ did not change significantly as compared to the reduction of the fresh LZ. The TPR profiles of the fresh catalyst and the spent catalysts are not shown here due to the similarity between them and lack of any additional insight.

The H_2 consumption for the reduction of the spent catalyst is in direct proportion to the temperature at which the DRM reforming reaction was conducted. Because the DRM reaction conditions are extremely reducing, this may have caused some of the Rh to destabilize from the bulk of the crystal and diffuse to the surface of the catalyst. As the Rh loading increased, the maximum capacity of the pyrochlore structure for Rh at the B site was exceeded, causing Rh atoms to break the coordination with the neighboring La, Zr, Rh, and O atoms in the bulk and move to the surface and form weakly bonded Rh, with a reducibility comparable to supported Rh catalysts.

XRD of the spent catalysts from DRM followed by TPO and TPR

After conducting TPR on the spent L5RhZ pyrochlore, the changes in their crystalline structure were studied by conducting XRD over these catalysts. We are not aware of any paper in the literature discussing these series of experiments over pyrochlores for DRM. The diffraction pattern

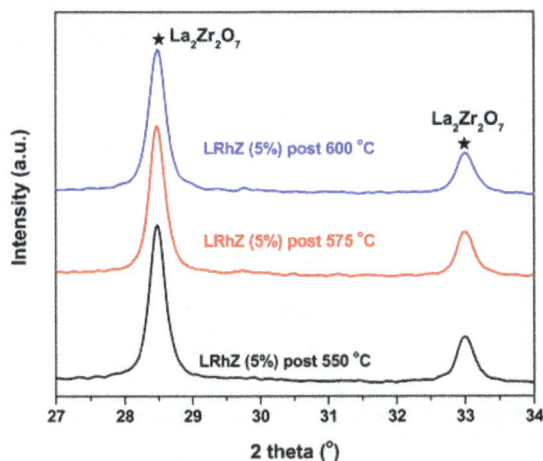

Fig. 15 Plot of the XRD pattern for the L5RhZ pyrochlores spent for DRM at 500, 575, and 600 °C with GHSV = 48,000 mL/g_{cat}/h at 1 atm followed by TPO and TPR up to 950 °C showing in particular that the LaRhO$_3$ perovskite peak vanishes after TPR

of these reduced L5RhZ pyrochlores did not show the presence of perovskite (LaRhO$_3$) phase (in Fig. 15). The observable perovskite phase (LaRhO$_3$) that was formed in the L5RhZ pyrochlore after the TPO of the spent catalysts, appears to be reduced by TPR to amorphous form which could not be detected by the X-rays during diffraction.

Conclusion

The XRD studies of the freshly calcined pyrochlores show a peak for LaRhO$_3$ which was observed for L5RhZ pyrochlore, possibly due to the higher Rh loading which led to separation of the excess Rh into the perovskite phase. XPS shows that rhodium is present primarily as Rh^{+3} species on the surface of L2RhZ and L5RhZ and that there was no major local surface aggregation due to higher concentration of Rh on the surface of L5RhZ. TPR results of the fresh catalysts show that the total reducibility of the pyrochlores (mg H$_2$ consumed/g_{cat}) increased with increasing Rh substitution. The conversion of CH$_4$ and CO$_2$ and the resultant product H$_2$/CO ratio over a series of substituted lanthanum zirconate pyrochlores increased with Rh loading and reaction temperature. The TPO of the catalysts after DRM shows that carbon formation decreases with an increase in Rh loading and increasing reaction temperature. The post reaction XRD plots show that there was no apparent change observed in the LZ and L2RhZ structures. However, for spent L5RhZ, the higher Rh loading could have caused the excess metal to separate out as a perovskite phase. When these spent L5RhZ pyrochlores were subjected to TPR after the TPO, the reduction temperature of the Rh was lower than that of the freshly calcined catalysts and there was also an observable

increase in the H$_2$ consumption. This was attributed to the diffusion of Rh metal from the bulk of the structure to the surface. However, the XRD pattern of the reduced spent L5RhZ pyrochlore did not show a perovskite phase, likely because the LaRhO$_3$ phase was reduced to some non-crystalline form. This result is novel and gives an insight into the behavior of Rh in the pyrochlore structure under alternating reducing (DRM and TPR) and oxidizing conditions. To our knowledge, this disappearance of perovskite (LaRhO$_3$) phase by alternative oxidation and reduction treatment has not been reported in the literature particularly for pyrochlores catalyzing DRM.

Acknowledgments This material is based upon work supported as part of the Center for Atomic Level Catalyst Design, an Energy Frontier Research Center funded by the U.S. Department of Energy, Office of Science, Office of Basic Energy Sciences under Award Number DE-SC0001058. We thank Ms. Kim Hutchison at North Carolina State University, Ms. Wanda LeBlanc at Louisiana State University and Dr. Nachal Subramanian at Georgia Tech (currently) for helping in getting the ICP, XRD, and XPS data, respectively.

References

1. Subramanian MA, Aravamudan G, Rao GVS (1983) Oxide pyrochlores—a review. Prog Solid State Chem 15(2):55–143.

2. Wilde PJ, Catlow CRA (1998) Defects and diffusion in pyrochlore structured oxides. Solid State Ionics 112(3–4):173–183.

3. Whittle KR, Cranswick LMD, Redfern SAT, Swainson IP, Lumpkin GR (2009) Lanthanum pyrochlores and the effect of yttrium addition in the systems La(2-x)Y(x)Zr(2)O(7) and La(2-x)Y(x)Hf(2)O(7). J Solid State Chem 182(3):442–450.

4. Schmalle HW, Williams T, Reller A, Linden A, Bednorz JG (1993) The twin structure of La$_2$Ti$_2$O$_7$—X-ray and transmission electron-microscopy studies. Acta Crystallogr Sect B Struct Commun 49:235–244.

5. Pakhare D, Haynes D, Shekhawat D, Spivey J (2012) Role of metal substitution in lanthanum zirconate pyrochlores (La$_2$Zr$_2$O$_7$) for dry (CO$_2$) reforming of methane (DRM). Appl Petrochem Res 2(1–2):27–35.

6. Iyer MV, Norcio LP, Kugler EL, Dadyburjor DB (2003) Kinetic modeling for methane reforming with carbon dioxide over a mixed-metal carbide catalyst. Ind Eng Chem Res 42(12):2712–2721.

7. Kumar N, Smith ML, Spivey JJ (2012) Characterization and testing of silica-supported cobalt-palladium catalysts for conversion of syngas to oxygenates. J Catal 289:218–226.

8. Bradford MCJ, Vannice MA (1996) Catalytic reforming of methane with carbon dioxide over nickel catalysts II. Reaction Appl Catal A 142(1):97–122.

9. Brungs AJ, York APE, Claridge JB, Marquez-Alvarez C, Green MLH (2000) Dry reforming of methane to synthesis gas over supported molybdenum carbide catalysts. Catal Lett 70(3–4): 117–122.

10. Coronel L, Munera JF, Lombardo EA, Cornaglia LM (2011) Pd based membrane reactor for ultra pure hydrogen production through the dry reforming of methane. Experimental and modeling studies. Appl Catal A 400(1–2):185–194.

11. de Arauj GC, de Lima SM, Assaf JM, Pena MA, Garcia Fierro JL, MdC Rangel (2008) Catalytic evaluation of perovskite-type oxide $LaNi_{1-x}Ru_xO_3$ in methane dry reforming. Catal Today 133:129–135.

12. Bradford MCJ, Vannice MA (1999) CO_2 reforming of CH_4. Catal Rev Sci Eng 41(1):1–42.

13. Subramanian ND, Gao J, Mo XH, Goodwin JG, Torres W, Spivey JJ (2010) La and/or V oxide promoted Rh/SiO2 catalysts: effect of temperature, H-2/CO ratio, space velocity, and pressure on ethanol selectivity from syngas. J Catal 272(2):204–209.

14. Schuurman Y, Mirodatos C, Ferreira-Aparicio P, Rodriguez-Ramos I, Guerrero-Ruiz A (2000) Bifunctional pathways in the carbon dioxide reforming of methane over MgO-promoted Ru/C catalysts. Catal Lett 66(1–2):33–37.

15. Ross JRH (2005) Natural gas reforming and CO_2 mitigation. Catal Today 100(1–2):151–158.

16. Gadalla AM, Bower B (1988) The role of catalyst support on the activity of nickel for reforming methane with CO_2. Chem Eng Sci 43(11):3049–3062.

17. Gallego GS, Batiot-Dupeyrat C, Barrault J, Florez E, Mondragon F (2008) Dry reforming of methane over LaNi1-yByO3 ± delta(B = Mg, Co) perovskites used as catalyst precursor. Appl Catal A 334(1–2):251–258.

18. Hou ZY, Zheng XM, Yashima T (2005) High coke-resistance of K-Ca-promoted Ni/alpha-Al_2O_3 catalyst for CH_4 reforming with CO_2. React Kinet Catal Lett 84(2):229–235

19. Haynes DJ, Berry DA, Shekhawat D, Spivey JJ (2008) Catalytic partial oxidation of n-tetradecane using pyrochlores: effect of Rh and Sr substitution. Catal Today 136(3–4):206–213.

20. Diaz-Guillen JA, Diaz-Guillen AR, Padmasree KP, Almanza JM, Fuentes AF, Santamaria J, Leon C (2008) Synthesis and electrical properties of the pyrochlore-type Gd(2-y)La(y)Zr(2)O(7) solid solution. Bol Soc Esp Ceram Vidr 47(3):159–164

21. Ashcroft AT, Cheetham AK, Jones RH, Natarajan S, Thomas JM, Waller D, Clark SM (1993) An insitu, energy-dispersive X-ray-diffraction study of natural-gas conversion by CO_2 reforming. J Phys Chem 97(13):3355–3358.

22. Haynes DJ, Campos A, Berry DA, Shekhawat D, Roy A, Spivey JJ (2010) Catalytic partial oxidation of a diesel surrogate fuel using an Ru-substituted pyrochlore. Catal Today 155(1–2):84–91.

23. Hayakawa I, Kamizono H (1993) Durability of an $La_2Zr_2O_7$ waste form containing various amounts of simulated HLW elements. J Nucl Mater 202(1–2):163–168.

24. Sheng PY, Chiu WW, Yee A, Morrison SJ, Idriss H (2007) Hydrogen production from ethanol over bimetallic Rh-M/CeO_2 (M = Pd or Pt). Catal Today 129(3–4):313–321.

25. Eriksson S, Rojas S, Boutonnet M, Fierro JLG (2007) Effect of Ce-doping on Rh/ZrO_2 catalysts for partial oxidation of methane. Appl Catal A 326(1):8–16.

26. Larichev YV, Netskina OV, Komova OV, Simagina VI (2010) Comparative XPS study of Rh/Al(2)O(3) and Rh/TiO(2) as catalysts for NaBH(4) hydrolysis. Int J Hydrogen Energy 35(13):6501–6507.

27. Wang Y, Song Z, Ma D, Luo HY, Liang DB, Bao XH (1999) Characterization of Rh-based catalysts with EPR, TPR, IR and XPS. J Mol Catal A Chem 149(1–2):51–61.

28. Sandell A, Libuda J, Bruhwiler PA, Andersson S, Maxwell AJ, Baumer M, Martensson N, Freund HJ (1996) Interaction of CO with Pd clusters supported on a thin alumina film. J Vac Sci Technol A Vac Surf Films 14(3):1546–1551.

29. Stoychev D, Valov I, Stefanov P, Atanasova G, Stoycheva M, Marinova T (2003) Electrochemical growth of thin La_2O_3 films on oxide and metal surfaces. Mater Sci Eng C Biomimetic Supramol Syst 23(1–2):123–128.

30. Wang HR, Chen YQ, Zhang QL, Zhu QC, Gong MC, Zhao M (2009) Catalytic methanol decomposition to carbon monoxide and hydrogen over Pd/CeO(2)-ZrO(2)-La(2)O(3) with different Ce/Zr molar ratios. J Nat Gas Chem 18(2):211–216.

31. Hoang DL, Lieske H (2000) Temperature-programmed reduction study of chromium oxide supported on zirconia and lanthana-zirconia. Thermochim Acta 345(1):93–99.

32. Wei JM, Iglesia E (2004) Isotopic and kinetic assessment of the mechanism of reactions of CH_4 with CO_2 or H_2O to form synthesis gas and carbon on nickel catalysts. J Catal 224(2):370–383.

33. Bitter JH, Seshan K, Lercher JA (2000) On the contribution of X-ray absorption spectroscopy to explore structure and activity relations of Pt/ZrO_2 catalysts for CO_2/CH_4 reforming. Top Catal 10(3–4):295–305.

34. Wei JM, Iglesia E (2004) Structural requirements and reaction pathways in methane activation and chemical conversion catalyzed by rhodium. J Catal 225(1):116–127.

35. Zhang ZL, Verykios XE (1996) Mechanistic aspects of carbon dioxide reforming of methane to synthesis gas over Ni catalysts. Catal Lett 38(3–4):175–179.

36. Fan M-S, Abdullah AZ, Bhatia S (2009) Catalytic technology for carbon dioxide reforming of methane to synthesis gas. ChemCatChem 1(2):192–208.

37. Nakamura J, Aikawa K, Sato K, Uchijima T (1994) Role of support in reforming of CH_4 with CO_2 over Rh catalysts. Catal Lett 25(3–4):265–270.

38. Zhang ZL, Tsipouriari VA, Efstathiou AM, Verykios XE (1996) Reforming of methane with carbon dioxide to synthesis gas over supported rhodium catalysts I. Effects of support and metal crystallite size on reaction activity and deactivation characteristics. J Catal 158(1):51–63.

39. Shamsi A, Baltrus JR, Spivey JJ (2005) Characterization of coke deposited on Pt/alumina catalyst during reforming of liquid hydrocarbons. Appl Catal A 293:145–152.

40. Barbier J (1986) Deactivation of reforming catalysts by coking— a review. Appl Catal 23(2):225–243.

41. Sarusi I, Fodor K, Baan K, Oszko A, Potari G, Erdohelyi A (2011) CO(2) reforming of CH(4) on doped Rh/Al(2)O(3) catalysts. Catal Today 171(1):132–139.

The environmental feasibility of algae biodiesel production

Tara Shirvani

Abstract Microalgae can grow in waste or seawater, have vastly superior biomass yields per hectare and, most importantly, the CO_2 removed from the atmosphere during photosynthetic growth of the plant offsets CO_2 released during fuel combustion. Algae-based fuel products are more promising than first-generation biofuels, as they exclude land use and food security issues, but require a mass production breakthrough to be viable. Through a life cycle approach, we evaluate whether algal biodiesel production can be a viable fuel source once the energy and carbon intensity of the process are managed accordingly. Currently, algae biodiesel production is 2.5 times as energy intensive as conventional diesel. Biodiesel from advanced biomass can only realize its inherent environmental advantages of GHG emissions reduction once every step of the production chain is fully optimized and decarbonized. In the case of Saudi Arabia which operates on a 100 % fossil-based electricity and heat grid, the inherent environmental advantages of producing algae biodiesel would be heavily overshadowed by the nation's carbon-intensive energy and power sector.

Keywords Algae biodiesel · Lifecycle analysis · Carbon footprint · Cumulative energy demand · Energy security · Renewable energy

T. Shirvani (✉)
Smith School of Enterprise and the Environment,
University of Oxford, Oxford, UK
e-mail: tara.shirvani@smithschool.ox.ac.uk

T. Shirvani
Department of Inorganic Chemistry, University of Oxford,
Oxford, UK

The search for alternative fuels for the transport industry has revived the interest in biofuels from sustainable cultivation and feedstock which is not in competition with food or animal feed [1]. While biofuels remain a viable alternative to fossil fuels, a full replacement of our 90 % hydrocarbon-based transport industry is unlikely in the near to mid-term future [2–8]. With the emphasis shifting towards the development of advanced biofuels, microalgae has proven to be a promising feedstock to overcome land use and food security issues while growing in waste or seawater [7, 9–11]. The oil-rich algae biomass, with its superior production yields [12–14] has attracted considerable attention as potential domestic and renewable substitute for imported fossil fuels. Given microalgae's high production yields, the required global land mass necessary to satisfy global fossil fuel consumption could be considerably reduced. For algae-derived biodiesel with a yield of 850 GJ/ha/year, to replace the total production of 748 million tons of petroleum-derived diesel in 2009, a land mass of around 57.3 million ha would be required. This land size approximated to an area somewhat smaller than Texas [7]. The inherent potential advantage of biodiesel production from algae is lower lifecycle Greenhouse Gases (GHG) emissions, as algae biomass converts atmospheric CO_2 through photosynthesis into bio-plant material which is eventually released back to the atmosphere via micro-organisms when used as a fuel, via engine tail pipe emissions [10, 15, 16]. Fossil fuel combustion releases additional carbon which took million of years to be removed from the atmosphere [17]. Moreover, compared to cultivation requirements for other advanced biofuel sources microalgae growth mainly requires solar radiation, carbon dioxide, water and nutrients in the form of inorganic salts [18]. Given an existing base of 50,000 species of known microalgae, only a fraction is appropriate for biodiesel

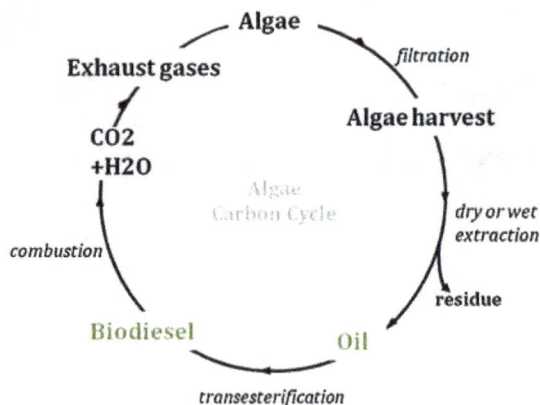

Fig. 1 Algae to biodiesel carbon cycle [7]

production, due to algae strains' varying lipid content and productivity levels [10, 19]. On average algal lipid content varies between 20 and 50 % by weight of dry biomass, although some strains can under certain optimally induced conditions accumulate as much as 90 % oil yield ratios [18, 20, 21].

Microalgae is cultivated in either open raceway ponds or closed photobioreactors (PBR), each of which has been designed in a variety of operating configurations [10]. Both farming approaches are still held back by substantial technical and economic hurdles. Although photobioreactors benefit from higher productivity rates, lower evaporative water losses or likelihood of culture collapse, the farming approach is still challenged by its significantly larger capital investment and operating cost base [9, 22]. While open pond algae cultivation is operated on a

60–100 % lower cost level, the technology is held back from its large-scale commercial breakthrough by problems of poor light utilization efficiency, lower volumetric productivity and higher risk of culture contamination or collapse [10, 19, 22, 23]. However, since this industry is not mature, there is ample space for optimization and we expect the future rise in oil prices to add to this effect.

The main stages of the algae biodiesel production process consist of algae farming, biomass harvesting, oil extraction, transesterification of algae oil, fuel distribution and combustion, see Fig. 1.

As part of our LCA study, microalgae batches are cultivated in open pond farming installations, harvested and dried in subsequent stages to generate 75 tonnes/ha/year of dry algae biomass. The extracted 30 % share of algae lipids (22.5 tonnes/ha/year) is further transesterified to yield 850 GJ/ha/year of biodiesel. 89 GJ/ha/year of glycerol is generated as a by-product and is exported as animal feed. The remaining 70 % residual algae biomass (689 GJ/ha/year) is used in various co-product utilization methods to offset the production process' substantial energy requirements. When we consider all energy inputs of the biodiesel production cycle as 100 % fossil fuel sourced, the production process is 2.5 times as energy intensive as conventional diesel from the United States and nearly equivalent to the high fuel-cycle energy use of oil shale diesel, see Fig. 2. The major disadvantage inherent in biodiesel production from microalgae is driven by the high energy input in the form of heat and electricity.

Biodiesel from algae biomass can only realize its inherent environmental advantages of GHG emissions

Fig. 2 Benchmarking the life cycle energy requirements of fossil fuels against algae biodiesel [7]

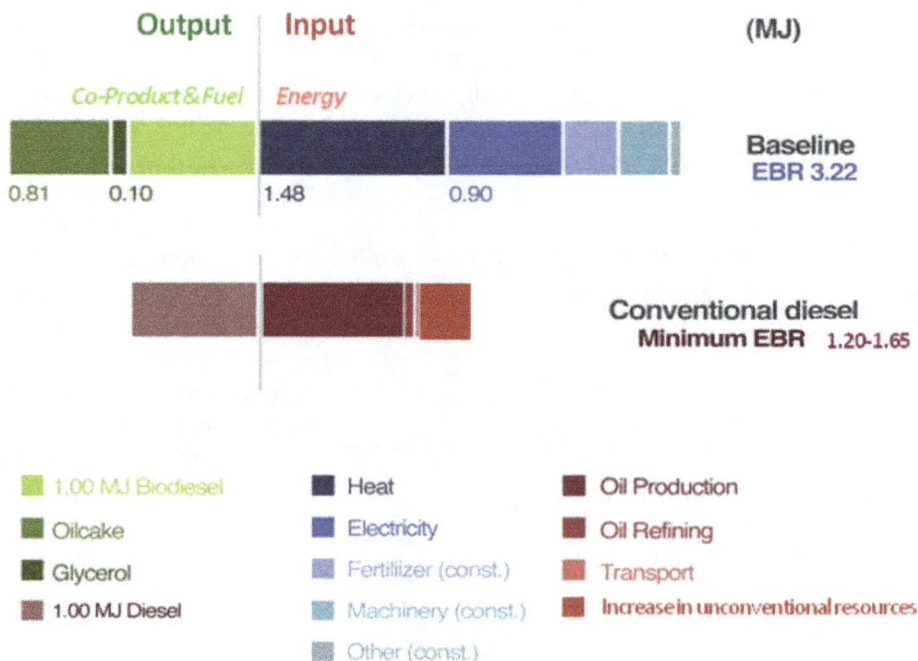

reduction, once every step of the production chain is fully optimized and decarbonized. This will entail the sourcing of all direct energy input in the form of heat and electricity, as well as indirect requirements for transport and building materials, from low-carbon energy sources. Moreover, the offset of carbon-intensive fertilizer requirements through the recycling of wastewater will be critical. Recent studies suggest that without the recycling of harvest water, the algae-to-biodiesel water footprint is as high as 3,726 kg water/kg biodiesel [11]. Further improvements in the production of cycle's carbon footprint can be achieved through the commercialization of new oil extraction technologies. By eliminating the need to dry algae biomass to a 90 % solid content [13] required for the subsequent oil extraction in vegetable oil mills, considerable energy savings could be achieved.

As a priority, countries will need to defossilize primary energy sources used by their electricity grids, as only then can the transport sector move towards low GHG emissions. In the case of Saudi Arabia which operates on a 100 % fossil-based electricity and heat grid, the inherent environmental advantages of producing algae biodiesel would be heavily overshadowed by the nation's carbon intensive energy and power sector. In contrary, Brazil and France, which essentially operate on a defossilized electricity grid, have the potential for biodiesel from algae to be a viable alternative to conventional diesel [7]. Ultimately with the transport fuel industry remaining to be fuelled by hydrocarbons for the foreseeable future, biofuels will need to contribute as well as complement our current fuel mix as part of a larger mission to diversify our global transport system within the mid-to long-term future.

References

1. Gallagher E (2008) Renewable fuels agency. East Sussex, UK
2. Alekkett K (2007) OECD/ITF Joint Transport Research Centre. Discussion Papers
3. Campbell JE, Lobell DB, Genova RC, Field CB (2008) The global potential of bioenergy on abandoned agriculture lands. Environ Sci Technol 42:5791–5794
4. Owen NA, Inderwildi OR, King DA (2010) The status of conventional world oil reserves—Hype or cause for concern? Energy Policy 38:4743–4749
5. Sperling D, Gordon D (2009) Two billion cars: driving toward sustainability. Oxford University Press, USA
6. Wells J (2007) Crude oil: uncertainty about future oil supply makes it important to develop a strategy for addressing a peak and decline in oil production. DIANE Publishing, Darby
7. Shirvani T, Yan X, Inderwildi OR, Edwards PP, King DA (2011) Life cycle energy and greenhouse gas analysis for algae-derived biodiesel. Energy Environ Sci 4:3773–3778
8. King D, Inderwildi O (2010) Future of mobility roadmaps, smith school of enterprise and the environment. University of Oxford, Oxford
9. Jorquera O, Kiperstok A, Sales EA, Embiruçu M, Ghirardi ML (2010) Comparative energy life-cycle analyses of microalgal biomass production in open ponds and photobioreactors. Bioresour Technol 101:1406–1413
10. Mata TM, Martins AA, Caetano N (2009) Microalgae for biodiesel production and other applications: a review. Renew Sustain Energy Rev 14(1):217–232
11. Yang J, Xu M, Zhang X, Hu Q, Sommerfeld M, Chen Y (2011) Life-cycle analysis on biodiesel production from microalgae: water footprint and nutrients balance. Bioresour Technol 102:159–165
12. Clarens AF, Resurreccion EP, White MA, Colosi LM (2010) Environ Sci Technol 44:1813–1819
13. Lardon L, He lias A, Sialve B, Steyer JP, Bernard O (2009) Life-cycle assessment of biodiesel production from microalgae. Environ Sci Technol 43:6475–6481
14. Schenk PM, Thomas-Hall SR, Stephens E, Marx UC, Mussgnug JH, Posten C, Kruse O, Hankamer B (2008) Bioenergy Res 1:20–43
15. Pickett J, Anderson D, Bowles D, Bridgwater T, Jarvis P, Mortimer N, Poliakoff M, Woods J (2008) Sustainable biofuels: prospects and challenges. The Royal Society, London
16. Wang MQ (1999) GREET 1.5-transportation fuel-cycle model: methodology, development, use, and results, ANL/ESD-39, vol 1. Argonne National Lab, IL
17. Sheehan J, Camobreco V, Duffield J, Graboski M, Shapouri H (1998) Life cycle inventory of biodiesel and petroleum diesel for use in an urban bus. Final report, National Renewable Energy Lab., Golden
18. Chisti Y (2007) Biotechnol Adv 25:294–306
19. Richmond A (2007) Handbook of microalgal culture: biotechnology and applied phycology. Blackwell, Oxford
20. Li Y, Horsman M, Wang B, Wu N, Lan CQ (2008) Effects of nitrogen sources on cell growth and lipid accumulation of green alga Neochloris oleoabundans. Appl Microbiol Biotechnol 81:629–636
21. Li Y, Horsman M, Wu N, Lan CQ, Dubois-Calero N (2008) Biofuels from microalgae. Biotechnol Prog 24:815–820
22. Pulz O (2001) Photobioreactors: production systems for phototrophic microorganisms. Appl Microbiol Biotechnol 57:287–293
23. Carvalho A, Malcata FX (2006) Microalgal reactors: a review of enclosed system designs and performances. Biotechnol Prog 22:1490–1506

5

Characterising carbon deposited during pyrolysis gasoline hydrogenation: enhanced TPO methodologies

Javed Ali · S. David Jackson

Abstract In this paper, the use of temperature-programmed oxidation (TPO) is highlighted as a means of interrogating the surface of catalysts. Most studies only concern themselves with the active portion of the catalyst bed; in this study, we examine both the active part and the part that experiences equilibrated gas after reaction. The extent and nature of carbonaceous deposits on both sections of catalyst have been characterised by TPO and reveal that there is significant deposition of low H:C ratio carbonaceous species on catalyst that has only seen equilibrated reaction gas. This has potential implications for the life of the catalyst bed and the activity and selectivity observed.

Keywords PyGas · Hydrogenation · Temperature programmed oxidation · Carbon deposition

Introduction

Pyrolysis gasoline (PyGas) is a by-product of high temperature naphtha cracking during ethylene and propylene production [1, 2]. It is a mixture of highly unsaturated hydrocarbons ranging from C_5 to C_{12} and contains considerable amounts of aromatics, typically 40–80 % (benzene, toluene and xylene), together with paraffins, olefins and diolefins [3]. Composition of the PyGas produced depends on the feedstock and operating conditions and varies from plant to plant; however, a typical PyGas composition is given in Table 1 [8].

The main reason for the deactivation of catalysts during PyGas hydrogenation is the deposition of carbonaceous residue (coke) on the surface of the catalyst [10, 11]. The nature of the coke depends on the process conditions and length of reaction with the carbonaceous residues deposited, varying from hydrocarbons and poly-aromatics to graphitic coke. The deposition of coke decreases accessibility for the reactants to the catalyst surface and can block pores. Continued heavy build up of coke in pores may result in fracturing of the support material which can cause plugging of reactor voids [7, 13].

Carbonaceous residue formation is mainly due to the polymerisation of coke precursors present in PyGas such as styrene, di-olefins, mono-olefins, cyclo-olefins and aromatic species [14]. The deposited residue can convert to polyaromatic hydrocarbons, which then further convert to highly condensed polyaromatics and graphitic type coke. This is summarised in Fig. 1 [7, 9].

The nature of the carbonaceous residues also changes with reaction time. The change in the nature of the coke is due to polymerisation, multi-side reactions and loss of hydrogen content. The deposits convert to more condensed and hydrogen deficient coke with the passage of time and then finally form graphitic type coke.

The deposition of coke is highly dependent on the reaction conditions, especially on reaction temperature and hydrogen partial pressure. Increases in the reaction temperature of PyGas hydrogenation not only increase the amount of coke deposition but also result in a more condensed hydrogen deficient coke on the surface. Carbon laydown can be decreased by increasing hydrogen partial pressure [1].

Temperature-programmed oxidation (TPO) is a simple and important technique for the investigation into the quantity and nature of coke deposited [5, 6]. The amount of

J. Ali · S. D. Jackson (✉)
Department of Chemistry, Centre for Catalysis Research,
WestCHEM, University of Glasgow, G12 8QQ Glasgow,
Scotland, UK
e-mail: david.jackson@glasgow.ac.uk

Table 1 Typical composition of PyGas

PyGas components	Weight percent (wt%)
Benzene, toluene and xylenes	50
Olefins and dienes	25
Styrene and other aromatics	15
Paraffins and naphthenes	10

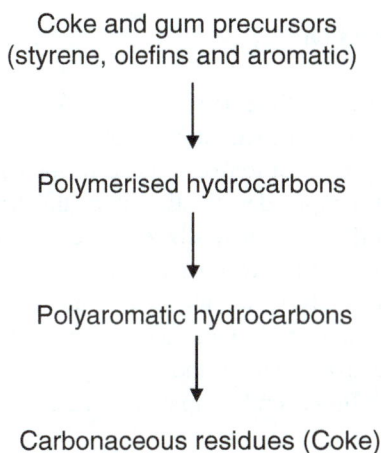

Table 2 Catalyst specifications

Catalyst	Metal loading (%wt)	Metal dispersion (%)	Surface area ($m^2 \, g^{-1}$)	Pore volume ($cm^3 \, g^{-1}$)
Ni/Al$_2$O$_3$	16.0	18.0	106	0.39

Coke and gum precursors
(styrene, olefins and aromatic)

↓

Polymerised hydrocarbons

↓

Polyaromatic hydrocarbons

↓

Carbonaceous residues (Coke)

Fig. 1 Formation of carbonaceous residues (Coke)

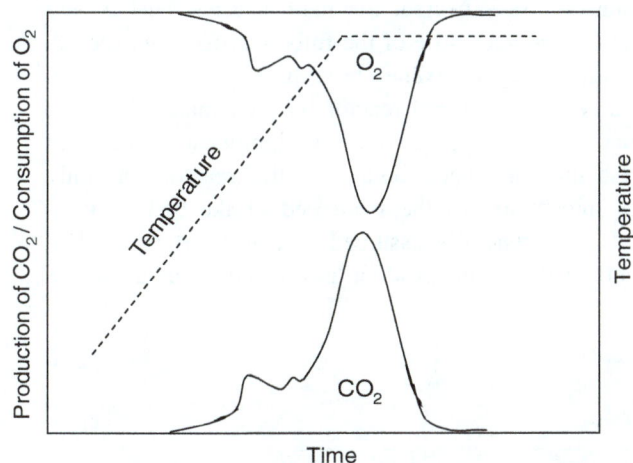

Fig. 2 Typical temperature programme oxidation profile [5, 6]

oxygen consumed and the amount of CO_2/CO produced indicate the amount of coke deposited on the surface of the catalyst. The oxygen consumption is commonly mirrored by the production of CO_2, as shown in Fig. 2.

The combustion of hydrogen rich coke, which is mostly present on the catalytic metal, occurs first. This is followed by the combustion of the hydrogen deficient coke present on the surface of the catalyst [4, 12]. The ratio of CO_2/H_2O and/or CO/H_2O can give insight into the carbon to hydrogen ratio present in the coke. A decrease in hydrogen

to carbon ratio represents more condensed polyaromatic type of coke [6].

In this paper, we use TPO to probe the carbon deposit in both the reaction zone and the rest of the catalyst bed that is subjected to equilibrated gas in the typical fixed bed configuration found in the most chemical plants.

Experimental

The Ni/Al$_2$O$_3$ (HTC 400) catalyst used in this project was provided by Johnson Matthey. The characterisation data of catalyst provided by catalyst manufacturer are shown in Table 2.

Hydrogenation of the synthetic pyrolysis gasoline (PyGas) was investigated over the Ni/Al$_2$O$_3$ catalyst in a fixed-bed, high-pressure, continuous flow reactor. The catalyst (particle size between 250 and 425 microns) was added to the reactor tube and positioned in the middle of reactor tube. The catalyst was reduced in situ with a flow of hydrogen gas. The Ni/Al$_2$O$_3$ catalyst was reduced for 16 h at a temperature of 723 K. Following this, the reactor temperature was decreased to the required reaction temperature (413 K). The reactor was then pressurised to the desired pressure (20 barg). Once the reactor had obtained the desired conditions, the reaction was started with a particular flow of PyGas feed of 0.042 ml min^{-1} (WHSV$_{PyGas}$ 8 h^{-1}) and a flow of H$_2$/N$_2$ gas (10 ml min^{-1}). The PyGas was vaporised and mixed with hydrogen and then passed through the catalyst bed. The hydrogenated PyGas was condensed in a knockout pot, which was chilled to 273 K. Liquid samples were collected every 2 h during daytime and analysed using a Thermo Finnigan Focus G.C. The hydrogenated PyGas mainly condensed in the knockout pot; however, very small amounts of the reaction products remained in the gaseous phase and, therefore, the effluent gas was also analysed at regular intervals by online Agilent G.C. All reactions were run for 76 h. After finishing each reaction, nitrogen gas was passed through the catalyst over night so as to remove any species present in the reactor. The reactor was then cooled to room temperature in the flow of nitrogen gas and then an in situ TPO was performed on the post reaction catalyst. The gas was switched to a flow of 2 % O$_2$/Ar (40 ml min^{-1}) and the temperature increased to 723 K at a ramp of 5 K min^{-1} and then the reactor temperature was held constant at 723 K until all evolution of gases other

Table 3 Weight percent (%) composition of PyGas

Components	1-pentene	Cyclopentene	1-octene	Toluene	Styrene	Heptane	Decane
PyGas wt%	10.0	10.0	10.0	55.0	10.0	2.5	2.5

than O_2 and Ar ceased. The TPO process was monitored by an online mass spectrometer (QMS) in which the ion current of main species of mass fragments with m/z 28 (CO), 44 (CO_2), 32 (O_2), 18 (H_2O) was monitored. Other species with m/z 68 (cyclopentene), 70 (1-pentene), 72 (pentane), 78 (benzene), 92 (toluene), 98 (methylcyclohexane), 104 (styrene), 106 (ethylcyclo-hexane), 112 (1-octene), 114 (octane), 15 (CH_3), 29 (C_2H_5) and 2 (H_2) were also followed for detailed investigation of deposited coke.

A synthetic PyGas was used in this study and the composition is shown in Table 3. No benzene was used in this mix due to health and safety concerns: toluene was used as a replacement. Heptane and decane were used as internal standards for GC analysis.

Results and discussion

During the course of the time on stream (TOS), the catalyst deactivated. A typical deactivation profile is shown in Fig. 3 with the hydrogenation of toluene to methylcyclohexane. There is a rapid deactivation over the first 24 h followed by a period of much slower deactivation. This behaviour was observed with all reactants. At the end of the 76-h TOS, the catalyst was subjected to an in situ TPO. The evolution of carbon dioxide and water is shown in Fig. 4, while the evolution of more complex species is shown in Fig. 5. In general, aliphatic species were not observed.

These graphs show that there has been considerable carbon deposition and that species such as benzene, which was not present in the feed, can be formed on the surface of the catalyst. As expected, the main species detected are aromatics, and very little of the fully hydrogenated species such as methylcyclohexane are seen.

In a typical fixed bed reactor on a chemical plant, the product flow will pass over a significant amount of catalyst that has been added to the reactor but only comes into action as the front end of the bed is deactivated. It is generally assumed that this part of the bed is unaffected by the product gases and is in a similar

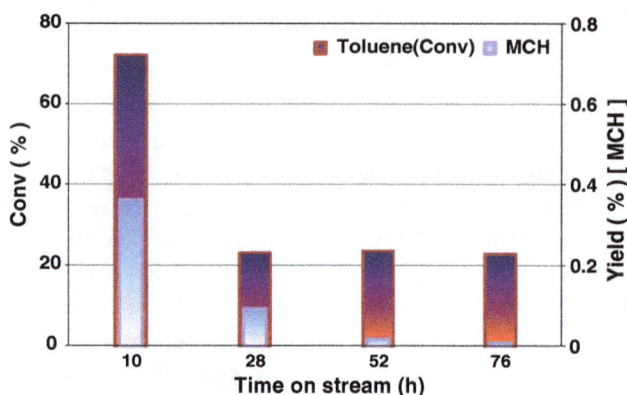

Fig. 3 Deactivation profile for toluene hydrogenation (component part of PyGas). Conditions: 413 K, WHSV 8 h^{-1}, H_2 pressure 5 barg, H_2:PyGas molar ratio = 5, overall pressure 20 barg

Fig. 4 Evolution of water and carbon dioxide as a function of temperature during TPO

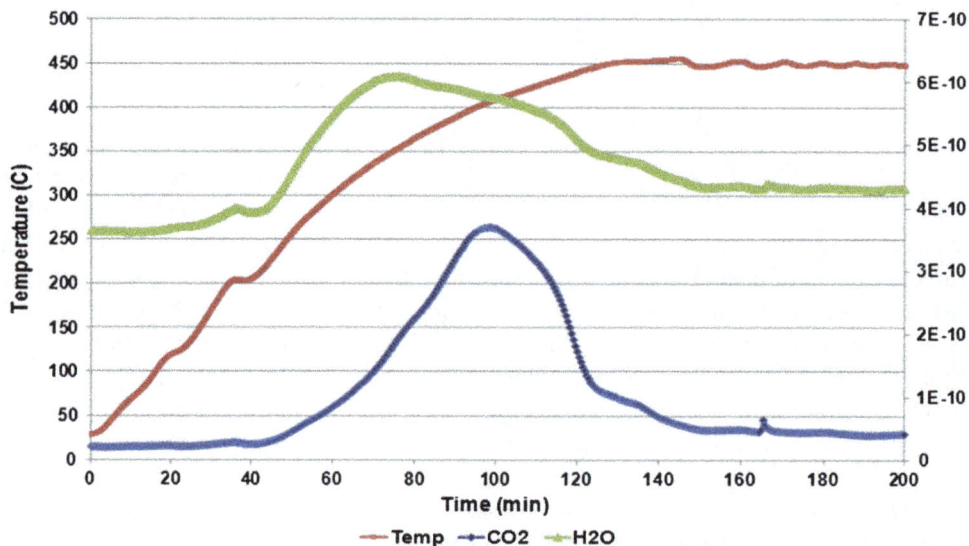

Fig. 5 Evolution of complex species during TPO

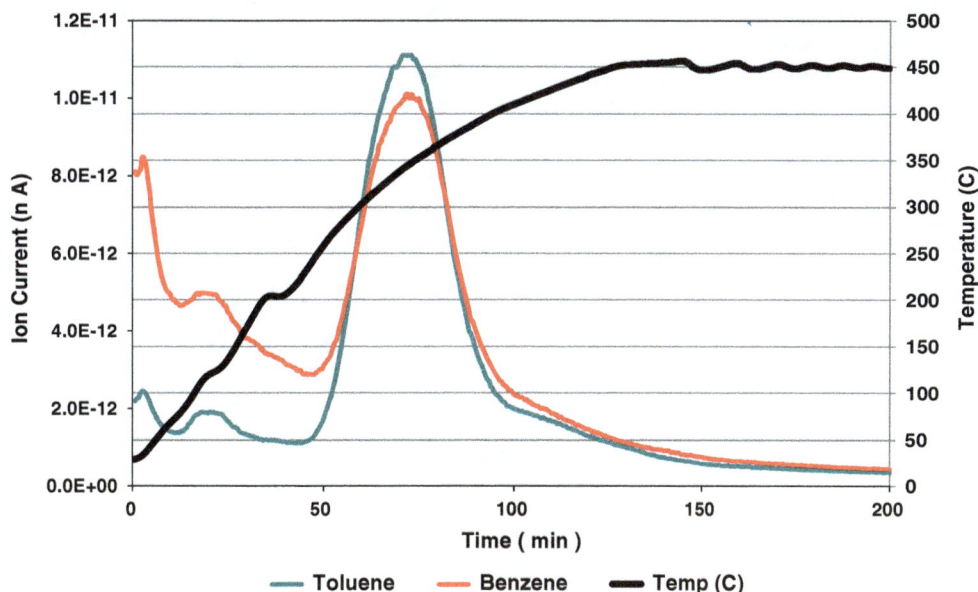

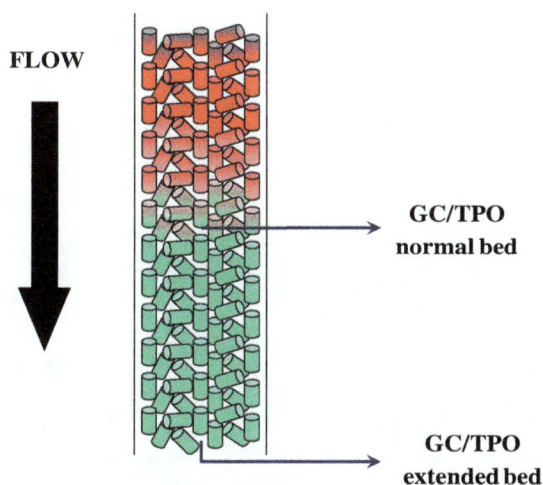

Fig. 6 Schematic of full-bed and half-bed reactor tests

Table 4 Conversion of PyGas components at normal-bed and extended-bed positions

Conversion (%)	Extended bed	Normal bed
1-pentene	90.6	87.3
1-octene	98.3	99.3
Cyclopentene	93.8	93.1
Toluene	24.1	23.1
Styrene	99.9	99.9

Conditions: 413 K, 20 barg pressure, 5 barg hydrogen, H_2:PyGas molar ratio = 5

state to fresh catalyst. To test this, the bed of catalyst was extended to double its length and used under the same conditions of flow, pressure and temperature. In this way, it would be possible to generate both GC and TPO data for both the normal bed and the extended bed (Fig. 6).

GC analysis was taken at the bottom of the normal bed and the extended bed to confirm whether the conversion was the same. The results are shown in Table 4 and confirm that the extent of hydrogenation changes very little between the normal-bed position and the extended-bed position as would be expected. In both cases, the catalyst was subjected to the same conditions and was on stream for the same length of time (76 h). TPOs were then run on both the normal and the extended bed. The

normal-bed profiles are shown in Figs. 4 and 5. The extended-bed profiles are shown in Figs. 7 and 8. It is immediately clear that the profiles for the carbon dioxide and water have significantly altered. In Fig. 9, both the carbon dioxide evolution profiles for the normal and extended bed are compared. The carbon dioxide profile for the extended bed reveals that there has been considerable carbon deposition on the "unreactive" part of the bed and this is especially noticeable at the higher temperatures. By 450 °C, carbon dioxide evolution has almost ceased with the normal bed, whereas there is still significant amount of carbon combusting with the extended bed. The water profile is also different. The water evolution from the normal bed covers the same temperature range as the carbon dioxide, although the specific shape is different indicating a variety of carbonaceous species on the surface with different H:C ratios. However, the water profile from the extended bed does not mirror the temperature range of the carbon dioxide. In this case, the water evolution peaks at 350 °C and is almost back to

Fig. 7 Evolution of water and carbon dioxide as a function of temperature during TPO with an extended bed

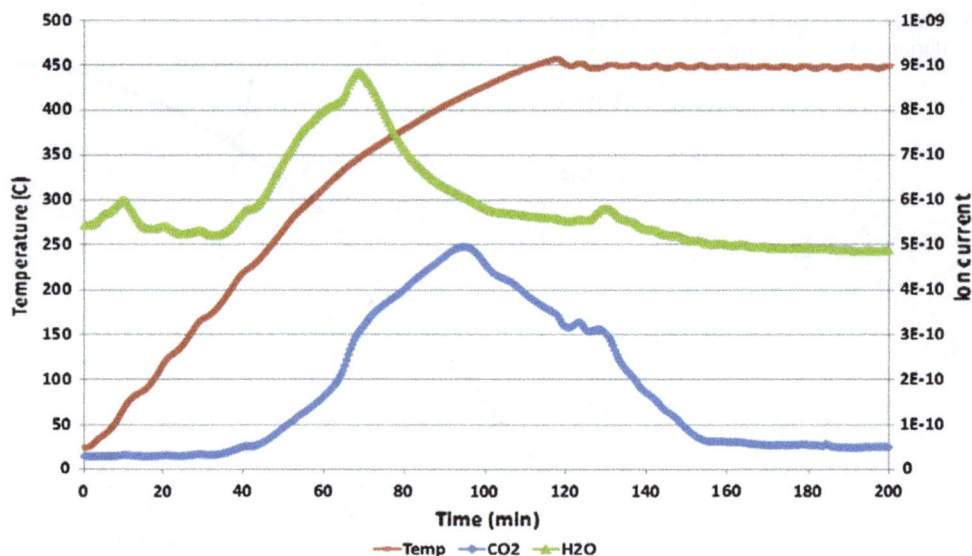

Fig. 8 Evolution of complex species during TPO with an extended bed

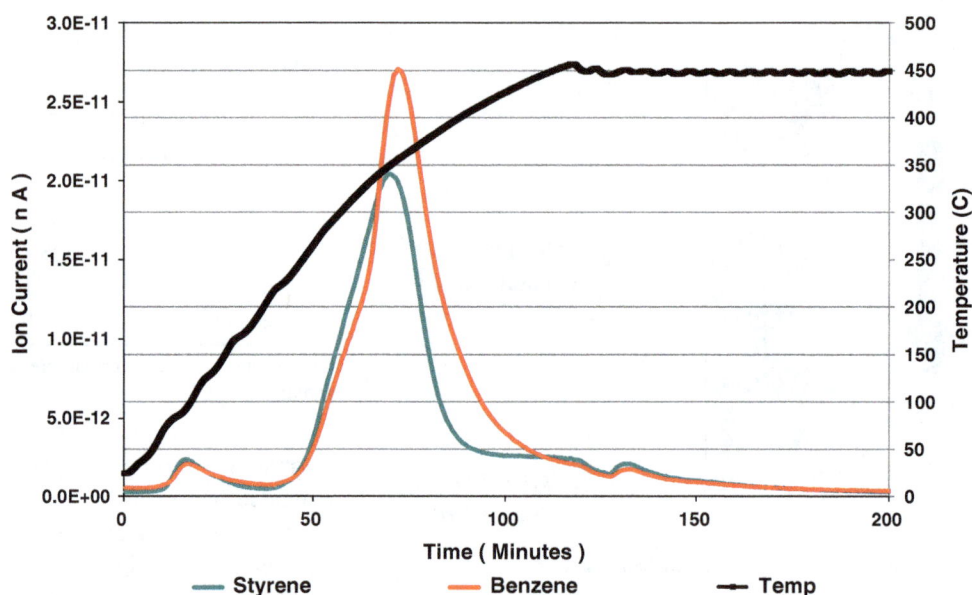

baseline by 450 °C. This change can give information related to the C:H ratio of the carbonaceous deposit. The carbonaceous deposit that is combusted at 450 °C on the extended bed produces carbon dioxide but very little water indicating that the deposit has a low H:C ratio, <1 suggesting the formation of polyaromatic or graphitic-like species. Note that this material is deposited in an atmosphere that has much less free hydrogen than that at the start of the bed. Even though much of the organic feed is now saturated, the lower gas phase hydrogen concentration does appear to affect the H:C ratio in a negative way. From the comparison of the water and carbon dioxide profiles, it is clear that on the normal bed the carbonaceous deposit has a much higher H:C ratio over the full profile and hence is easier to combust at lower temperatures.

In Fig. 10, the profiles obtained from benzene and toluene are compared and it can be seen that the extended bed has significantly more aromatic species than the normal bed but there is no evidence for desorption at higher temperatures.

Conclusions

Therefore using this enhanced TPO methodology, we have been able to show that in a standard fixed bed reactor, passing reaction equilibrated gas over fresh catalyst does not necessarily inhibit carbonaceous deposition and that the species deposited may be more deleterious to the catalyst activity than that deposited under reaction chemistry. This has potential implications for catalyst life in the plant and

Fig. 9 Comparison of carbon dioxide and water evolution from the normal (NB) and extended (EB) beds

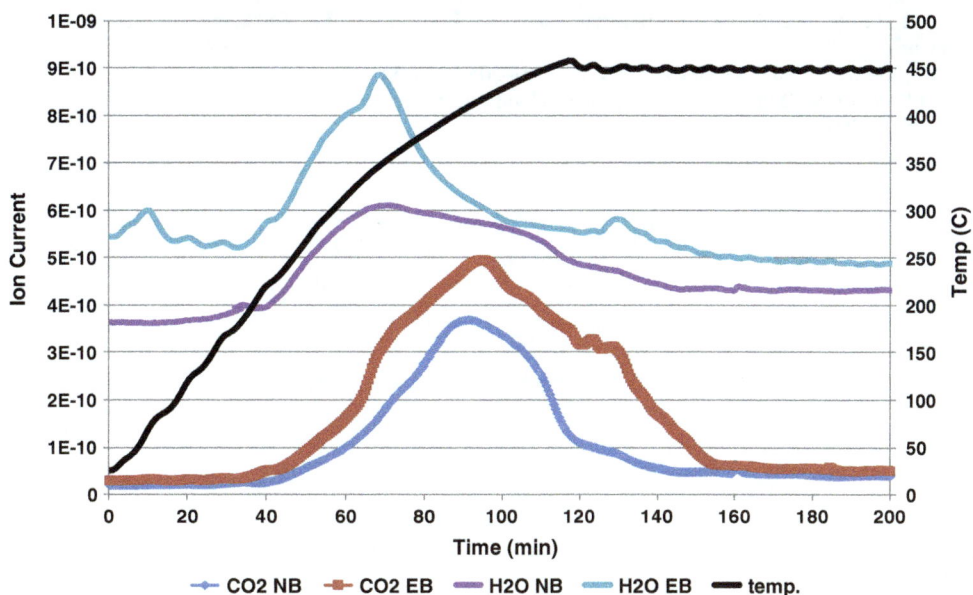

Fig. 10 Comparison of benzene and toluene evolution from the normal (NB) and extended (EB) beds

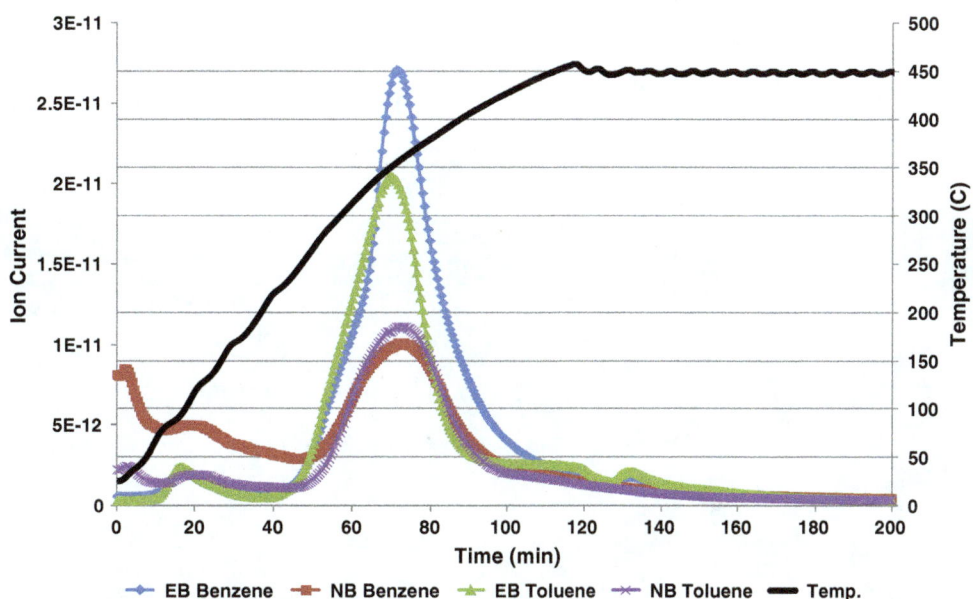

on the expected activity and selectivity. Using these techniques, a better understanding of the catalyst operating in the plant can be obtained.

References

1. Cheng YM, Chang JR, Wu JC (1986) Appl Catal 24:273–285
2. Reddy KM, Pokhriyal SK, Ratnasamy P, Sivasanker S (1992) Appl Catal A 83:1–13
3. (1979) US Patent 4150061, Process for converting pyrolysis gasoline to benzene and ethylbenzene-lean xylenes; assigned to Standard Oil Company (Indiana)
4. Anderson JA, Garcia MF (2005) Supported metals in catalysis. Imperial College Press, London
5. Barbier J (1986) Appl Catal 23:225–243
6. Barbier J, Churin E, Parera JM, Riviere J (1985) React Kinet Catal Lett 29:323–330
7. Bartholomew CH (2001) Appl Catal A 212:17–60
8. de Medeiros JL, Araujao OQF, Gaspar AB, Silva MAP, Britto JM (2007) Braz J Chem Eng 24:119–133
9. Hagen J (2006) Industrial catalysis: a practical approach, 2nd edn. Wiley-VCH, Mannheim
10. Le Page JF, Cosyns J, Courty EFP, Marcilly C, Martino JMG, Montarnal R, Sugier A, Landeghem HV (1987) Applied heterogeneous catalysis: design, manufacture, use of solid catalyst. Technip, Paris

11. Lloyd L (2008) Handbook of industrial catalysts. Springer, London
12. Porta F, Rossi M (2003) J Mol Catal A Chem 204–205:553–559
13. Richardson JT (1989) Principles of catalyst development. Plenum Press, New York
14. Weitkamp J, Raichle A, Traa Y (2001) Appl Catal A 222:277–297

6

CO₂ recycling using microalgae for the production of fuels

M. H. Wilson · J. Groppo · A. Placido · S. Graham ·
S. A. Morton III · E. Santillan-Jimenez · A. Shea ·
M. Crocker · C. Crofcheck · R. Andrews

Abstract CO_2 capture and recycle using microalgae was demonstrated at a coal-fired power plant (Duke Energy's East Bend Station, Kentucky). Using an in-house designed closed loop, vertical tube photobioreactor, *Scenedesmus acutus* was cultured using flue gas as the CO_2 source. Algae productivity of 39 g/(m² day) in June–July was achieved at significant scale (18,000 L), while average daily productivity slightly in excess of 10 g/(m² day) was demonstrated in the month of December. A protocol for low-cost algae harvesting and dewatering was developed, and the conversion of algal lipids—extracted from the harvested biomass—to diesel-range hydrocarbons via catalytic deoxygenation was demonstrated. Assuming an amortization period of 10 years, calculations suggest that the current cost of capturing and recycling CO_2 using this approach will fall close to \$1,600/ton CO_2, the main expense corresponding to the capital cost of the photobioreactor system and the associated installation cost. From this it follows that future cost reduction measures should focus on the design of a culturing system which is less expensive to build and install. In even the most optimistic scenario, the cost of algae-based CO_2 capture is unlikely to fall below \$225/ton, corresponding to a production cost of ∼ \$400/ton biomass. Hence, the value of the algal biomass produced will be critical in determining the overall economics of CO_2 capture and recycle.

Keywords Microalgae · Carbon dioxide · Flue gas · Capture · Techno-economic analysis · Biofuels

Introduction

Despite concerns surrounding the contribution of fossil fuel combustion to global warming, the need for fossil fuels will remain significant for the foreseeable future. With direct replacement unlikely, strategies to reduce the emitted CO_2 are in high demand. The current array of options encompasses four main areas: (1) modifications to existing power plants to increase the efficiency of combustion, (2) improvements to the efficiency of energy use by consumers, (3) chemical carbon capture with subsequent sequestration, and (4) bio-mitigation with carbon recycling. All of these methods have promise, yet they are beset with significant technical and non-technical challenges. Alterations to the methods of use and production involve issues related to expensive plant modifications or changes to the behavioral patterns of consumers. Despite technological gains for CO_2 capture and sequestration, the costs associated with energy-intensive CO_2 concentration and compression are significant and anticipated to result in a parasitic power plant load on the order of 30–40 %. In addition, the uncertainty surrounding risk and liability issues related to long-term geologic sequestration is potentially strong enough to deter investment and adoption.

M. H. Wilson · J. Groppo · A. Placido · S. Graham ·
E. Santillan-Jimenez · M. Crocker (✉) · R. Andrews
Center for Applied Energy Research, University of Kentucky,
Lexington, KY 40511, USA
e-mail: mark.crocker@uky.edu

S. A. Morton III
Department of Engineering, James Madison University,
Harrisonburg, VA 22807, USA

A. Shea · C. Crofcheck
Department of Biosystems and Agricultural Engineering,
University of Kentucky, Lexington, KY 40506, USA

R. Andrews
Department of Chemical and Materials Engineering, University
of Kentucky, Lexington, KY 40506, USA

As is frequently stated, microalgae are the fastest growing photosynthetic organisms with growth rates and CO_2 bio-fixation potentials generally in excess of terrestrial plants [21]. Depending on growth conditions (light intensity, temperature, and physical nature of the environment), the levels of available CO_2, and the nutrient needs of the organism, a combination of carbohydrates, proteins, and lipids are produced. From these metabolites a range of fuel and chemical feedstocks/resources can be produced [4, 29]. It is this combination of growth rate and lipid productivity that has led to algae being touted as an ideal source of bio-derived oil [28].

The use of microalgae-based CO_2 mitigation suffers from two principal disadvantages: (1) in bio-mitigation, CO_2 is captured and subsequently recycled (effectively, the carbon is used twice); in other words, CO_2 is not permanently removed from the carbon cycle; and (2) a range of challenges exist which are primarily related to system complexity and scale-up issues that are driven more by economic constraints than technical issues. However, biological carbon capture and recycling has the potential to generate a revenue stream to offset, at least in part, the overall cost of implementation [12]. Indeed, the use of algae as a carbon dioxide bio-mitigation strategy and as a potential source of renewable fuels has long been a focus of research and development [2, 5, 7, 14, 16, 20]. A primary concern is that the cultivation strategy selected (typically a large shallow open pond) requires vast amounts of water and land. This is further complicated by the need to keep the algae cultivation system close to the carbon dioxide source, where a primary limiting factor becomes the suitability of the available land. To resolve this problem, most studies have inferred that the optimal approach would be to construct a fossil fuel power facility where land availability is not a problem for large-scale cultivation; however, this solution does not resolve the issue of carbon dioxide mitigation from existing fossil fuel energy production facilities, or, for that matter, other types of CO_2 point sources.

A few recent reports have attempted to address these challenges through the development of closed loop photobioreactors, which on an areal basis are typically more productive than open ponds [33]. Indeed, a study by Doucha et al. [10] employing flue gas from a natural gas fired boiler fed to a thin layer photobioreactor containing *Chlorella* sp. found that up to 50 % CO_2 removal could be attained. A study reported by Vunjak-Novakovic et al. [31] placed this figure as high as 82 % on sunny days for *Dunaliella* strains grown in an airlift reactor, with 50 % CO_2 removal on cloudy days. An older study by Laws and Berning [17] performed in Hawaii using the marine chlorophyte *Tetraselmis suecica* (grown in outdoor flumes) indicated that CO_2 emitted from an oil-fired electric plant could be successfully substituted for pure CO_2 for the cultivation of algae. More limited than the number of these studies are the commercially viable microalgae production processes that utilize flue gas from power plants. In fact, only Seambiotic in Israel has produced significant quantities of algae in this manner, using the flue gas from the Israel Electric Corporation's coal-fired Ashkelon power plant (Seambiotic [27]).

Against this background, we set out to determine the technical and economic feasibility of algae-based carbon capture in Kentucky. This required designing and demonstrating a process capable of utilizing flue gas through operation of a continuous microalgae culture, and evaluating the economics of CO_2 capture using this approach. This paper summarizes the results of our initial work, including the development and scale-up of component technologies, and demonstration of the integrated process at Duke Energy's East Bend Station, situated in northern Kentucky.

Experimental

Algae culturing

Scenedesmus acutus was obtained from the University of Texas Culture Collection (UTEX B72) and was used for all experiments. Cultures were grown in urea medium previously optimized for this *S.* strain (Crofcheck et al. [8]). Initial cultures were grown in 500 mL Erlenmeyer flasks under warm (Philips F32T8/TL741 Alto, 32 Watts) and cool white (Philips F32T8/TL735 Alto, 32 Watts) fluorescent lights [70 $\mu mol/(m^2\ s)$] in a 16:8 h light:dark illumination period. Flasks were bubbled with 3 % CO_2 (from gas cylinders) and kept at room temperature (22 °C). The flask cultures were eventually transferred to 7.5 L airlift photobioreactors (PBRs). These airlifts also received a constant supply of 3 % CO_2, but were grown under natural light conditions in a greenhouse. A number of airlift PBRs were used to inoculate a 650 L Varicon BioFence PBR which, in turn, was used to seed a 1,000 L PBR. Both of these large greenhouse reactors were needed to produce enough algae to inoculate the East Bend Station PBR. The larger PBRs were constantly monitored by probes for pH (Hach DPD1R1), dO_2 (Hach 5740DOB), temperature, dCO_2 (Mettler Toledo InPro 5000i), and photosynthetically active radiation (PAR, Apogee Instruments SQ-215). The large greenhouse PBRs were fed CO_2 whenever the culture pH rose above a certain set point (usually pH 7.0). The East Bend Station PBR operated the same way using flue gas as its CO_2 source.

Culture growth was monitored by means of dry mass (g/L) (Crofcheck et al. [8]) and qualitative microscopy analyses. In addition, ultraviolet–visible spectrophotometry

(Thermo Scientific Evolution 60) was used to monitor the density of algal cultures, absorbance being measured at 680 nm. Typically, one 50 mL sample was taken daily from the PBRs for analysis. In addition, ion and urea concentrations in the cultures in the large PBRs were monitored on a regular basis by ion chromatography (IC) and high performance liquid chromatography (HPLC), respectively. The concentrations of urea and specific nutrient ions were tracked to determine the rate of nutrient consumption. Elemental analysis of harvested algal biomass was conducted using inductively coupled plasma-optical emission spectroscopy (ICP-OES).

Lipid extraction and purification

The algae used in all experiments were *S. acutus* (UTEX B72) autotrophically cultured at East Bend Station using a urea-based medium (Crofcheck et al. [8]). After harvesting and dewatering, the algae (10–15 % solids) were dried in an oven at 60 °C for 24 h. The oven-dried *S.* algae were ground up in a coffee grinder until the algae particles were reduced to a size of <1 mm. After grinding and before all extractions, the algae used in the extractions were heated to 100 °C for 20 min to remove residual water (moisture content <3 wt%). Extractions were performed according to the Bligh–Dyer method [3] with one modification, namely, a biomass to total solvent ratio of 10 g/180 g was used. After removal of solvent, the crude lipid was weighed, re-dissolved in $CHCl_3$ and filtered through a plug of K10 montmorillonite (Sigma-Aldrich) to remove the chlorophyll present (which remained strongly adsorbed on the clay). Solvent was then removed under vacuum to afford the purified lipid as a colorless waxy solid.

Lipid conversion to fatty acid methyl esters

Algal lipids were converted to the corresponding methyl esters using a two-step process of esterification (to convert free fatty acids) and transesterification (to convert triacylglycerides) (Canakci and Van Gerpen [6]). 1 g of extracted, purified algal lipids was mixed with 1.5 mL anhydrous methanol containing 2 % H_2SO_4 (wt/wt) and refluxed at 65 °C for 2 h. The reaction was then cooled in an ice bath and the contents were mixed with approximately 3 mL of a 1:1 (v/v) mixture of water:cyclohexane. The organic layer was extracted, dried over sodium sulfate and the cyclohexane was removed by a rotary evaporator. The remaining lipid was then reacted with methanol containing potassium methoxide (0.5 wt% of lipid feedstock). This mixture was refluxed at 65 °C for 30 min. The reaction contents were then cooled, mixed with a water:cyclohexane mixture and the organic layer was separated, dried over sodium sulfate and the cyclohexane was removed on a rotary evaporator.

FAME analysis was performed using an HP6890 GC equipped with a J&W Scientific HP-88 capillary column (30 m × 250 μm × 0.2 μm). The inlet was set to 250 °C. The split ratio was 20:1 with a constant flow of 1 mL/min. The oven began at 50 °C and was ramped at 20 °C/min to 140 °C and held for 5 min prior to a second ramp of 3 °C/min to 240 °C. The detector was held at 300 °C. The GC was calibrated using a 37-component FAME GC standard (Sigma-Aldrich). Samples were diluted 1,000:1 in cyclohexane and toluene was used as an internal standard.

Lipid conversion to hydrocarbons

Lipid deoxygenation experiments were performed in a fixed bed stainless steel tubular reactor (1/2 in. o.d.) equipped with an HPLC pump. 0.5 g of Ni–Al LDH catalyst (particle size 150–300 μm) was first reduced under H_2 at 400 °C for 3 h. Details of the catalyst preparation and characterization have been reported elsewhere [25]. After reduction of the catalyst, the system was taken to the reaction temperature (300 °C) and pressurized with H_2 to 580 psi. A 1.33 wt% solution of the algal lipids dissolved in dodecane was introduced to the system at a rate of 0.1 mL/min along with a flow of H_2 (50 mL/min). Samples were collected from a liquid/gas separator placed downstream from the catalyst bed. The liquid feed and reaction products were analyzed using an Agilent 7890A GC equipped with an Agilent J&W DB-5HT column (30 m × 250 μm × 0.1 μm), an Agilent Multimode inlet, a deactivated open ended helix liner and a flame ionization detector (FID). Data acquired using the GC-FID were processed using SimDis Expert 9 software purchased from Separation Systems, Inc. The dodecane solvent was subtracted and/or quenched from the chromatogram prior to processing the chromatographic data. Further details can be found in [25].

Results and discussion

Photobioreactor development

The cultivation of an autotrophic organism requires the provision of a controlled growth environment, which involves exposure of the organism to appropriate levels of sunlight, CO_2, and nutrients [30]. The mass cultivation of algae can be realized in either an open culture system (pond), or a closed loop system (photobioreactor). The selection of an open or closed culture system revolves around a number of system parameters: (1) the microalgae to be cultured, (2) the anticipated carbon source, (3) the accessibility to required resources, and (4) the cost of construction, operation, and maintenance of the culture system. Based upon simple mass balance calculations, an algae unit size to reduce the CO_2 output of a power plant would need to be of an enormous scale. Photobioreactors (PBRs) were chosen as the cultivation method in this study on the basis of their higher areal productivities [33] and limited water loss [30] due to evaporation. A number of prototype reactors of different configurations were first constructed in an effort to incorporate the lessons learned into larger scale reactors. Specifically, variations in construction materials, tube orientation and spacing, as well as flow patterns, were examined. The most important factor in designing photobioreactors is to allow exposure of the algal culture to sunlight to drive photosynthesis. Given that a vertical system typically enables a higher surface to footprint ratio than other configurations, a design based on a tubular photobioreactor was selected, oriented vertically, and constructed from low-cost, off-the-shelf parts.

The hydrodynamics of the reactor is another area of specific concern. Having a good understanding of the flow characteristics of the PBR is an important step toward enabling process control. Having a well-mixed system, with even flow and limited dead zones, ensures that measurements taken at a centralized point are descriptive of the entire system. In addition, any stagnant areas or zones with lower flow (with the potential to collect biomass) should be limited. If the biomass remains trapped in the reactor it can degrade and release compounds that affect culture health. Moreover, poor mixing can lead to anaerobic conditions which favor microbial denitrification, resulting in N_2O emissions [11, 15]. Given that N_2O is a potent greenhouse gas (GHG), this negatively affects the GHG balance of the system.

Different methods for circulating the algal culture and keeping it well mixed were, therefore, evaluated before developing a series flow, serpentine-style PBR (Fig. 1). This style of reactor most closely resembles a plug flow reactor, and is constructed by connecting multiple tubes in series to provide the algae access to the solar radiation needed to drive growth. Initial concerns over oxygen accumulation and low carbon dioxide levels were set aside due to the relatively low kinetic rate of photosynthesis (Grima et al. [13]). In order for detrimental concentrations of dissolved O_2 to accumulate, the liquid path, and thereby the residence time, would have to be extremely long. If care is taken in overall reactor design and operation, this issue can be resolved.

Historically, one of the main challenges in algae cultivation is CO_2 limitation. Carbon constitutes almost 50 % by weight of the elemental composition of algal biomass, with CO_2 representing the most significant nutrient requirement. One of the inherent benefits of working with coal flue gas is the high percentage of CO_2 in the gas (10–15 %) as compared to atmospheric conditions (0.04 %). Introducing CO_2-rich gas to the system can often be energy intensive, as is the case in systems requiring gas compression and bubbling. To minimize the costs associated with CO_2 entrainment, while maximizing mass transfer, a liquid driven vacuum pump (i.e., venturi or eductor) was employed in this study. An eductor uses the Bernoulli principle to entrain gas in a driven liquid flow. The extremely turbulent nature of the biphasic flow encourages good mixing and mass transfer, thereby facilitating CO_2 dissolution in the growth medium.

Photobioreactor operating strategy

Daily productivity rates of algal cultures are dependent on multiple factors, including the nature of the organism being cultured, nutrient concentrations, the concentration of dissolved carbon, temperature, light intensity (i.e., photosynthetically active radiation, PAR), and pH. In this study

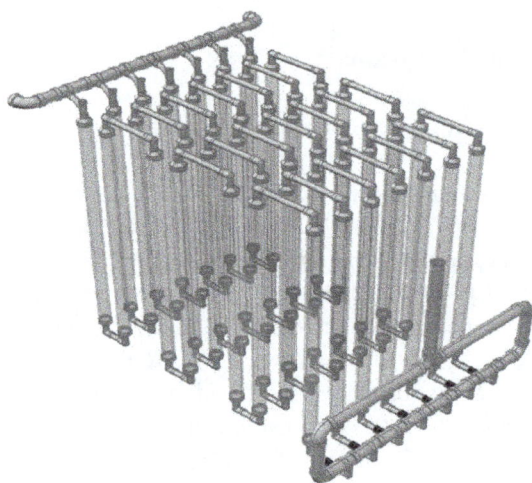

Fig. 1 CAD image showing PBR design and photograph of the PBR installed in a greenhouse

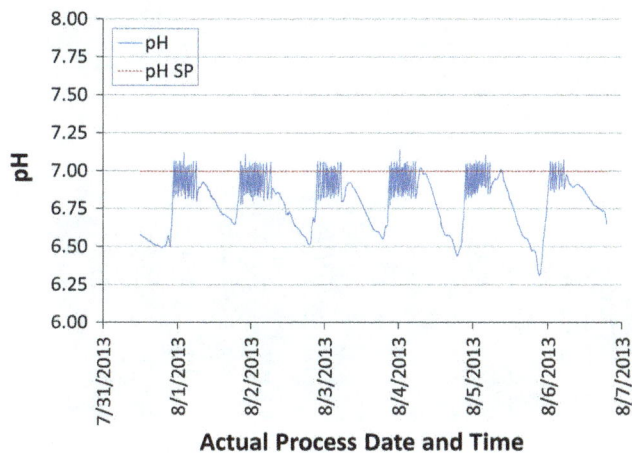

Fig. 2 Photobioreactor pH control. pH SP refers to the pH set point

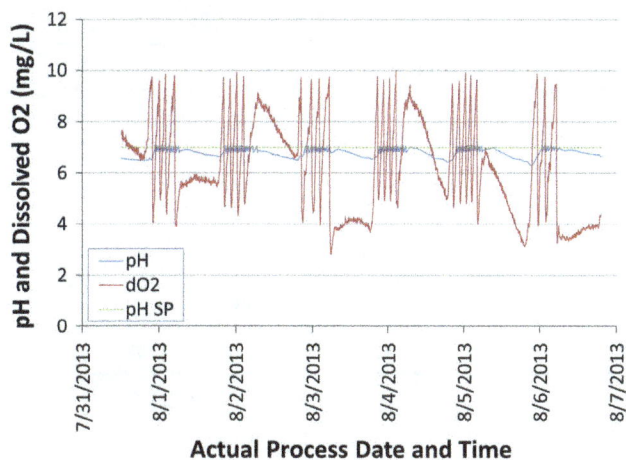

Fig. 3 O_2 production and system pH. The oscillations of the dO_2 signal are due to automated sparging of the culture with N_2 to maintain the dO_2 concentration below the set point value of 10 mg/L

S. acutus was cultured, a prior screening study having shown it to be appropriate for CO_2 capture based on its robust growth, tolerance to a wide range of pH values (Crofcheck et al. [9]), as well as ease of harvesting. Combustion flue gases can be a rich source of CO_2 for algae cultivation; however, the addition of excess flue gas must be avoided as this can result in over-acidification of the culture medium. This can be achieved by feeding CO_2 (as flue gas) on demand based upon the pH of the system. As algae grow, consuming CO_2, the pH of the solution is increased. The introduction of a CO_2-rich gas increases the concentration of carbonic acid and other dissolved carbon species, thereby lowering the pH. This approach maintains the system pH within an optimum pH range for algal growth while providing enough CO_2 to sustain growth. This is particularly important if other acidic flue gas components such as SOx and NOx are present. The dissolution of SOx in particular, which forms H_2SO_3/H_2SO_4, can result in over-acidification of the culture medium, thereby inhibiting growth (Crofcheck et al. [9]). For this reason, it is important that SOx is not added to the cultivation system faster than its dissolution products can be utilized by the algae.

Figure 2 illustrates the pH control method used to regulate CO_2 flow to the reactor during 6 days of algal cultivation in a 650 L PBR. The horizontal line indicates the pH set point of the reactor, while the trace shows the measured pH of the system. This graph also captures the occurrence of respiration, which produces CO_2, thereby lowering the pH during the night hours.

Appropriate reactor design can eliminate the buildup of O_2 in the photoactive portion of the reactor, but dissolved O_2 accumulation can still occur in a closed system over time. Elevated levels of dissolved O_2 can inhibit photosynthesis so it is important to have a method to remove excess O_2 from the system (Weissman et al. [32]). One method is to periodically sparge the main process tank with

an oxygen-lean gas (such as post-combustion flue gas or nitrogen) which will remove dissolved O_2 preferentially over dissolved CO_2. Figure 3 shows the response of a PBR with N_2 sparging to remove excessive O_2 concentrations (i.e., >100 % atmospheric saturation or ~ 9 ppm). As anticipated, a high frequency of pH oscillation (corresponding to strong CO_2 consumption) is paired with a strong response in dissolved O_2. The effects of respiration on the pH and dissolved O_2 concentration during the night are also illustrated.

Another important variable that must be controlled is the culture density of the reactor. As a culture increases its number of cells, the increased chlorophyll concentration of the culture attenuates light much more quickly, starving some cells of required levels of solar radiation. Regular harvesting and dilution of the culture are, therefore, required to maintain a stable system capable of operating for an extended period of time.

Demonstration facility

Field testing of the system described above was conducted at Duke Power's East Bend Station (650 MW) located in Boone County, Kentucky. This single unit plant burns high sulfur coal as the fuel source and utilizes a wet limestone scrubber for SOx control and selective catalytic reduction (SCR) with ammonia injection for NOx control. Flue gas used for algae growth studies was obtained after the scrubber and SCR treatments with typical composition summarized in Table 1.

The site layout consisted of a PBR tube array located on an embankment situated approximately 7.5 m above a lower level where the 19,000 L feed tank, 5,700 L harvest tank and system control enclosure were located. The PBR assembly was constructed on a concrete pad poured above

Table 1 East Bend flue gas analysis (3/1/11–3/1/12)

	CO_2 (%)	NOx (ppm)	SO_2 (ppm)
Average:	8.9	53.4	28.0
Minimum:	7.2	14.5	6.5
Maximum:	9.6	97.2	84.3

Table 2 Analysis of source water at East Bend Station (average value of duplicate measurements ± standard deviation)

Analyte	Concentration, ppm
Chloride	3.79 ± 0.01
Nitrate–N	3.89 ± 1.17
Sulfate	25.15 ± 0.21
Phosphorus, total	<0.04
Calcium	89.45 ± 2.90
Magnesium	28.10 ± 0
Hardness by calculation	335.5 ± 12
Potassium	1.17 ± 0
Sodium	4.11 ± 0.78

Fig. 4 Photobioreactor installed at East Bend Station

a gravel drainage bed lined with a geomembrane below a French drain to collect all surface run-off and potential tube leakages. The drain flowed down the 7.5 m embankment to another concrete pad poured to provide a stable foundation for the feed and harvest tanks.

Water used to fill the PBR was drawn from several wells located on the property; typical analyses are shown in Table 2. Before water was fed into the PBR, it was passed through a UV sterilizer to minimize potential contamination by any organisms that may be present in this otherwise untreated water.

The PBR was constructed of clear PET (polyethylene terephthalate) tubes (8.9 cm diameter × 244 cm high) connected by 7.5 cm diameter schedule 40 PVC (polyvinyl chloride) pipe. Reactor tubes were arranged in 10 parallel flow paths, each consisting of 51 tubes connected in a serpentine path extending linearly for 18.3 m (Fig. 4). Feed was introduced by a centrifugal pump via a manifold where flow velocity through each tube was maintained at 16 cm/s, providing a residence time of approximately 13 min in the photosynthetically active volume of the clear tubes for each pass through the PBR. At the end of the PBR, the flow from each parallel flow path was combined in a common manifold and returned to the feed tank. As the return volume flowed back to the top of the 19,000 L feed tank, flow was directed to fall through T-pipe fittings to create suction, which was used as a means to introduce flue gas into the system (see Fig. 5 for a schematic of the PBR system).

The suction end of the return piping arrangement was connected to an air manifold with three automated control valves. When slurry pH rose above the desired set point (pH 7.0) the control valve connected to the flue gas would open, allowing flue gas to be introduced to the feed tank. When the pH dropped below the set point, the flue gas control valve would close and another control valve would open, allowing air in the head space to be recirculated, preventing CO_2 in the head space from venting from the system. The third control valve was used as a relief valve to prevent the feed tank from becoming pressurized as flue gas was added to the system. In this manner, as CO_2 was consumed by the algae, additional CO_2 was automatically fed into the system as needed to maintain the desired operating pH.

The PBR at East Bend Station was seeded on 7 December 2012 and operated continuously until 31 December 2012. During this time, flue gas was added as needed to maintain the pH at 7.0. Summary results (Fig. 6) show that productivity as high as 23 g/(m^2 day) was achieved during this period. These data also illustrate that productivity is related to available sunlight (i.e., PAR) as growth rate increased following periods of increased available PAR. The fact that the productivity data do not align perfectly with the PAR values requires comment, albeit that a clear trend is evident. This can be explained on the basis that (1) productivity is dependent on a combination of PAR and temperature (indeed, little growth was observed below 10 °C), and (2) the time at which the reactor is sampled during the day (am or pm) can introduce a lag into the data, i.e., samples taken in the early morning reflect growth the previous day (as well as night losses due to respiration), while samples taken in late afternoon reflect growth on the same day. During this time period, average daily temperature ranged from 4 to 20.5 °C. While this particular organism is not known to be particularly well suited for winter growth, reasonable growth rates were achieved, provided that sufficient PAR was available.

Fig. 5 Schematic of photobioreactor system at East Bend Station

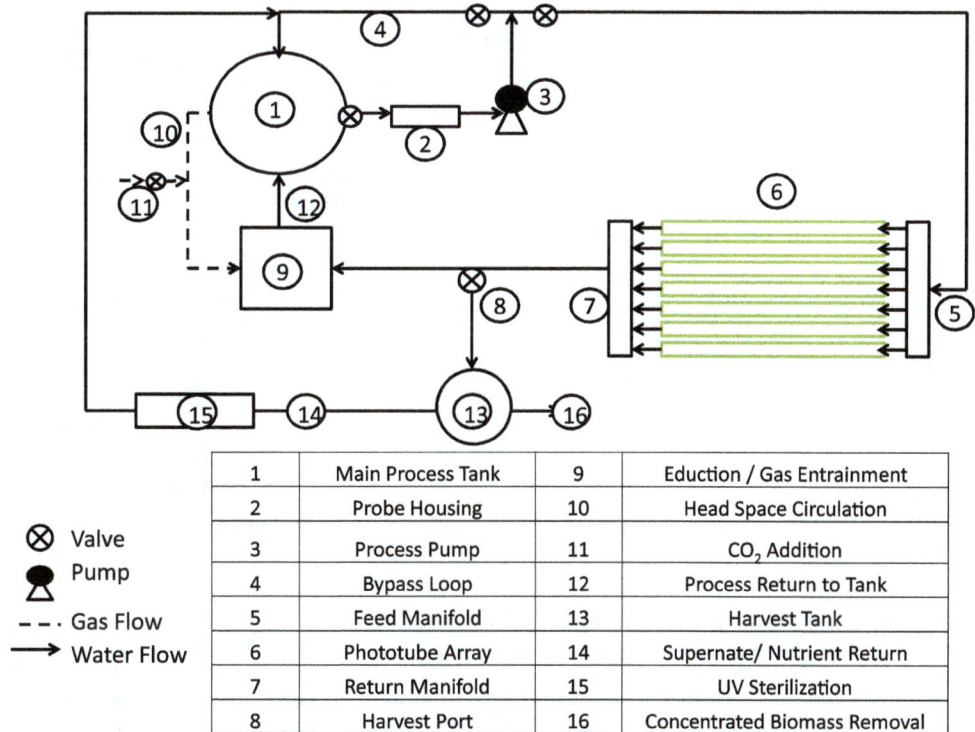

1	Main Process Tank	9	Eduction / Gas Entrainment
2	Probe Housing	10	Head Space Circulation
3	Process Pump	11	CO_2 Addition
4	Bypass Loop	12	Process Return to Tank
5	Feed Manifold	13	Harvest Tank
6	Phototube Array	14	Supernate/ Nutrient Return
7	Return Manifold	15	UV Sterilization
8	Harvest Port	16	Concentrated Biomass Removal

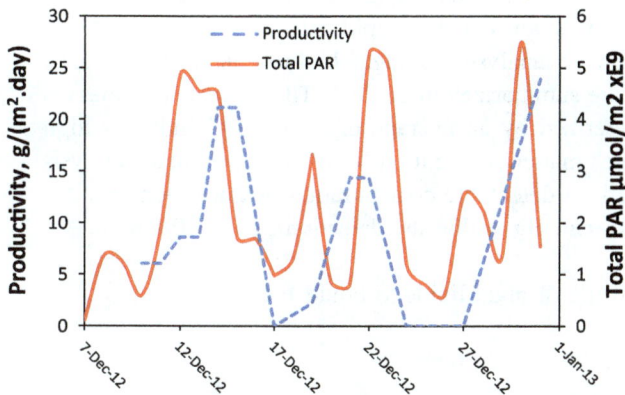

Fig. 6 Algal productivity and PAR during December growth study

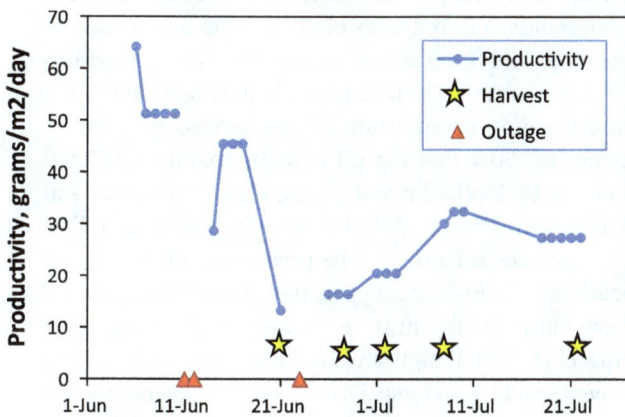

Fig. 7 Algal productivity during June–July growth study

Summer growth studies were conducted during June and July 2013 (see Fig. 7). As was observed with the winter growth study, productivity essentially followed periods of available sunlight (not shown). This study was complicated by several unforced electrical outages and an electrical failure to the feed system power supply caused by a lightning strike. Nevertheless, during periods of operation a mean growth rate of 32.9 g/(m² day) [with a standard deviation of 14.2 g/(m² day)] was recorded.

Harvesting and dewatering

During continuous PBR operation, it is necessary to periodically remove algae to control culture density to minimize self-shading and maintain culture health. Development of a suitable harvesting/dewatering strategy also addressed other system needs such as recycling clarified water to minimize water consumption and recycle unused soluble nutrients. Since the algae culture is very dilute (0.4–1.0 g dry mass/L), a cost effective, high capacity solid/liquid separation strategy was warranted. After considering a number of options, including dissolved air flotation and centrifugation, it was decided to pursue the use of sedimentation, thickening and filtration, an approach commonly used for treatment of industrial waste water.

Harvesting cycles were conducted as deemed necessary to maintain culture density, typically on a cycle of two to three times per week. A schematic diagram of the process is shown in Fig. 8. While the system continued normal

Fig. 8 Schematic diagram of harvesting/dewatering process

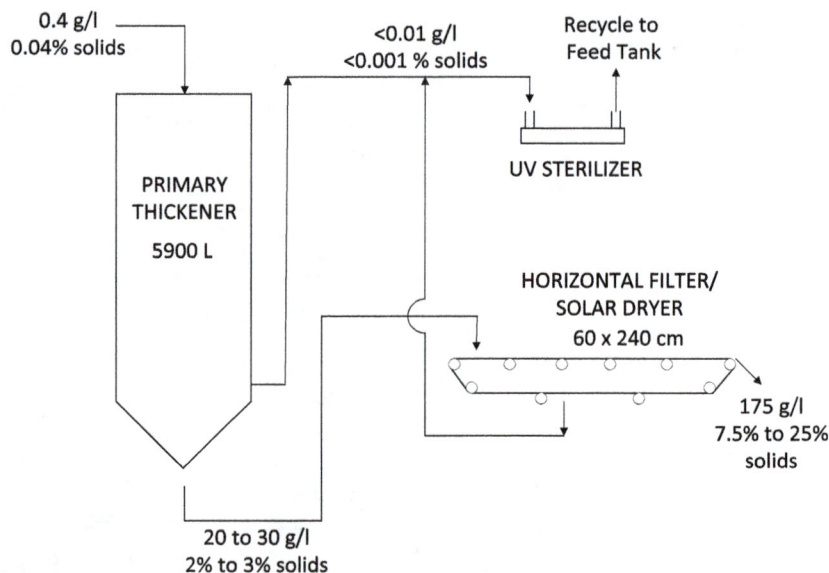

Table 3 Ultimate analyses of algal biomass harvested at East Bend Station (average of seven separate algae harvests ± standard deviation)

Carbon (%)	42.47 ± 4.18
Hydrogen (%)	6.50 ± 0.55
Nitrogen (%)	6.77 ± 0.70
Total sulfur (%)	0.52 ± 0.07
Oxygen (%)	24.38 ± 1.60
Ash (%)	19.36 ± 6.65
Volatile matter (%)	66.54 ± 4.25
Fixed carbon (%)	9.09 ± 2.40
As (ppm)	<0.1
Se (ppm)	<0.1
Cd (ppm)	<0.1
Hg (ppm)	<0.1

operation, approximately 4,000 L of culture was diverted into a cylindrical, cone bottomed harvesting tank. Moderate molecular weight cationic polyacrylamide flocculant was added at a dosage of 3–5 ppm, dependent upon harvest culture density, and mixed using a recirculating centrifugal pump. The flocculated system was allowed to settle for 4–8 h, after which settled biomass (20–30 g/L) was removed from the thickener and clarified water was pumped back into the feed tank to recycle water along with remaining soluble nutrients. As water was recycled, it was passed through a UV sterilizer. The concentrated algae slurry was then transferred to a horizontal gravity filter/solar dryer and allowed to drain on a multifilament filter fabric. Clear filtrate was passed through the UV sterilizer and returned to the feed tank and dewatered biomass cake was removed from the filter/dryer. If adequate sunlight was available, the filter cake was dry (≤3 % moisture) after

approximately 24 h. If inadequate sunlight was available, the filter cake typically contained 7.5–25 % solids and was transferred to an oven and dried at 100 °C. After the biomass was thoroughly dried, a representative portion was characterized and the remainder stored in a freezer for utilization studies such as lipid extraction and upgrading.

Ultimate analyses of algal biomass harvested at East Bend are summarized in Table 3. The harvested biomass is characterized by an average of 42.47 % C and very high volatile matter content (66.54 %). Elemental analysis showed no detectable concentration of trace elements As, Se, Cd, and Hg within the detection limit of 0.1 ppm.

Upgrading of algal lipids to liquid fuels

Fatty acid methyl esters

To analyze the fatty acid profile of the lipids present in the harvested algae, lipids were extracted by the Bligh–Dyer method and converted to fatty acid methyl esters (FAME, more commonly referred to as biodiesel) via a sequence of esterification and transesterification. The gas chromatogram of the resulting FAME mixture is shown in Fig. 9 and the corresponding composition is summarized in Table 4. These results show that the oil consists mainly of C16:0 (palmitic), C18:1 (elaidic and oleic), C18:2 (linoleic) and C18:3 (linolenic) fatty acid chains. While C16 and C18 chain lengths are suitable for the production of diesel fuel hydrocarbons via hydrodeoxygenation or decarboxylation/decarbonylation (vide infra), or indeed for the production of biodiesel (FAME), higher value fatty acids such as EPA (eicosapentaenoic acid) and DHA (docosahexaenoic acid) are not present. Consequently, the value of the oil for nutraceutical purposes would be low.

Fig. 9 Gas chromatogram of fatty acid methyl esters obtained from lipids extracted from *Scenedesmus acutus*

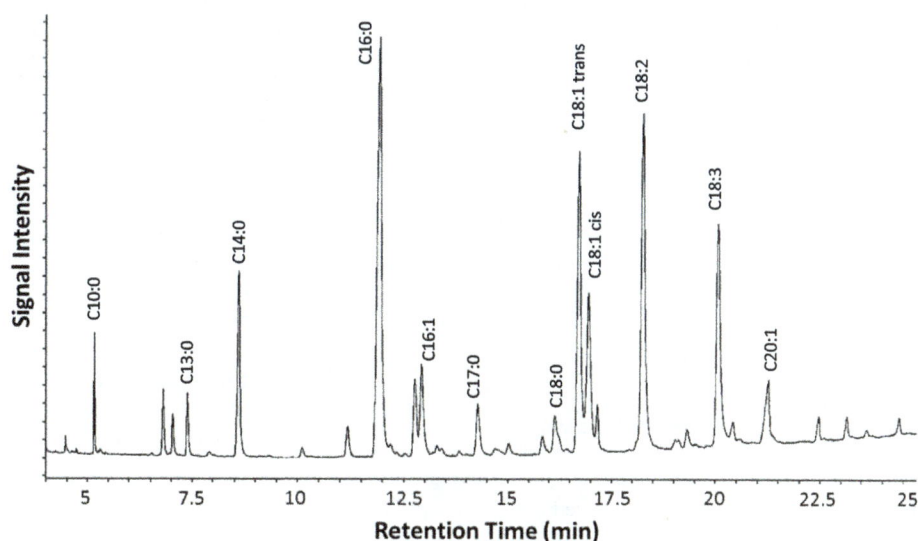

Table 4 Distribution of fatty acid chains in lipids extracted from *Scenedesmus acutus*

Fatty acid chain (X:Y)[a]	Algal lipid (GC area %)
Capric (10:0)	1.2
Tridecanoic (13:0)	1.3
Myristic (14:0)	5.3
Palmitic (16:0)	20.6
Palmitoleic (16:1)	3.9
Heptadecanoic (17:1)	2.0
Stearic (18:0)	2.1
Elaidic (18:1n9t)	12.1
Oleic (18:1n9c)	7.9
Linoleic (18:2)	15.3
γ-Linolenic (18:3n6)	10.5
Cis-11-eicosenoic (20:1)	3.3
Other	14.5

[a] $X{:}Y$ = carbon number: number of double bonds

Table 5 Conversion of algae oil to diesel-range hydrocarbons

Catalyst	Conversion	Selectivity to C10–C17 (%)[a]	Selectivity to C17 (%)
Ni–Al LDH	95	73	7

Conditions: fixed bed reactor, 300 °C, 580 psi H_2, feed = 1.33 wt% algae oil in dodecane, feed rate = 6 mL/h

[a] Note that this value underestimates actual C10–C17 selectivity due to the fact that any C12 produced is not included in the calculation (given that C12 is used as the reaction solvent)

Diesel-range hydrocarbons

Hydrodeoxygenation (-H_2O) via hydrotreating forms the basis of a number of commercial or semi-commercial processes for the production of high quality drop-in hydrocarbon fuels from the lipids in vegetable oils and animal fats. Unfortunately, these processes require sulfided catalysts that risk contaminating the products with sulfur; in addition, they are constrained to use high pressures of H_2 that are typically only available in centralized facilities. An alternative lies in the deoxygenation of lipids via decarboxylation/decarbonylation (deCO$_x$), an approach that proceeds under considerably lower H_2 pressures and uses simple metal catalysts [24]. In recent work, we have shown that Ni-based catalysts are highly active for the upgrading of soybean oil and model triglycerides via deCO$_x$ [18, 19,

26]. Similarly, Lercher and co-workers have demonstrated that Ni-containing bifunctional catalysts can be employed to convert algal lipids to diesel-range alkanes in both batch and continuous modes (Peng et al. [22]; Peng et al. [23]; Zhao et al. [34]).

Building on the above studies, oil extracted from *Scenedesmus* microalgae harvested from the East Bend facility was subjected to upgrading via catalytic decarboxylation/decarbonylation. Prior to upgrading, the crude lipids were purified by filtration through K10 montmorillonite (an acid-treated clay) to remove chlorophyll (the presence of which might lead to the formation of deposits such as coke and Mg^{2+} on the catalyst during reaction). The purified oil was then upgraded as a solution in dodecane over a Ni/Al layered double hydroxide (LDH) catalyst [25] in fixed bed mode under H_2 (580 psi). Results are summarized in Table 5 and Figs. 10 and 11.

These results confirm that catalytic deCO$_x$ is a viable process for the conversion of microalgal lipids to diesel/jet fuel range hydrocarbons. Notably, some C18 is obtained (with 5 % selectivity), indicating that hydrodeoxygenation occurs in parallel with decarbonylation/decarboxylation, although together the latter processes constitute the major pathway given the higher selectivity of C17 observed. It is

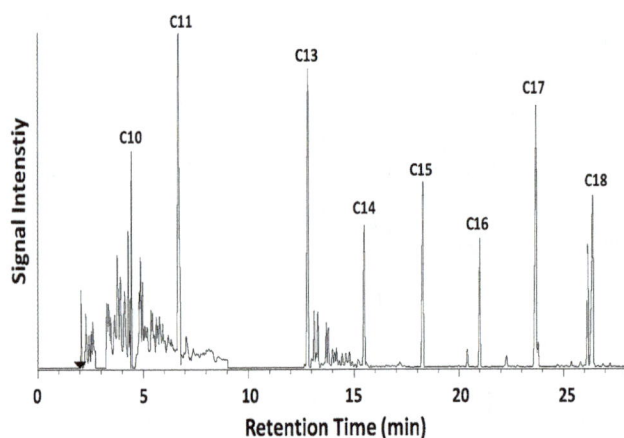

Fig. 10 Gas chromatogram of liquid product sampled after 4 h on stream during decarbonylation/decarboxylation of algal lipids. Note that the dodecane solvent (C12) has been subtracted from the chromatogram

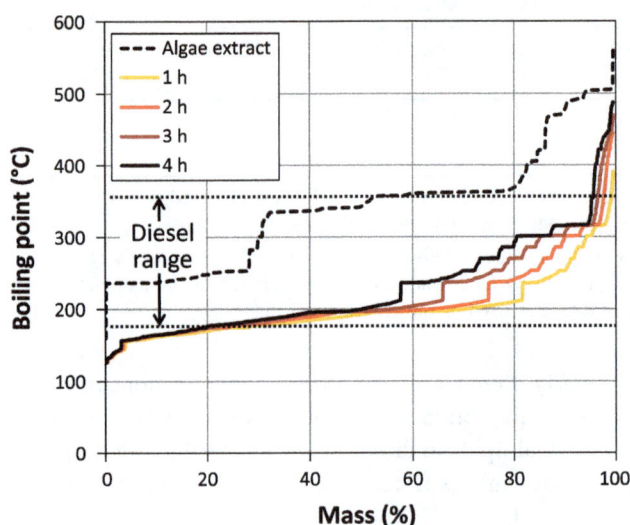

Fig. 11 Simulated-distillation boiling point distribution plots of algal lipid feed and liquid product sampled at 1, 2, 3, and 4 h during decarboxylation/decarbonylation. Note that the dodecane solvent (C12) has been removed from the plots

also evident that cracking of the unsaturated C18 fatty acid chains occurs, given the significant amounts of C10–C13 hydrocarbons obtained. Such cracking may or may not be beneficial depending on whether hydrocarbons are being targeted for jet fuel/lighter diesel-range applications or not. In other work [18, 19], we have found that less highly unsaturated fatty acid chains produce comparatively higher yields of the longer chain hydrocarbons (e.g., C15 and C17), which are well suited for diesel fuel blending.

Preliminary techno-economic analysis

The baseline scenario considered for this study was a 1,000 MW power plant, requiring 30 % CO_2 capture. Key

Table 6 Summary of inputs and assumptions used in the techno-economic analysis (base case)

Input	Value	Comment
Required CO_2 capture efficiency	30 %	Required capture efficiency if CO_2 emissions are to be maintained at 1,990 value
PBR cost, $/L (raw materials)	0.55	Custom design, PETG and PVC parts. Current cost, discounted by 55 % for bulk manufacture of parts
PBR installation cost, $/L	100	Assumed to be 100 % of raw material costs
PBR tube useful life, years	5	UV degradation limits tube life
Operation and maintenance costs, $/L	5 %	Labor + minor consumables, 5 % of PBR material costs
Areal productivity, g/(m^2 day) (year average)	30	Value based on data collected at East Bend Station and at the University of Kentucky
Nutrient costs, $/kg algae	0.14	Nutrient recipe utilizes bulk grade fertilizer
Nutrient recycle	97 %	Water + nutrients from algae harvesting and dewatering are recycled
Flocculant concentration, ppm	3	Commercial cationic flocculant
Flocculant cost, $/kg	4.40	Commercial cationic flocculant
Electricity cost, $/kWh	0.02	Discounted rate at utility site
Water cost, $/L	–	Water at site is free
Operating days per year	300	Estimated power plant operation (allowing for plant maintenance and unscheduled outages)
Payback period, years	10	Assumed payback time

inputs and assumptions used are collected in Table 6, while Table 7 summarizes the calculated costs associated with CO_2 capture. The techno-economic model is based on the capital and operating costs of a microalgae cultivation system sized to consume a given amount of CO_2. On a stoichiometric basis, algae consume ~1.76 tons of CO_2 to produce 1 ton of algal biomass, the exact figure depending on the elemental composition of the biomass produced [1]. The CO_2 emissions are based on the average rating of a coal-burning power plant, which relates the BTU content of the coal to the CO_2 emission. Algal productivity is expressed in grams per meters squared per day, a value of 30 g/(m^2 day) being used in this analysis. This number is derived from East Bend Station data (collected in the months of December, June and July) and data collected at the University of Kentucky (UK) over an approximately 12 month period; the East Bend data followed the trends previously observed at UK with respect to algae productivity as a function of the time of year. Relating algal

Table 7 Summary of costs associated with CO_2 mitigation using microalgae

	Cost in $ per ton of CO_2 removed
Growing system	
PBR capital	775
PBR installation	775
PBR operation and maintenance	40
Energy	1
Nutrients	15
Growth subtotal	1,606
Dewatering	
1st stage dewatering	15
2nd stage dewatering	0.50
Dewatering subtotal	15.50
Total cost	1,621

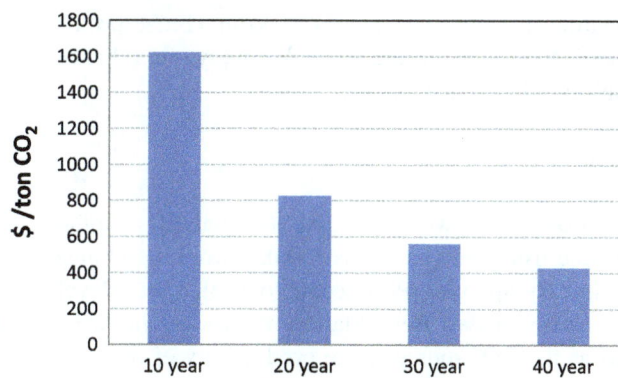

Fig. 12 Effect of amortization period on the cost of CO_2 capture

productivity to the carbon emissions of a typical coal-fired power plant results in the total land area required for the cultivation system, i.e., an algae farm big enough to consume the requisite amount of CO_2. Based on this productivity, a cultivation system equivalent to 26,200 acres (40.9 square miles) would be required for the baseline scenario (30 % CO_2 mitigation for a 1,000 MW capacity plant). The costs associated with this process (capital, operating, dewatering) are then normalized by land area ($/m^2) to be compared with areal productivity, and thereby CO_2 consumption. As with capital costs, the energy consumption of the PBR system is extrapolated from current system values. The pressure drop associated with adding additional tubes in series is negligible, resulting in a lower energy cost per unit area (Watts/m^2) of installed and operating PBR. [Note that each tube pair has a calculated pressure drop of 0.00134 psi based on the frictional resistance; a module of 500 tubes as for the East Bend PBR has 25 tube pairs (10 rows) and would, therefore, have a pressure drop of 0.034 psi]. In this way, we are able to estimate the costs associated with consuming a ton of CO_2.

The key assumptions start with the size of the power plant, combined with the percentage of flue gas to be consumed, which together set the CO_2 emission rate of the system. In this case, we based our system on a 300 MW slip stream from a 1 GW plant, i.e., a 30 % slipstream. Operating days per year (300) and an amortization period are chosen to calculate the total amount of CO_2 that would be emitted over the lifetime of the algae-based mitigation system. The overall capital costs are estimated based on the current design of the demonstration facility at East Bend Station and are normalized based on the land area that the system would occupy. The most important assumption is areal productivity, which controls the size and thereby cost of the overall system.

Although significant progress was made in reducing both the capital and recurring operating costs of the PBR employed, according to the analysis in Table 7, the current cost of capturing CO_2 falls close to $1,600/ton. Moreover, the capital cost and installation of the algae growth facility constitute 98 % of the overall cost of CO_2 capture. Further progress is clearly needed to reduce the overall capital cost of the system and thus reduce the cost of CO_2 capture. In addition, it should be noted that an amortization period of 10 years was used in the analysis. A less conservative approach would involve increasing this 10 year period to longer periods, bearing in mind that the contribution of operations and maintenance should be increased accordingly. Specifically, allowance has to be made for replacement of the PET tubes every 5 years, these comprising 8 % of the total capital cost; other parts are fabricated from PVC and are assumed to have a lifetime of >30 years. As shown in Fig. 12, extending the operating life of the facility improves the cost per ton of CO_2 mitigated considerably, although further cost reductions would require a decrease in the various cost elements associated with the PBR. A second option would be to increase the areal productivity although the scope for this seems limited given climatic constraints.

An important consideration is that this analysis takes no account of the value of the algal biomass produced. From this, it follows that there is a strong incentive to maximize this value, e.g., by conversion of the biomass to valuable products (nutraceuticals, animal food additive, premium organic fertilizer, etc.), to generate a revenue stream which can help to defray the costs of CO_2 capture/recycle. The size of the markets for algae-derived products is inversely related to the product price. Lower value products such as liquid fuels may be less attractive from a profitability perspective, but a utility-scale installation would produce a quantity of algal biomass that would inevitably oversupply lower volume, higher value product markets. However, there is no reason that utility-scale biomass utilization could not focus on developing both markets; lower value markets to utilize significant volumes, along with limited participation in higher value markets to maximize profits. By doing so, higher value market supply, and hence, profitability will be maintained.

Conclusions

Based on our initial work, we conclude that CO_2 capture and recycle using microalgae is feasible from a technical standpoint. By applying PBR technology, that was developed in-house, at East Bend Station, Kentucky, and using flue gas as the CO_2 source, algae productivity of routinely ≥ 30 g/(m^2 day) in the summer months was achieved at significant scale (18,000 L). These values compare favorably with values reported in the literature for both pond- and PBR-based cultivation studies. Moreover, average daily productivity slightly in excess of 10 g/(m^2 day) was demonstrated in the month of December and 39 g/(m^2 day) in June–July. To harvest and dewater the produced algal biomass, a protocol was developed based on flocculation and sedimentation, followed by filtration. Extraction of lipids from the harvested biomass was also demonstrated, followed by their conversion to diesel-range hydrocarbons via catalytic deoxygenation.

Conservative estimates suggest that the current cost of capturing and recycling CO_2 using this approach will fall close to \$1,600/ton CO_2 (assuming an amortization period of 10 years). The largest sources of cost reside in the algae culturing stage of the process, corresponding mainly to the capital cost of the photobioreactor system and the associated installation cost. From this it follows that future cost reduction measures should focus on the design of a culturing system which is less expensive to build and install. Even in the most optimistic scenario, the cost of algae-based CO_2 capture is unlikely to fall below \$225/ton, corresponding to a production cost of $\sim$ \$400/ton biomass. Clearly, the economics of CO_2 capture and recycle can be significantly improved if the algal biomass produced can be sold. In view of the fact that the markets for algae are in their infancy, the value of algal biomass is at present hard to quantify with the exception of its fuel value. That said, the literature suggests that several large-volume markets do exist, such as animal feed and organic fertilizer, albeit that these applications require the absence of bioaccumulated heavy metals in the biomass. In this regard, the absence of heavy metals in the algae grown at East Bend, at a detection level of 0.1 ppm, is encouraging.

Acknowledgments The authors would like to thank Dr. Anne Ware, Tonya Morgan, Landon Mills, Wade Bailey and Neil McCarthy for assistance with lipid extraction and upgrading experiments, together with Nick Rhea and many undergraduate students who contributed to the construction of the photobioreactor. We would also like to acknowledge the considerable assistance of Doug Durst (Duke Energy) with the activities at East Bend Station, and Prof. James Dawson (Pittsburg State University) for preliminary experiments. The Kentucky Department of Energy Development and Independence, Duke Energy and the University of Kentucky are thanked for financial support.

References

1. Bayless DJ, Kremer G, Vis M, Stuart B, Shi L, Ono E, Cuello JL (2006) Photosynthetic CO_2 mitigation using a novel membrane-based photobioreactor. J Environ Eng Manag 16(4):209–215
2. Benemann JR, Oswald WJ (1996) Systems and economic analysis of microalgae ponds for conversion of CO_2 to biomass. Final report. DOE/PC/93204–T5; p 214
3. Bligh EG, Dyer WJ (1959) A rapid method of total lipid extraction and purification. Can J Biochem Phys 37:911–917
4. Brennan L, Owende P (2010) Biofuels from microalgae—a review of technologies for production, processing, and extractions of biofuels and co-products. Renew Sustain Energy Rev 14:557–577
5. Campbell PK, Beer T, Batten D (2009) Greenhouse gas sequestration by algae—energy and greenhouse gas life cycle studies. CSIRO, Australia
6. Canakci M, Van Gerpen J (2001) Biodiesel production from oils and fats with high free fatty acids. ASAE Trans 4(6):1429–1436
7. Copeland CH, Paul P, Samantha W, Paul E, Richard S, David B (2003) Chemical fixation of CO_2 in coal combustion products and recycling through biosystems. TVA/DE-FC26-00NT40933; p 70
8. Crofcheck C, Shea A, E X, Montross M, Crocker M, Andrews R (2012) Influence of media composition on the growth rate of *Chlorella vulgaris* and *Scenedesmus acutus* utilized for CO2 mitigation. J Biochem Technol 4(2):589–594
9. Crofcheck C, Shea A, Montross M, Crocker M, Andrews R (2013) Influence of flue gas components on the growth rate of *Chlorella vulgaris* and *Scenedesmus acutus*. Trans ASABE 56(6):1421–1429
10. Doucha J, Straka F, Livansky K (2005) Utilization of flue gas for cultivation of microalgae (*Chlorella* sp.) in an outdoor open thin-layer photobioreactor. J Appl Phycol 17(5):403–412
11. Fagerstone KD, Quinn JC, Bradley TH, De Long SK, Marchese AJ (2011) Quantitative measurement of direct nitrous oxide emissions from microalgae cultivation. Environ Sci Technol 45:9449–9456
12. Farrelly DJ, Everard CD, Fagan CC, McDonnell KP (2013) Carbon sequestration and the role of biological mitigation: a review. Renew Sustain Energy Rev 21:712–727
13. Grima EM, Fernandez A, Camacho FG, Chisti Y (1999) Photobioreactors: light regime, mass transfer, and scaleup. J Biotechnol 70:231–247
14. Johnson DA, Sprague S (1987). FY 1987 Aquatic species program: annual report. SERI/SP-231-3206; p 269
15. Harter T, Bossier P, Verreth J, Bodé S, Van der Ha D, Debeer A, Boon N, Boeckx P, Vyverman W, Nevejan N (2013) Carbon and nitrogen mass balances during flue gas treatment with *Dunaliella salina* cultures. J Appl Phycol 25:359–368
16. Kadam KL (2001) Microalgae production from power plant flue gas: environmental implications on a life cycle basis. NREL/TP-510-29417; p 55
17. Laws EA, Berning JL (1991) A study of the energetics and economics of microalgal mass culture with the marine

chlorophyte *Tetraselmis suecica*: implications for use of power plant stack gases. Biotechnol Bioeng 37:936–947

18. Morgan T, Grubb D, Santillan-Jimenez E, Crocker M (2010) Conversion of triglycerides to hydrocarbons over supported metal catalysts. Top Catal 53:820–829

19. Morgan T, Santillan-Jimenez E, Harman-Ware AE, Ji Y, Grubb D, Crocker M (2012) Catalytic deoxygenation of triglycerides to hydrocarbons over supported nickel catalysts. Chem Eng J 189–190:346–355

20. Nakamura T, Senior CL (2005) Recovery and sequestration of CO_2 from stationary combustion systems by photosynthesis of microalgae. DOE/PSI-1356/TR-2016; p 200

21. Nielsen SL, Enriquez S, Duarte CM, Sand-Jensen K (1996) Scaling maximum growth rates across photosynthetic organisms. Funct Ecol 10(2):167–175

22. Peng B, Yao Y, Zhao C, Lercher JA (2012) Towards quantitative conversion of microalgae oil to diesel-range alkanes with bifunctional catalysts. Angew Chem Int Ed 51:2072–2075

23. Peng B, Yuan X, Zhao C, Lercher JA (2012) Stabilizing catalytic pathways via redundancy: selective reduction of microalgae oil to alkanes. J Am Chem Soc 134:9400–9405

24. Santillan-Jimenez E, Crocker M (2012) Catalytic deoxygenation of fatty acids and their derivatives to hydrocarbon fuels via decarboxylation/decarbonylation. J Chem Technol Biotechnol 87:1041–1050

25. Santillan-Jimenez E, Morgan T, Shoup J, Harman-Ware AE, Crocker M (2013) Catalytic deoxygenation of triglycerides and fatty acids to hydrocarbons over Ni–Al layered double hydroxide. Catal Today, (in press)

26. Santillan-Jimenez E, Morgan T, Lacny J, Mohapatra S, Crocker M (2012) Catalytic deoxygenation of triglycerides and fatty acids over carbon-supported nickel. Fuel 103:1010–1017

27. Seambiotic http://www.seambiotic.com/. Accessed June 2013

28. Sheehan J, Dunahay T, Benemann J, Roessler P (1998) Look back at the US Department of Energy's Aquatic Species Program: biodiesel from algae; close-out report. NREL/TP-580-24190; p 325

29. Spolaore P, Joannis-Cassan C, Durn E, Isambert A (2006) Commercial applications of microalgae. J Biosci Bioeng 101(2):87–96

30. Tredici MR (2004) In: Richmond A (ed) Handbook of microalgal culture: biotechnology and applied phycology. Blackwell Science Ltd, Oxford, pp 178–214

31. Vunjak-Novakovic G, Kim Y, Wu XX, Berzin I, Merchuk JC (2005) Air-lift bioreactors for algal growth on flue gas: mathematical modeling and pilot-plant studies. Ind Eng Chem Res 44(16):6154–6163

32. Weissman JC, Goebel RP, Benemann JR (1988) Photobioreactor design: mixing, carbon, utilization, and oxygen accumulation. Biotechnol Bioeng 31:336–344

33. Williams PJB, Laurens LML (2010) Microalgae as biodiesel and biomass feedstocks: review and analysis of the biochemistry, energetics and economics. Energy Environ Sci 3(5):554–590

34. Zhao C, Brück T, Lercher JA (2013) Catalytic deoxygenation of microalgae oil to green hydrocarbons. Green Chem 15(1720): 1739

7

A kinetic model for ethylene oligomerization using zirconium/aluminum- and nickel/zinc-based catalyst systems in a batch reactor

Adil A. Mohammed · Seif-Eddeen K. Fateen ·
Tamer S. Ahmed · Tarek M. Moustafa

Abstract The aim of this work is to develop a kinetic model of the oligomerization of ethylene to linear alpha olefins (LAOs) for zirconium/aluminum and nickel/zinc catalyst systems. The development of such model helps in the study of the behavior of industrial LAOs reactors as well as in the optimization of their operation. The kinetic model was developed based on a four-step mechanism: site activation, initiation and propagation, chain transfer and site deactivation. A novel stochastic optimization algorithm, Intelligent Firefly Algorithm, was used to obtain the kinetic model parameters that best fit the available experimental data that were obtained from published sources. The values of the kinetic parameters were obtained for the developed kinetic models for two catalyst systems. The performance of the model with the estimated parameters was tested against the experimental data. The proposed kinetic model predicts the product distribution for the zirconium/aluminum catalyst system with suitable accuracy. The model can also predict the product distribution for the nickel/zinc catalyst system with good accuracy for all products. As expected, the accuracy of the model to predict the concentration of the higher carbon products decreases with the carbon number.

Keywords Ethylene · Oligomerization · Modeling · Zirconium/aluminum catalyst · Nickel/zinc catalyst

Abbreviation

Notation

A	Pre-exponential factor
C_{CAT}	Catalyst concentration, mol/l
C_{CAT}^k	Active catalyst concentration, mol/l
$C_{CAT^k \cdot M}$	Complex active catalyst/ethylene concentration, mol/l
C_{decy}	Moles of deactivated catalysts, mol
C_M	Concentration of ethylene monomer, mol/l
C_M^k	Concentration of active ethylene monomer, mol/l
$C_{M^k \cdot TEA}$	Concentration of complex active ethylene monomer/co-catalyst, mol/l
C_{P_0}	Concentration of active site, mol/l
C_{P_i}	Concentration of living polymers, mol/l
$C_{P_i^k}$	Concentration of active living polymers, mol/l
$C_{P_i^k \cdot TEA}$	Concentration of complex active living polymers/co-catalyst, mol/l
C_{TEA}	Co-catalyst concentration, mol/l
C_{TEA}^k	Active co-catalyst concentration, mol/l
$C_{TEA^k \cdot CAT}$	Complex active co-catalyst/catalyst concentration, mol/l
$C_{TEA_1^k \cdot CAT}$	Complex active co-catalyst/catalyst concentration-catalyst, mol/l
D	Moles of dead oligomer, mol
E	Activation energy, cal/mol
k_1	Rate constant of active site
k_2	Rate constant of chain initiation
k_3	Rate constant of chain propagation
k_4	Rate constant of chain transfer
k_5	Rate constant of deactivation
k_{+1}	Rate constant of attachment of the catalyst in the site activation
k_{-1}	Rate constant of detachment of the catalyst in the site activation

A. A. Mohammed · S.-E. K. Fateen · T. S. Ahmed (✉) ·
T. M. Moustafa
Chemical Engineering Department, Faculty of Engineering,
Cairo University, Giza 12613, Egypt
e-mail: tamer.s.ahmed@eng1.cu.edu.eg

k_{+2} — Rate constant of attachment of the monomer in the site activation

k_{-2} — Rate constant of detachment of the monomer in the site activation

k_{+4} — Rate constant of attachment in the chain propagation reaction

k_{-4} — Rate constant of detachment in the chain propagation reaction

k_{+5} — Rate constant of attachment in the chain transfer reaction

k_{-5} — Rate constant of detachment in the chain transfer reaction

k'_c — Rate constant of the first reaction of co-catalyst in the site activation

k''_c — Rate constant of the second reaction of co-catalyst in the site activation

K_A — Equilibrium constant of the catalyst in the site activation

K_B — Equilibrium constant of the monomer in the site activation

K_C — Equilibrium constant of the co-catalyst in the site activation

K_D — Equilibrium constant in the chain propagation reaction

K_E — Equilibrium constant in the chain transfer reaction

M — Ethylene monomer

M^k — Active ethylene monomer

P_i — Living oligomers

P_i^k — Active living oligomers

P_0 — Active site

R — Gas constant, mol/cal K

T — Temperature, K

T_r — Reference temperature, K

v — Volume, l

Greek letters

α Alpha position

Subscripts

c   Calculated mole fractions of product

e   Experimental mole fractions of product

CAT   Catalyst

CAT^k Active catalyst

TEA   Co-catalyst

TEA^k Active co-catalyst

Introduction

Linear alpha olefins (LAOs) are linear hydrocarbons having a double bond between the first and second carbon atoms. Linear α-olefins have found wide application in various areas of petrochemical synthesis. They are used for copolymerization with ethylene for the purpose of obtaining low- and high-density linear polyethylene and for the preparation of detergents and synthetic lubricants [12, 14, 41, 48, 54, 61, 62]. They can also be used in the production of surfactants, agents for enhanced oil recovery, corrosion inhibitors, high-performance lubricating oils, linear alkyl-benzenes, oxo-synthesis alcohols, α-olefin sulfonates, oil additives, and as drilling fluids [1, 3, 6, 28, 35, 51, 59]. The worldwide growth rate of LAOs production between 1980 and 1990 amounted to 10–15 wt% per year. From 1990 to 2000, the annual increase in LAO production came to 4.5–5 % wt%. Their average annual demand over the next 10–15 years is expected to increase by 3.5 wt %. The total amounts of LAOs manufactured worldwide of years 1999, 2003, and 2005 were 2.6, 3.5, and 4.2 million tons, respectively [7].

Ethylene oligomerization is one of the vital industrial processes in which linear alpha olefins can be produced. The products of Ethylene oligomerization can be further separated into different LAOs based the carbon number of the product [9, 38]. Different industrial processes are used to produce LAOs and extensive research studies are under way on the oligomerization of ethylene in different catalysts and processes conditions. The companies that produce alpha olefins via ethylene oligomerization include Ethyl, Chevron, Shell, Idemitsu, Mitsubishi, IFP-Axens, UOP, and Sabic-Linde [7, 15, 38, 45]. These processes are catalyzed by different Ziegler–Natta catalysts. Investigations of catalytic oligomerization of ethylene have been mainly focused on the complexes of titanium, zirconium, chromium, and nickel for decades. Complexes of titanium and nickel are the most often used catalysts for ethylene oligomerization [47, 49, 58]. Zirconium complexes have also been found to be very active catalyst in this reaction [5, 33, 34, 42]. The use of other transition metal complexes in ethylene oligomerization is uncommon. [3, 4, 17, 23, 30, 33, 43, 44, 53, 63]

Several authors proposed models to describe the kinetics of the polymerization processes on different catalysts. Zacca and Ray [57] discussed a new model for the polymerization of olefins in a loop reactor. Their reactor was modeled as a two-tubular interconnected reactor. They studied the effects of recycle rate, axial dispersion and heat transfer. Sharma and coworkers [8] developed a mathematical model for the isothermal, slurry polymerization of ethylene using solid Ziegler–Natta catalysts. They discussed the effect of gas liquid mass transfer limitations on the overall rate and polymer properties. Kiparissides [36] produced synthetic polymers via a multitude of reaction mechanisms and processes, including addition and step growth polymerizations. He provided an overview of the different polymerization processes and the mathematical modeling approaches and addressed the problems related with the computer-aided design, monitoring, optimization and control of polymerization reactors.

Dube et al. [19] developed a practical methodology for the computer modeling of multi-component chain growth polymerizations, which is applicable to many multi-component systems. Various co-monomer systems were used to illustrate the development of practical semi-batch and continuous reactor operational policies for the manufacture of copolymers with high quality and productivity. Hatzantonis et al. [24] reviewed recent developments in modeling gas phase olefin polymerization fluidized-bed (FBR) reactors in the presence of a multi-site Ziegler–Natta catalyst. In addition, they developed an FBR model to account for the effect of varying of bubble size with the bed height on the reactor dynamics and the molecular properties of the polymer product. They studied the effect of important reactor parameters on the dynamic and steady-state behavior of the FBR. Alhumaizi [2] developed a dynamic mathematical model for the ethylene dimerization reactor to simulate the distribution of butene-1 and other oligomers in the products under different operating conditions. He studied the effects of the recycle flow rate and the cooler heat transfer coefficient for optimum conditions for production of butene-1. Soares [46] reviewed the principal of mathematical modeling techniques for describing the microstructure of polyolefins produced by coordination polymerization. Fernandes and Lona [21, 22] developed a heterogeneous model describing the behavior of the three-phase fluidized-bed reactor in polymer production. The model incorporated the interactions between phases and also predicted the physicochemical properties of the polymer.

Yoon et al. [56] reviewed recent developments in modeling techniques for the calculation of polymer properties and the application of process models to the design of model-based reactor optimizations and controls. Jin et al. [29] developed a mathematical model to simulate the ethylene polymerization in continuous stirred tank reactor. The molecular properties of various polymers were calculated based on the method of moment. Valencia and Soares [52] developed a simulation model for the polymerization of ethylene in a process with (n) reactors working in series, which could predict raw material conversions and product properties. Their model parameters were obtained from laboratory data. Zhang et al. [60] developed a mathematical model to describe the effect of different polymerization conditions on the preparation of ethylene-1-hexene copolymerization with a tandem catalysis system. The general feature of the model simulation agreed well with both their experimentation data and other results reported in the literature. Ghasem and coworkers [26, 27] developed a modified dynamic three-phase structure model that took into account the presence of particles participating in the reaction with emulsion and catalyst phases. The control system was discussed using of neural network controller. In addition, they modified the model to include heat and mass transfer between the bubble and the cloud as well as between the cloud and the emulsion phases. Touloupides et al. [50] developed a comprehensive mathematical model, which was capable of simulating the dynamic operation of an industrial slurry phase cascade loop reactor series under different plant operating conditions. They used multi-site ZN kinetic and method of moments for calculation of polymer molecular properties.

However, to the best of our knowledge, there has been no published study that proposed kinetic models for the oligomerization of ethylene with zirconium-based catalyst with aluminum-based co-catalyst or with DPA(Na)/$NiCl_2 \cdot 6H_2O$/Zn catalyst system. The objective of this work is to develop two kinetic models for the said catalyst systems. The remainder of this paper is divided as follows. "Model formulation" introduces the model formulation. The experimental data are introduced briefly in "Experimental data". "Parameters estimation" describes the methodology of estimating the parameters of the kinetic model using the Intelligent Firefly Algorithm. "Result and discussion" presents the results and discusses its significance. The conclusions of this study are summarized in "Conclusion".

Model formulation

Many studies focused on the kinetics of olefin polymerization using Ziegler–Natta catalysts. The polymerization reactions occur at several reactive sites on the catalyst. Due to the complexity of the process of polymerization and oligomerization of olefins with Ziegler–Natta catalysts, the mechanism of the reaction is usually divided into a number of reaction steps [10, 11, 13, 16, 25, 31, 32, 37, 39, 57]. These steps are:

Site activation

The active site is formed by the reaction between the catalyst with co-catalyst. The co-catalyst acts as an alkylating and reducing agent.

Initiation and propagation

The active site having a coordination vacancy attracts the electrons in the olefin π-bond. Coordination is followed by the insertion into the polymer chain (R) and the re-establishment of the coordination vacancy for further monomer insertion.

Chain transfer

Several chain transfer mechanisms are applied in Z–N catalyst such as transfer by β-hydride elimination, transfer to monomer, and transfer to co-catalyst.

Site deactivation

For some polymerizations, the rate of reaction decreases with time due to the catalyst deactivation reaction. The two active sites form the deactivated catalyst, which is inactive for monomer oligomerization. Another deactivation reaction forms a dead polymer chain and a dead catalyst site.

Model mechanism

The proposed model is limited to the production of α-olefins up to C_{20}. The oligomerization reaction can be broken into five parts; catalyst activation, chain initiation, chain propagation, chain transfer and catalyst deactivation.

Model features

The following points summarize the main features of the model.

1. The concentrations of the catalyst (CAT), ethylene (M), co-catalyst (TEA), and oligomers (P_i^k) are divided into active and inactive stages.
2. The catalyst, ethylene, and co-catalyst are involved in the reactions as follows:

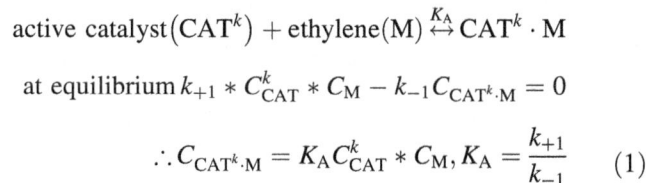

$$\text{active catalyst}(\text{CAT}^k) + \text{ethylene}(\text{M}) \overset{K_A}{\leftrightarrow} \text{CAT}^k \cdot \text{M}$$

at equilibrium $k_{+1} * C_{\text{CAT}}^k * C_M - k_{-1} C_{\text{CAT}^k \cdot M} = 0$

$$\therefore C_{\text{CAT}^k \cdot M} = K_A C_{\text{CAT}}^k * C_M, K_A = \frac{k_{+1}}{k_{-1}} \quad (1)$$

balance of $C_{\text{CAT}}: C_{\text{CAT}} = C_{\text{CAT}}^k + C_{\text{CAT}^k \cdot M}$

$$C_{\text{CAT}}^k = \frac{C_{\text{CAT}}}{K_A C_M + 1}$$

$$\text{active ethylene}(\text{M}^k) + \text{cocatalyst (TEA)} \overset{K_B}{\leftrightarrow} \text{M}^k \cdot \text{TEA}$$

at equilibrium $k_{+2} * C_M^k * C_{\text{TEA}} - k_{-2} C_{\text{M}^k \cdot \text{TEA}} = 0$

$$\therefore C_{\text{M}^k \cdot \text{TEA}} = K_B C_M^k * C_{\text{TEA}}, K_B = \frac{k_{+2}}{k_{-2}}$$

balance of $C_M: C_M = C_M^k + C_{\text{M}^k \cdot \text{TEA}}$

$$C_M^k = \frac{C_M}{K_B C_{\text{TEA}} + 1}$$

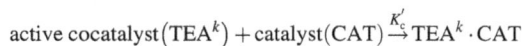

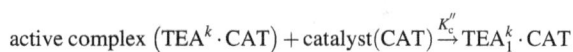

$$(2)$$

$$\text{active cocatalyst}(\text{TEA}^k) + \text{catalyst}(\text{CAT}) \overset{K_c'}{\to} \text{TEA}^k \cdot \text{CAT}$$

$$\text{active complex } (\text{TEA}^k \cdot \text{CAT}) + \text{catalyst}(\text{CAT}) \overset{K_c''}{\to} \text{TEA}_1^k \cdot \text{CAT}$$

$$C_{\text{TEA}^k \cdot \text{CAT}} = k_c' C_{\text{TEA}}^k C_{\text{CAT}}$$

$$C_{\text{TEA}_1^k \cdot \text{CAT}} = k_c'' C_{\text{TEA}^k \cdot \text{CAT}} C_{\text{CAT}} = k_c'' k_c' C_{\text{TEA}}^k C_{\text{CAT}}^2$$

balance of $C_{\text{TEA}}: C_{\text{TEA}} = C_{\text{TEA}}^k + C_{\text{TEA}^k \cdot \text{CAT}} + C_{\text{TEA}_1^k \cdot \text{CAT}}$ $\quad (3)$

$$= C_{\text{TEA}}^k + k_c' C_{\text{TEA}}^k C_{\text{CAT}} + k_c'' k_c' C_{\text{TEA}}^k C_{\text{CAT}}^2$$

$$= C_{\text{TEA}}^k \left(1 + k_c' C_{\text{CAT}} + k_c'' k_c' C_{\text{CAT}}^2\right)$$

Assuming that $k_c' = 2K_c, k_c'' = \dfrac{k_c'}{4}$

$$C_{\text{TEA}}^k = \frac{C_{\text{TEA}}}{(K_C C_{\text{CAT}} + 1)^2}$$

3. The oligomers of propagation proceed as in the following reaction

$$\text{TEA} + \text{active oligomers}(P_i^k) \overset{K_D}{\leftrightarrow} P_i^k \cdot \text{TEA}$$

at equilibrium $k_{+4} * C_{P_i}^k * C_{\text{TEA}} - k_{-4} C_{P_i^k \cdot \text{TEA}} = 0$

$$\therefore C_{P_i^k \cdot \text{TEA}} = K_D C_{P_i}^k * C_{\text{TEA}}, K_D = \frac{k_{+4}}{k_{-4}}$$

balance of $C_{P_i}: C_{P_i} = C_{P_i}^k + C_{P_i^k \cdot \text{TEA}} C_{P_i}^k = \frac{C_{P_i}}{K_D C_{\text{TEA}} + 1}$

$$(4)$$

4. The oligomers of chain transfer proceed as in the following reaction:

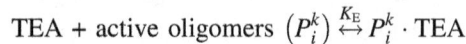

$$\text{TEA} + \text{active oligomers } (P_i^k) \overset{K_E}{\leftrightarrow} P_i^k \cdot \text{TEA}$$

at equilibrium $k_{+5} * C_{P_i}^k * C_{\text{TEA}} - k_{-5} C_{P_i^k \cdot \text{TEA}} = 0$

$$\therefore C_{P_i^k \cdot \text{TEA}} = K_E C_{P_i}^k * C_{\text{TEA}}, K_E = \frac{k_{+5}}{k_{-5}} \quad (5)$$

balance of $C_{P_i}: C_{P_i} = C_{P_i}^k + C_{P_i^k \cdot \text{TEA}}$

$$C_{P_i^k} = \frac{C_{P_i}}{K_E C_{\text{TEA}} + 1}$$

5. The activator affects the rate equations of the chain propagation and chain transfer by rate law of C_{TEA}^a and C_{TEA}^b, respectively.
6. Only one single site is involved in the catalytic process.
7. The co-catalyst does not deactivate.
8. The only transfer mechanism included is the transfer to monomer. The transfer reactions are assumed to produce monomer (M) and catalyst (C_{CAT}).

The mechanism of the reaction is as follows:

Site activation

The active catalyst (CAT^k), active ethylene (M^k), and active co-catalyst (TEA^k) react to form active site (P_0). The active center for olefins in Ziegler–Natta catalyst systems is a complex acting as an alkylating and reducing agent.

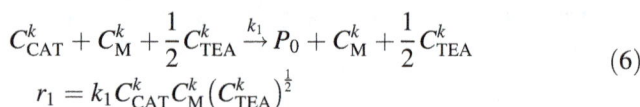

$$C_{CAT}^k + C_M^k + \frac{1}{2}C_{TEA}^k \xrightarrow{k_1} P_0 + C_M^k + \frac{1}{2}C_{TEA}^k$$
$$r_1 = k_1 C_{CAT}^k C_M^k \left(C_{TEA}^k\right)^{\frac{1}{2}} \tag{6}$$

Chain initiation

After the catalyst is activated, it is ready to polymerize with monomer. The active site that has a coordination vacancy reacts with an ethylene monomer to form a live oligomer chain of length one (P_1^k).

$$P_0 + M \xrightarrow{k_2} P_1^k \quad r_2 = k_2 C_{P_0} C_M \tag{7}$$

Chain propagation

Coordination is followed by the insertion of a monomer into the chain, which increases the chain length by one monomer unit until chain transfer takes place.

$$P_i^k + M \xrightarrow{k_3} P_{i+1} \quad r_3 = k_3 C_{P_i^k} C_M C_{TEA}^a \tag{8}$$

Chain transfer

Chain transfer to monomer may take place if the growing chain abstracts an atom from an unreacted monomer existing in the reaction medium. This leads to the death of the active chain (P_i^k) to form a dead chain (De_i) and begins a new one.

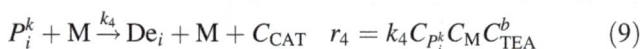

$$P_i^k + M \xrightarrow{k_4} De_i + M + C_{CAT} \quad r_4 = k_4 C_{P_i^k} C_M C_{TEA}^b \tag{9}$$

Site deactivation

Catalyst deactivation happens when active sites form a stable complex that is inactive for monomer oligomerization. Most Ziegler–Natta catalysts deactivate to form deactivation site (C_d) and/or dead chain (De_i).

$$P_0 \xrightarrow{k_5} C_d \quad r_{5,0} = k_5 C_{P_0} \tag{10}$$

$$P_i \xrightarrow{k_5} C_d + De_i \quad r_{5,i} = k_5 P_i \tag{11}$$

Kinetic rate constants are known to be temperature dependent; this dependence is usually described using Arrhenius relationships. The comparative form of the Arrhenius equation was used to estimate kinetic rate constants to improve the convergence in parameter estimation; the rate constants take the form:

$$k_i = A_i \exp\left(\frac{E_i}{R}\left(\frac{1}{T} - \frac{1}{T_r}\right)\right) \tag{12}$$

Population balance equations in batch reactors

Population balances are defined for all active species in a batch reactor. Population balance equations describe how the concentrations of living and dead chains of different lengths vary in time during the polymerization in a batch reactor. The material balance for driving population balances in batch reactor for the volume reactor (v) is:

rate of accumulation = rate of generation by chemical reaction

Population balance for active site:

$$\frac{1}{v}\frac{dP_0}{dt} = \left(r_1 - r_2 - r_{5,0}\right) \tag{13}$$

For living oligomer chains with $i \geq 2$, the following population balance can be derived:

$$\frac{1}{v}\frac{dP_i}{dt} = \left(r_{3,i-1} - r_{3,i} - r_{5,i}\right) \tag{14}$$

For living oligomer chains with $i = 1$, a slightly different equation is applied:

$$\frac{1}{v}\frac{dP_1}{dt} = \left(r_2 - r_{3,1} - r_{5,1}\right) \tag{15}$$

Similarly, the equations can be derived for dead oligomer formation for $i = 2$–10

$$\frac{1}{v}\frac{dD_i}{dt} = \left(r_{4,i} + r_{5,i}\right) \tag{16}$$

The concentration of the catalyst deactivation can be derived as follows:

$$\frac{1}{v}\frac{dC_{decy}}{dt} = \sum_{i=1}^{10} r_{5i} \tag{17}$$

Experimental data

Kinetic modeling in the current study was performed using experimental data obtained from the literature. All experimental data were done in slurry batch reactors. Two sets of experimental data were used in this study to estimate the model parameters for two different catalyst systems. Experimental data (1) are for zirconium-based catalyst and aluminum-based co-catalyst system with molar ratio of 17–45 between catalyst and co-catalyst, range of temperature 65–95 °C and pressure 2–4 MPa [40]. Experimental data (2) are for DPA(Na)/NiCl$_2$·6H$_2$O/Zn catalyst system with ratio of 1.5 between DPA

(Na) and $NiCl_2 \cdot 6H_2O$ and a molar ratio of 2–20 between zinc and $NiCl_2 \cdot 6H_2O$ at 95 °C and 5.5–6.0 MPa [18].

Parameters estimation

Regression was performed to obtain the parameters that best fit the model to the experimental data through minimization of the difference between experimental points and model predictions. The objective function for estimation of the model parameters is based on the difference in mole fractions of the products, and it takes the form

$$\text{Min} \left(\sum_{i=1}^{n} \sum_{j=1}^{m} \left(y_{e,i,j} - y_{c,i,j} \right)^2 \right) \qquad (18)$$

The minimization routine used was a stochastic global optimization routine named the Intelligent Firefly Algorithm (IFA) [20]. Stochastic global optimization algorithms have significant advantages over deterministic optimization methods and thereby deemed more appropriate for our application. IFA is a metaheuristic algorithm, inspired by the flashing behavior of fireflies. The primary purpose for a firefly's flash is to act as a signal system to attract other fireflies, a metaheuristic is a procedure designed to find a good solution to a difficult optimization problem. The Firefly algorithm was developed by Xin-She Yang [55] and it is based on idealized behavior of the flashing characteristics of fireflies. All fireflies are unisex, so that one firefly is attracted to other fireflies regardless of their sex. Attractiveness is proportional to their brightness, thus for any two flashing fireflies, the less bright one will move towards the brighter one. The attractiveness is proportional to the brightness and they both decrease as their distance increases. If no one is brighter than a particular firefly, it will move randomly. The brightness or light intensity of a firefly is determined by the value of the objective function of a given problem.

Results and discussion

The values of the kinetic parameters for both experimental data used are tabulated and discussed as follows.

Results of experimental data for zirconium/aluminum catalyst system

Values of the proposed model kinetic parameters of experimental data for zirconium/aluminum catalyst system are shown in Table (1).

Table 1 Parameters estimated value for experimental data (1)

Parameter	Estimated value	Unit	Expression
A1	4.525×10^5	l/(mol s)	A factor of activation
E1	3.985×10^3	cal/mol	Activation energy of activation and deactivation
A2	5.101×10^8	l/(mol s)	A factor of initiation
E2	5.267×10^4	cal/mol	Activation energy of initiation
A3	4.949×10^8	l/(mol s)	A factor of propagation
E3	1.514×10^4	cal/mol	Activation energy of propagation
A4	4.650×10^5	l/(mol s)	A factor of Chain transfer
E4	1.433×10^4	cal/mol	Activation energy of Chain transfer
A5	1.255	1/s	A factor of deactivation
K_A	4.812×10^2	l/mol	Equilibrium constant of the catalyst in the site activation
K_B	2.771×10^2	l/mol	Equilibrium constant of the monomer in the site activation
K_C	6.906×10^3	l/mol	Equilibrium constant of the co-catalyst in the site activation
K_D	4.083×10^4	l/mol	Equilibrium constant in the chain propagation reaction
K_E	5.955×10^1	l/mol	Equilibrium constant in the chain transfer reaction
a	1.116		
b	0.954		

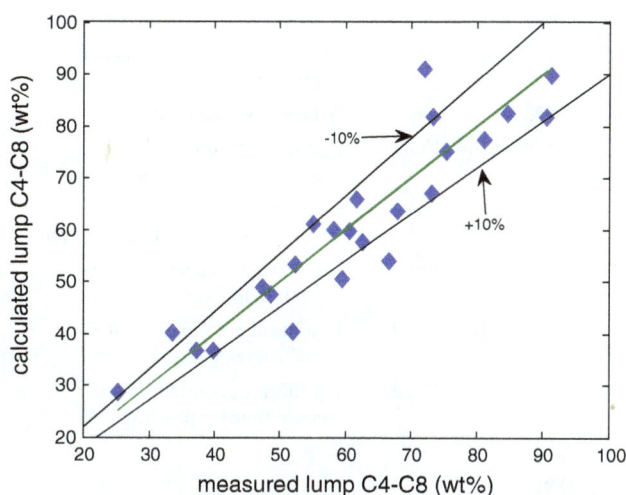

Fig. 1 Comparison calculated and measured data of lumped C_4, C_6 and C_8 product

Figure (1) shows the parity plot for lumped butane, hexane and octane using the estimated parameters. Each point represents one experimental run. Almost 80 % of the points are predicted with <10 % error lines. Figure (2) shows the experimental and calculated points for C_{10+}. It is clear that many runs are predicted outside the 20 % error boundary lines.

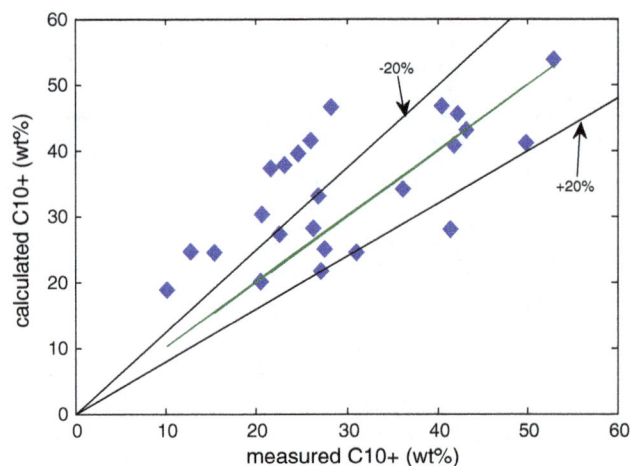

Fig. 2 Comparison calculated and measured data of C_{10+} product

Table 2 Kinetic parameters of experimental data (2)

Parameter	Estimated value	Unit	Expression
A1	4.447×10^6	l/(mol s)	A factor of activation
E1	1.455×10^3	cal/mol	Activation energy of activation and deactivation
A2	6.842×10^8	l/(mol s)	A factor of initiation
E2	4.757×10^4	cal/mol	Activation energy of initiation
A3	5.928×10^5	l/(mol s)	A factor of propagation
E3	1.237×10^4	cal/mol	Activation energy of propagation
A4	2.634×10^5	l/(mol s)	A factor of Chain transfer
E4	1.137×10^4	cal/mol	Activation energy of Chain transfer
A5	0.973	1/s	A factor of deactivation
K_A	8.546×10^4	l/mol	Equilibrium constant of the catalyst in the site activation
K_B	1.464×10^4	l/mol	Equilibrium constant of the monomer in the site activation
K_C	9.750×10^3	l/mol	Equilibrium constant of the co-catalyst in the site activation
K_D	7.946×10^4	l/mol	Equilibrium constant in the chain propagation reaction
K_E	9.604×10^3	l/mol	Equilibrium constant in the chain transfer reaction
a	0.845		
b	0.919		

Results for experimental data for nickel/zinc catalyst system

Values of the proposed models' kinetic parameters for experimental data for nickel/zinc catalyst system are shown in Table (2).

Figure (3) shows the parity plot of lumped butane, hexane and octane using the estimated parameters. All the

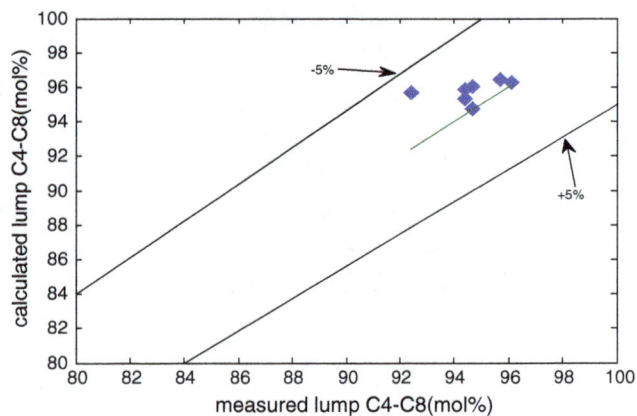

Fig. 3 Comparison calculated and measured data of lumped C_4, C_6 and C_8 product

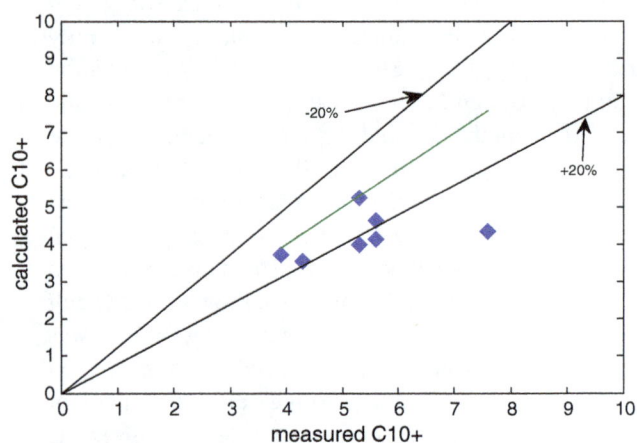

Fig. 4 Comparison calculated and measured data of C_{10+} product

points lay between the 5 % error lines for all three products. Figure (4) shows the experimental and calculated points for C_{10+} production. Some of the runs are predicted outside the 20 % error boundary lines.

The results show that the proposed model predicts both experimental data for zirconium/aluminum and nickel/zinc catalyst systems in a different way. The prediction of the model is better for lump C_4–C_8 compared to C_{10+}. For lump C_4–C_8, the model predictions for the zirconium/aluminum catalyst system lie mostly between the 10 % confidence limits. On the other hand, for the nickel/zinc catalyst system, the model predictions for lump C_4–C_8 lie between the 5 % confidence limits. As for the C_{10+}, the prediction of the model is lower for both systems. In both catalyst systems, most of the points lie between the 20 % confidence limits.

The results show that the proposed model fits the experimental data for nickel/zinc catalyst system with higher accuracy as compared to the experimental data for zirconium/aluminum catalyst system. This may be attributed to that for the zirconium/aluminum catalyst system, the experimental data for the reaction are available over a wide temperature range, while for the nickel/zinc catalyst system, the experimental data are only available at a constant temperature.

Finally, the model prediction is better for low carbon atoms. As the number of carbon atoms increases, the concentration of the polymer decreases and the model fails to predict the product concentration. The formation of higher carbon number polymer is a complex process that may not be well represented by this simple model. Relaxing some of the model assumptions may positively affect the results of its prediction.

Conclusions

A kinetic model of the oligomerization of ethylene was proposed in this study. Two sets of experimental data were used for the estimation of the model parameters. The performance of the model with estimated parameters has been tested against the experimental data. The proposed model predicts the product distribution for experimental data for zirconium/aluminum catalyst system with a suitable accuracy. The model predictions of the concentration for the components of C_4 up to C_8 fit within the 10 % confidence limits, while the absolute errors for C_{10+} were higher between the 20 % confidence limits. Similarly, the model could predict the experimental data for nickel/zinc catalyst system well. The prediction accuracy was better in comparison to the prediction of the zirconium/aluminum catalyst kinetic model. This may be attributed to that for the zirconium/aluminum catalyst system, the experimental data for the reaction are available over a wide temperature range, while for the nickel/zinc catalyst system, the experimental data are only available at a constant temperature. Finally, the model prediction is better for low carbon atoms. Relaxing some of the model assumptions may positively affect the results of its prediction. In all cases, the importance of the model lies in its ability to predict the concentration of the low carbon number oligomers.

References

1. Agapie T (2011) Selective ethylene oligomerization: recent advances in chromium catalysis and mechanistic investigations. Coord Chem Rev 255:861–880

2. Alhumaizi KI (2000) Stability Analysis of the Ethylene Dimerization Reactor for the Selective Production of Butene-1. Chem Eng Res Des 78:492–497

3. Aliyev V, Mosa F, Al-Hazmi M (2010) Catalyst composition for oligomerization of ethylene oligomerization process and method for its preparation. US 2010/0292423 A1 1–5

4. Al-Jarallah AM, Anabtawi JA, Siddiqui MAB, Aitani AM, Al-Sa'doun AW (1992) Ethylene dimerization and oligomerization to butene-1 and linear α-olefins: a review of catalytic systems and processes. Catal Today 14:1–121

5. Attridger J, Jackson R, Maddock J, Thompson T (1973) Ethylene oligomerisation with zirconium arylalkyls and alkenyls. J Chem Soc Chem Commun 1973:132–133

6. Bahuleyan BK, Ahn IY, Appukuttan V, Lee SH, Ha C-S, Kim I (2010) Ethylene oligomerization by tridentate cobalt complexes bearing pendant donor modified α-diimine ligands. Macromol Res 18:701–704

7. Belov GP, Matkovsky PE (2010) Processes for the production of higher linear α-olefins. Pet Chem 50:283–289

8. Bhagwat MS, Bhagwat SS, Sharma MM (1994) Mathematical modeling of the slurry polymerization of ethylene: gas-liquid mass transfer limitations. Ind Eng Chem Res 33:2322–2330

9. Bianchini C, Giambastiani G, Rios IG, Mantovani G, Meli A, Segarra AM (2006) Ethylene oligomerization, homopolymerization and copolymerization by iron and cobalt catalysts with 2,6-(bis-organylimino)pyridyl ligands. Coord Chem Rev 250:1391–1418

10. Bohm LL (1978) Ethylene polymerization process with a highly active Ziegler–Natta catalyst: 1. Kinet Polym (Guildf) 19:553–561

11. Bohm LL (1978) Reaction model for Ziegler–Natta polymerization processes. Polym (Guildf) 19:545–552

12. Cai T (1999) Studies of a new alkene oligomerization catalyst derived from nickel sulfate. Catal Today 51:153–160

13. De Carvalho AB, De Gloor PE, Hamielec AE (1989) A kinetic mathematical model for heterogeneous Ziegler–Natta copolymerization. Polym (Guildf) 30:280–296

14. Chandran D, Kwak CH, Oh JM, Ahn IY, Ha C-S, Kim I (2008) Ethylene oligomerizations by sterically modulated salicylaldimine cobalt(II) complexes combined with various alkyl aluminum cocatalysts. Catal Lett 125:27–34

15. Choo H, Kevan L (2001) Catalytic study of ethylene dimerization on Ni (II) -exchanged clinoptilolite. J Phys Chem B 105:6353–6360

16. Deffieux A, Fontanille M, Ribeiro MR, Portela MF (1996) Kinetic investigation of parameters governing the high-temperature polymerization of ethylene initiated by supported VCl3 catalytic systems. Eur Polym J 32:811–819

17. Dixon JT, Green MJ, Hess FM, Morgan DH (2004) Advances in selective ethylene trimerisation—a critical overview. J Organomet Chem 689:3641–3668

18. Dong-bing Liu DL (1998) Catalytic oligomerization of ethylene to lower α-olefins by the catalyst system DPA(Na)/NiC12.6H2O/Zn. Appl Catal A Gen 166:L255–L258

19. Dube MA, Soares JBP, Penlidis A, Hamielec AE (1997) Mathematical modeling of multicomponent chain-growth polymerizations in batch, semibatch, and continuous reactors: a review. Ind Eng Chem Res 36:966–1015

20. Fateen S-EK, Bonilla-Petriciolet A (2013) Intelligent firefly algorithm for global optimization. In: X. Yang (ed) Springer, Germany

21. Fernandes FAN, Lona LMF (2002) Heterogeneous modeling of fluidized bed polymerization reactors. Influence of mass diffusion into the polymer particle. Chem Eng Sci 26:841–848

22. Fernandes FAN, Lona LMF (2001) Heterogeneous modeling for fluidized-bed polymerization reactor. Chem Eng Sci 56:963–969

23. Hang W, Weidong Y, Tao J, Binbin L, Wenqing X, Jianjiang M, Youliang H (2002) Ethylene oligomerization by novel iron (II) diimine complexes/MAO. Chin Sci Bull 47:1616–1618

24. Hatzantonis H, Yiannoulakis H, Yiagopoulos A, Kiparissides C (2000) Recent developments in modeling gas-phase catalyzed olefin polymerization fluidized-bed reactors: the effect of bubble size variation on the reactor's performance. Chem Eng Sci 55:3237–3259

25. Huang J, Rempel GL (1995) Ziegler–Natta catalysts for olefin polymerization: mechanistic insights from metallocene systems. Prog Polym Sci 20:459–526

26. Ibrehem AS, Hussain MA, Ghasem NM (2009) Modified mathematical model for gas phase olefin polymerization in fluidized-bed catalytic reactor. Chem Eng J 149:353–362

27. Ibrehem PS, Hussain MA, Ghasem NM (2008) Mathematical model and advanced control for gas-phase olefin polymerization in fluidized-bed catalytic reactors. Chin J Chem Eng 16:84–89

28. Jiang T, Chen H, Ning Y, Chen W (2006) Preparation of 1-octene by ethylene tetramerization with high selectivity. Chin Sci Bull 51:521–523

29. Jin X, Hu P, Dong S, Yuan P (2004) steady-state modeling of ethylene polymerization. 5Ih World Congr. Intell Control Autom China pp 3493–3497

30. Jones D, Cavell K, Keim W (1999) Zirconium complexes as catalysts for the oligomerisation of ethylene: the role of chelate ligands and the Lewis acid cocatalyst in the generation of the active species. J Mol Catal A Chem 138:37–52

31. Asua José M (2007) Polymer reaction engineering. Blackwell Publishing Ltd, UK

32. Kazuo Soga TS (1997) Ziegler–Natta catalysts polymerizations. Prog Polym Sci 22:1503–1546

33. Khanmetov AA, Azizov AG, Ibragimova MD, Kuliev BV, Alieva RV, Kalbalieva ES, Bagirova SR, Mamedova RZ (2007) The multiple-site nature of ethylene polymerization catalysts based on zirconyl carboxylates. Pet Chem 47:176–183

34. Khanmetov AA, Azizov AG, Piraliev AG (2006) Ethylene oligomerization in the presence of catalyst systems based on mixed-ligand β-diketonatozirconium chlorides. Pet Chem 46:338–342

35. Kim I, Kwak CH, Kim JS, Ha C-S (2005) Ethylene oligomerizations to low-carbon linear α-olefins by structure modulated phenoxy-imine nickel(II) complexes combined with aluminum sesquichloride. Appl Catal A Gen 287:98–107

36. Kiparissides CA (1996) Polymerization reactor modeling: a review of recent developments and future directions. Chem Eng Sci 51:1637–1659

37. Kissin V, Mink I, Company MC, October R, Manuscript R, January R (1993) Ethylene oligomerization and chain growth mechanisms with Ziegler–Natta catalysts. Macromolecules 26:2151–2158

38. Fan-Hua Kong, Ming-He Xie (1999) Catalysts for 1-hexene Synthesis from Ethylene Oligomerization. J Nat Gas Chem 8:256–263

39. Meyer T, Keurentjes J (2005) Handbook of polymer reaction engineering, 1st edn. Wiley–VCH

40. Moustafa TM (2005) Unpublished data

41. Nelkenbaum E, Kapon M, Eisen MS (2005) Synthesis and molecular structures of neutral nickel complexes. Catalytic activity of the oligomerization of ethylene, and the dimerization of propylene. Organometallics 24:2645–2659

42. Oouchi K, Mitani M, Hayakawa M, Yamada T (1996) Ethylene oligomerization catalyzed with dichlorobis- (P-diketonato) zirconium/organoaluminium chloride systems. Macromol Chem Phys 1551:1545–1551

43. Shiraki Y, Nakamoto Y, Souma Y (2002) ZrCl 4 -TEA-EASC three-component catalyst for the oligomerization of ethylene: the role of organoaluminum co-catalysts and additives. J Mol Catal A Chem 187:283–294

44. Siedle AR, Lamanna WM, Newmark RA, Schroepfer JN (1998) Mechanism of olefin polymerization by a soluble zirconium. J Mol Catal A Chem 128:257–271

45. Skupiňsk J (1991) Oligomerization of α-olefins to higher oligomers. Chem Rev 91:613–648

46. Soares BP (2001) Mathematical modelling of the microstructure of polyoleÿns made by coordination polymerization: a review. Chem Eng Sci 56:4131–4153

47. de Souza RF, Bernardo-Gusmão K, Cunha GA, Loup C, Leca F, Réau R (2004) Ethylene dimerization into 1-butene using 2-pyridylphosphole nickel catalysts. J Catal 226:235–239

48. Su B, Feng G (2010) Influence of the metal centers of 2,6-bis(imino)pyridyl transition metal complexes on ethylene polymerization/oligomerization catalytic activities. Polym Int 59:1058–1063

49. Tembe GL, Ravindranathan M (1991) Oligomerization of ethylene to linear a-olefins by a titanium aryl oxide-alkylaluminum catalyst. Ind Eng Chem Res 30:2247–2252

50. Touloupides V, Kanellopoulos V, Pladis P, Kiparissides C, Mignon D, Van-Grambezen P (2010) Modeling and simulation of an industrial slurry-phase catalytic olefin polymerization reactor series. Chem Eng Sci 65:3208–3222

51. Turner AH (1983) Purity aspects of higher alpha olefins. J Am Oil Chem Soc 60:623–627

52. Valencia FP, Soares JBP (2007) Steady state simulation of ethylene polymerization using multiple-site coordination catalysts. Macromol Symp 259:110–115

53. Wang M, Li R, Qian M, Yu X, He R (2000) The effect of co-catalysts on the oligomerization and cyclization of ethylene catalyzed by zirconocene complexes. J Catal A Chem 160:337–341

54. Yang P, Yang Y, Zhang C, Yang X-J, Hu H-M, Gao Y, Wu B (2009) Synthesis, structure, and catalytic ethylene oligomerization of nickel(II) and cobalt(II) complexes with symmetrical and unsymmetrical 2,9-diaryl-1,10-phenanthroline ligands. Inorganica Chim Acta 362:89–96

55. Yang XS (2008) Firefly algorithm. Nature-inspired metaheuristic algorithms. Luniver Press, Frome, pp 79–90

56. Yoon WJ, Kim YS, Kim IS, Choi KY (2004) Recent advances in polymer reaction engineering: modeling and control of polymer properties. Korean J Chem Eng 21:147–167

57. Zacca JJ, Ray H (1993) Modelling of the liquid phase polymerization of olefins in loop reactors. Chem Eng Sci 48:3743–3765

58. Zaidman AV, Kayumov RR, Belov GP, Pervova IG, Lipunov IN, Kharlampidi KE (2010) Ethylene oligomerization in the presence of catalytic systems based on nickel(II) formazanates. Pet Chem 50:450–454

59. Zhang B, Wang Y, Wang G, Cao J, Sun S, Xing L, Sun Y, Han Y (2007) Oligomerization and polymerization of ethylene initiated by a novel Ni (II) -based acetyliminopyridine complexes as single-site catalysts. J Nat Gas Chem 16:64–69

60. Zhang J, Fan H, Li B-G, Zhu S (2008) Modeling and kinetics of tandem polymerization of ethylene catalyzed by bis(2-dodecylsulfanyl-ethyl)amine- and. Chem Eng Sci 63:2057–2065

61. Zhang Z, Chen S, Zhang X, Li H, Ke Y, Lu Y, Hu Y (2005) A series of novel 2,6-bis(imino)pyridyl iron catalysts: synthesis, characterization and ethylene oligomerization. J Mol Catal A Chem 230:1–8

62. Zhang Z, Zou J, Cui N, Ke Y, Hu Y (2004) Ethylene oligomerization catalyzed by a novel iron complex containing fluoro and methyl substituents. J Mol Catal A Chem 219:249–254

63. Zhukov VI, Val'kovich GV, Skorik IN, Petrov YM, Belov GP (2007) Ethylene oligomerization in the presence of ZrO(O-COR)2-Al(C2H5)2Cl-Modifier catalytic system. Pet Chem 47:49–54

8

Characterization of single-chain polymer folding using size exclusion chromatography with multiple modes of detection

Peter Frank · Alka Prasher · Bryan Tuten ·
Danming Chao · Erik Berda

Abstract We highlight here recent work from our laboratory on the subject of fabricating nanostructures from single polymer chains. These so-called single-chain nanoparticles are synthesized by inducing intra-molecular cross-linking on discrete macromolecules in dilute solution. Among the biggest challenges in this rapidly expanding area of research is reliable and accurate means to characterize this process. In this paper, we review our preferred method of characterization: size exclusion chromatography featuring multiple modes of detection. Multi-angle light scattering in conjunction with a concentration detector can provide absolute molecular weight data; viscometric detection can provide information about solution size and conformation. Correlation of these data provides a simple and robust way to quantify the process by which we fold single polymer coils into architecturally defined unimolecular nanostructures.

Keywords Single-chain nanoparticles · Polymer folding · TEM · SEC

P. Frank · A. Prasher · E. Berda (✉)
Department of Chemistry, University of New Hampshire,
Durham, USA
e-mail: Erik.Berda@unh.edu

B. Tuten · E. Berda
Materials Science Program, University of New Hampshire,
Durham, USA

D. Chao
Alan G. MacDiarmid Institute, College of Chemistry, Jilin
University, Changchun, People's Republic of China

Introduction

A fast growing research area, single-chain nanoparticles (SCNP) have demonstrated promise for use in catalysis [1], nanoreactors [2], controlled release [3], chemical sensors [4], enzyme mimetics [5–7], diagnostics [8] and optics [9]. This breadth of application is attributed to the precise control over the composition, functionality, geometry and topology this technology permits for accessing architectures in the sub-20 nm size regime. These features are programmed into SCNP a priori by means of robust living polymerization techniques including ring-opening metathesis polymerization [10] (ROMP), reversible addition-fragmentation chain-transfer [11] (RAFT), and atom transfer radical polymerization [12] (ATRP). Such techniques afford parent polymers with narrow molecular weight distributions ($Đ$), targeted molecular weights, tunable functionality and diverse architectures [13–17]. These parent chains are then subjected to intra-molecular cross-linking under ultra-dilute conditions, thus folding the chains into architecturally defined nanoparticles with dimensions smaller than the original dimensions of the solvated coil (Fig. 1) [18].

Although young, this field is developing rapidly; a number of important contributions have appeared in the recent literature. [1, 12, 19–21] The Meijer group showed that the molecular weight and backbone rigidity of the parent polymer does not significantly influence the SCNP fabrication induced by supramolecular interactions. Instead, the solvent plays a key role in mediating intra-molecular folding [21]. Meijer and coworkers also expanded the scope of possible architectures by successfully synthesizing SCNP from cylindrical brush block polymers as well as an orthogonally collapsible ABA triblock copolymer [6, 22]. Additionally, this group studied

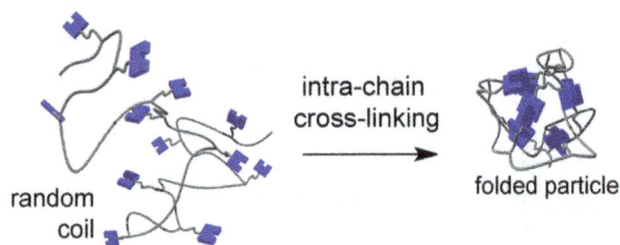

Fig. 1 Schematic representation of SCNP synthesis

temperature and co-solvent induced changes in the secondary structures of benzene-1,3,5-tricarboxamides (BTA)-based SCNPs, and developed a compartmentalized metal ion SCNP sensor that shows impressive binding affinities [4, 23]. The Barner-Kowollik group demonstrated a photo-induced Diels–Alder ligation to create SCNP in addition to using metal–ligand complexation to induce controlled folding by linking chain ends [1, 24]. Other work has demonstrated that SCNP can be formed from a host of reactions including benzocylcobutene coupling [25], olefin cross-metathesis [26], azide–alkyne click chemistry [19, 27–30], thiol-ene click chemistry [31, 32], Bergman cyclization [33–35], Curtius rearrangement [36], Glasser–Hay coupling [37], nitrene-mediated C-alkylation (sulfonyl azide) [38], oxidative polymerization through pendant moieties [12], benzosulfone chemistry [39], carbodiimide coupling [40], benzoxazine chemistry [11], amide bond formation [10, 31], cinnamoyl [40] and coumarin [2] photodimerization, guest–host interactions [41], hydrogen bonding [20, 42–44], and vulcanization [45].

These methods draw synthetic parallel to nature's elegant biomacromolecules, albeit in a crude and simplistic fashion [3, 5]. Given that synthetic polymers are always subject to distributions in molecular weights and heterogeneities in microstructure [46, 47], it is important to develop convenient and consistent methods for characterizing SCNP fabrication. We discuss here the analysis of SCNPs using size exclusion chromatography (SEC), specifically the application of multi-detection SEC. We have recently shown that when applied in conjunction with microscopy and dynamic light scattering (DLS), multi-detection SEC is a powerful tool for studying the folding of individual polymer chains into architectural defined nanostructures.

Polymer design and SCNP synthesis

There are a variety of synthetics strategies that can be used to fabricate SCNP [48]. The first methodology successfully employed in our labs involves the addition of an external, bifunctional cross-linker to a chain decorated with complementary reactive handles (Scheme 1) [10]. Polymer chains decorated with cyclic anhydride units were folded by the introduction of p-aminophenyl disulfide as the bifunctional cross-linker. We chose to work with poly(norbornene-exo-anhydride), as the requisite reactive monomer is commercially available and polymerization (or copolymerization with cyclooctene) via ROMP proceeds rapidly with excellent control. In series **1a**, the number of intra-chain linkages can be controlled by modulating the amount of bifunctional diamine cross-linker added to a solution of

Scheme 1 Synthetic approach to series **1** via addition of an external bifunctional cross-linker

Series **1a**

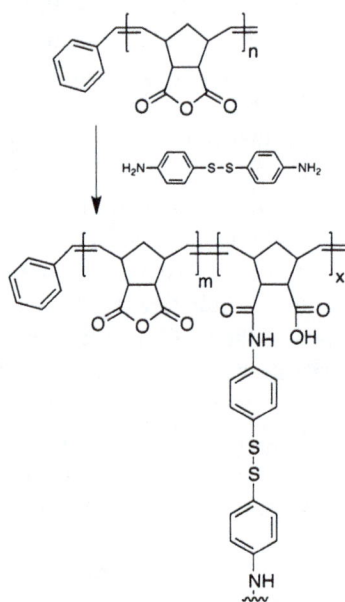

Series **1b**

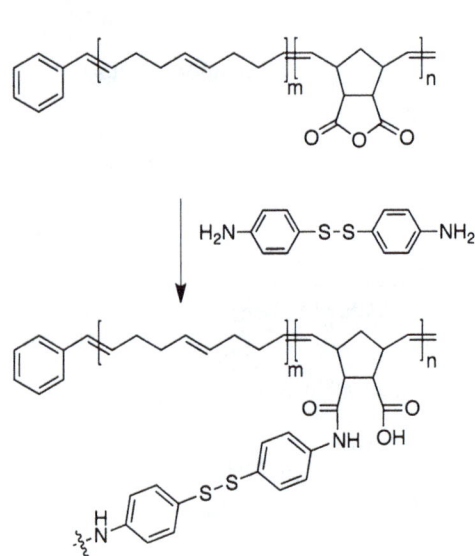

Series 2

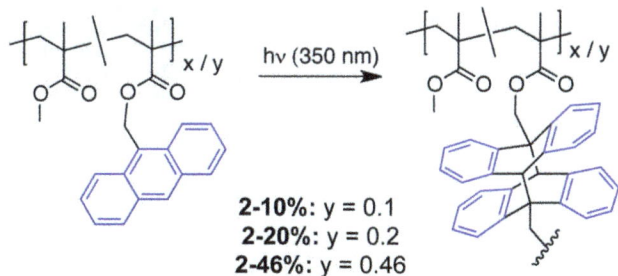

2-10%: y = 0.1
2-20%: y = 0.2
2-46%: y = 0.46

Scheme 2 Synthetic approach to series **2** using photo cross-linkable pendant groups

the polyanhydride. In series **1b**, the extent of intra-chain cross-links is controlled by changing the feed ratio of anhydride co-monomer relative to cyclooctene, followed by the addition of stoichiometric amounts of diamine. Using a cross-linker with a reversible covalent bond permits the unfolding and refolding of the chain and allows us to better characterize this coil to globule transition [10].

The second route we are using to fabricate SCNPs involves the homocoupling of a built in functional unit [49]. For series **2**, foldable methacrylic chains bearing pendant anthracene units were synthesized by RAFT polymerization (Scheme 2). Anthracene is well known for its ability to undergo a photo-

Scheme 3 Synthetic approach to series **3** using sequentially activated orthogonal intra-chain interactions

Series 3

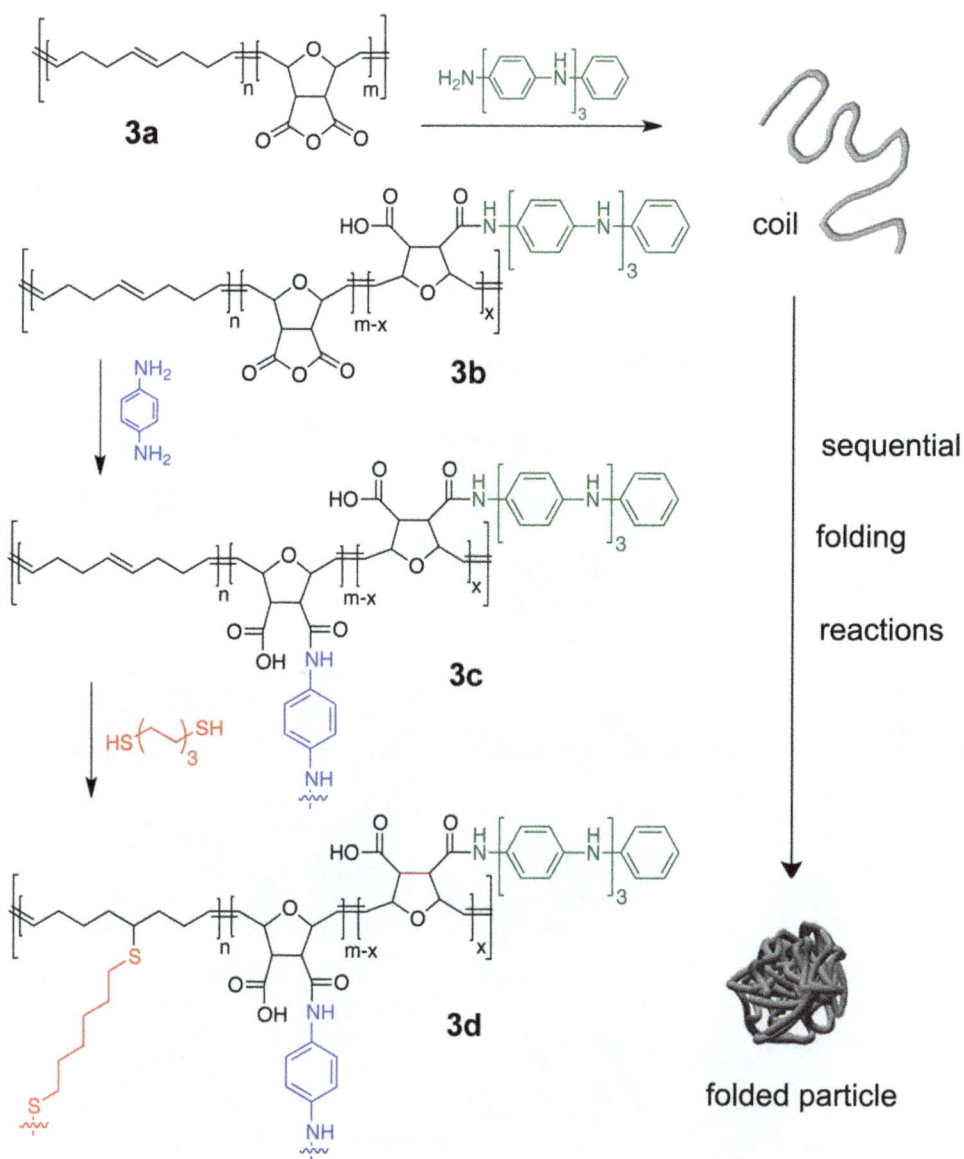

Fig. 2 SEC traces for series **1a**.
a Addition of varying amounts
of cross-linker, **b** after unfolding
via disulfide reduction,
c reversible folding and
unfolding via redox chemistry
(Tuten et al. [10]—adapted with
permission of The Royal
Society of Chemistry)

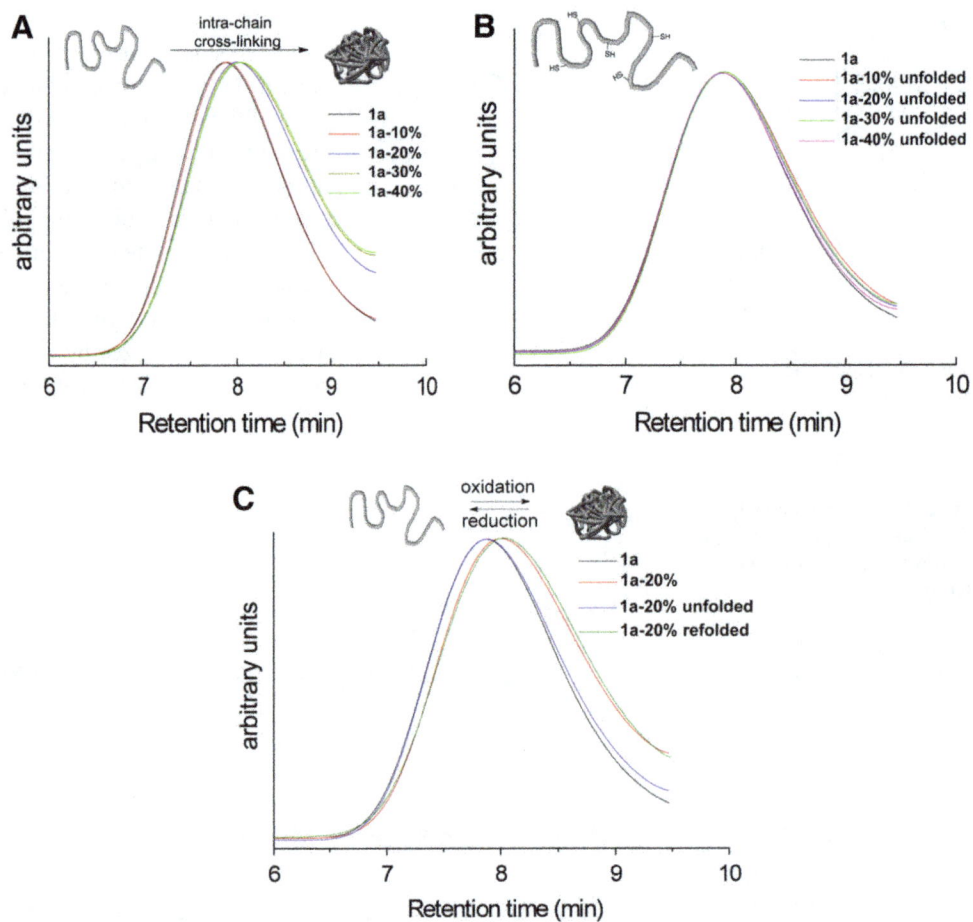

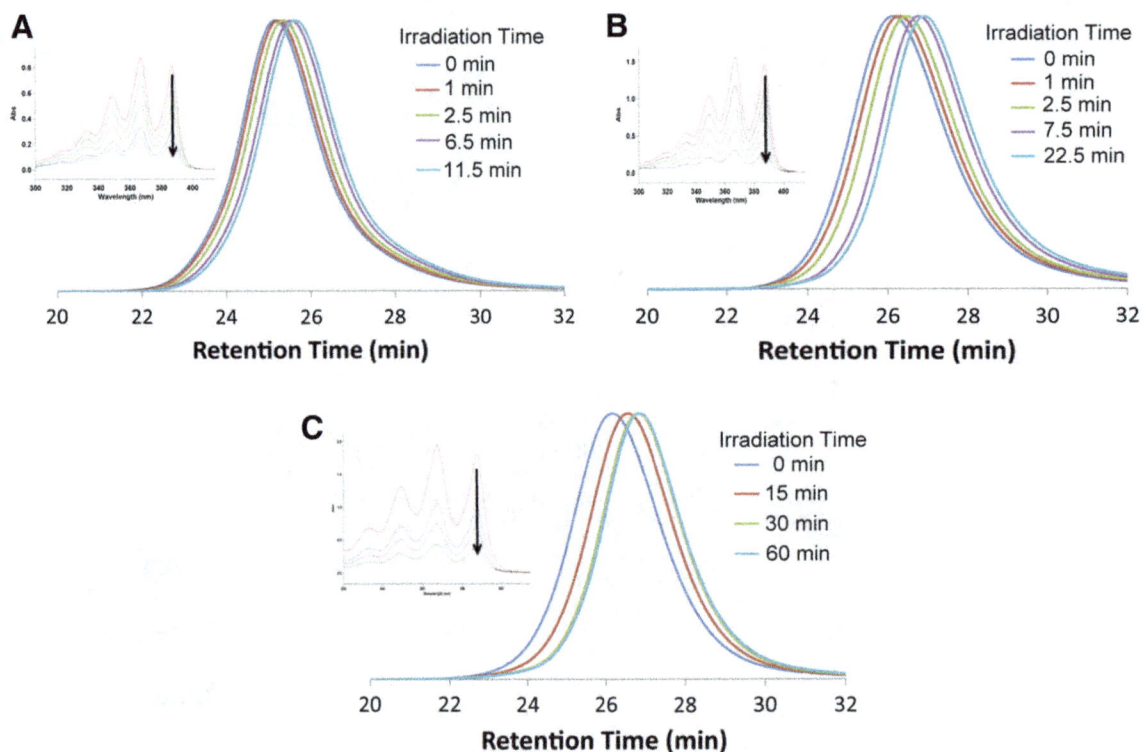

Fig. 3 SEC traces and corresponding UV–Vis spectra for series **2** photodimerization studies (Frank et al. [49]—adapted with permission of John Wiley And Sons Inc)

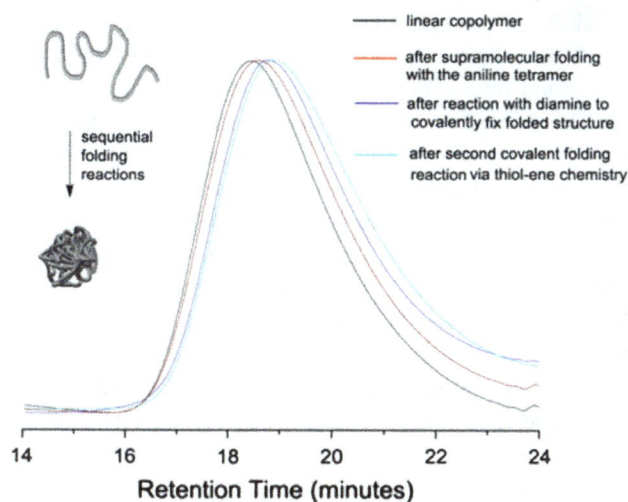

Legend:
— linear copolymer
— after supramolecular folding with the aniline tetramer
— after reaction with diamine to covalently fix folded structure
— after second covalent folding reaction via thiol-ene chemistry

sequential folding reactions

Retention Time (minutes)

Fig. 4 SEC traces displaying the decrease in hydrodynamic volume with each sequential folding reaction for series 3 (Chao et al. [31]—adapted with permission of The Royal Society of Chemistry)

Table 1 SEC data for series **1b**

	Peak retention time (min)[a]	M_w (kg/mol)[b]	Đ	$[\eta]$ (mL/g)	R_h (nm)
Parent polymer (**1b**)	17.8	50.1	1.22	13.9	4.5
Folded nanoparticle	18.8	59.3	1.32	8.7	3.9

[a] Retention times are from MALS detector trace

[b] Absolute molecule weight calcultated by MALS

Table 2 SEC data for series **2** before and after complete photo-induced folding

Polymer	Irradiation time (min)	Peak retention time (min)	M_w (kg/mol)	Đ	R_h (nm)	$[\eta]$
2a-10	0	24.9	30.8	1.17	4	13.3
	11.5	25.4	33.5	1.16	3.4	7.5
2b-20	0	25.6	30.6	1.14	3.6	9.5
	22.5	26.5	29.9	1.14	2.9	5.1
2c-46	0	25.7	42.9	1.16	3.5	6.6
	60	26.4	36.7	1.14	2.3	2.4

Table 3 Comparison of predicted vs. actual absolute M_w for series **3**

Polymer	Predicted molecular weights M_w (kg/mol)	Actual molecular weights M_w (kg/mol)	R_h (nm)
3a	16.1	15.0	6.5
3b	19.7	19.9	5.7
3c	20.5	22.6	4.9
3d	23.9	23.3	4.3

R_h values from DLS confirm the efficacy of each sequential folding reaction

induced 4 + 4 cycloaddition when irradiated with 350 nm-centered UV light, resulting in photodimers that can be reversed via irradiation with wavelengths <300 nm or thermally at temperatures >180 °C [50], providing an attractive route to synthesizing thermally responsive and photosensitive nanomaterials. In addition, using light as stimulus for SCNP fabrication offers additional functional group tolerance that other cross-linking strategies lack [49].

A third method for SCNP fabrication takes advantage of multiple orthogonal chemistries triggered sequentially to fold synthetic chains into particles [31]. Scheme 3 shows a series of poly(oxanorbornene anhydride-*co*-cyclooctadiene) materials synthesized using ROMP. Analogous to series **1**, the polymer structures contained in series **3** feature anhydride groups as reactive handles that can be easily modified with a variety of nucleophiles, simultaneously inducing folding and installing functionality. In this system, aniline tetramer (AT) was used to fold the chains via supramolecular interactions (π–π and hydrogen bonding) between these pendant groups. The chains were then reacted with *p*-aminoaniline, a bifunctional covalent cross-linker that pulls in the remaining anhydride units and further collapses the chain. Finally, internal olefin units installed within the backbones during ROMP were connected using photoinitiated thiol-ene click chemistry with 1,6-hexanedithiol as a bifunctional cross-linker. This sequential folding approach demonstrates that the application of orthogonal chemistries in a synergistic fashion results in more tightly collapsed structures with the goal of more closely mimicking natural folded polymeric constructs. [31]

SCNP characterization

Conventional SEC

SEC with conventional calibration is the standard method used to characterize SCNPs formation. Comparing the retention time of the parent polymer chains to the retention time of the folded nanoparticles after intramolecular cross-linking is induced typically reveals shifts to longer retention times. This indicates a decrease in hydrodynamic volume based on the inherent principles of size exclusion chromatography [51–53]. This is clearly demonstrated in all of the polymer series discussed here [10, 31, 49].

SEC traces for series **1a** below show the effect of adding increasing amounts of bifunctional cross-linker: as more intra-chain linkages are formed the SEC traces shift to longer retention times as expected (Fig. 2a). To ensure this is truly an effect of conformational changes and not due to other factors, such as interactions between the SEC columns and the analyte, we take advantage of the cleavable disulfide linkage that was installed during single-chain nanoparticles formation. Cleaving this linkage after the initial folding reaction takes place allows us to determine

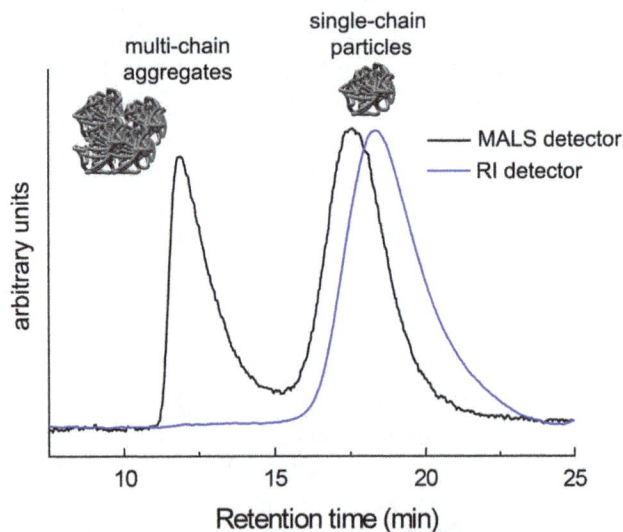

Fig. 5 MALS and RI detection traces for SEC of **1b** highlighting the ability of MALS to detect large, multi-chain aggregates (Tuten et al. [10]—Reproduced by permission of The Royal Society of Chemistry)

that the shifts in the retention time is actually due to the folding of the chain and not other behavior that might affect retention time. This is indeed the case, as shown in Fig. 2b. The reduction of the disulfide bridges via dithiothreitol (DTT) results in the unfolding of the SCNPs, confirmed by SEC traces showing that retention times (and therefore hydrodynamic volume) revert to those of the parent polymer chain, regardless of the amount of cross-linker added. Furthermore, refolding of the unfolded chains via thiol oxidation in the presence of catalytic $FeCl_3$ revealed shifts to longer retention times, indicating the reduction in the solvent dynamic volume (refolding), confirming the utility of this characterization method as well as demonstrating this chemistry as a route to responsive nanostructures (Fig. 2c).

Analyzing the SEC traces from series **2** shows that the extent of intra-molecular folding or collapse can be controlled very easily using built-in reactive co-monomer. Both reaction time and the amount of co-monomer incorporation can be used to control the extent of intra-chain cross-links when employing this method. Figure 3a–c highlights this behavior. All samples show steady shifts to longer retention times with increasing irradiation times, an

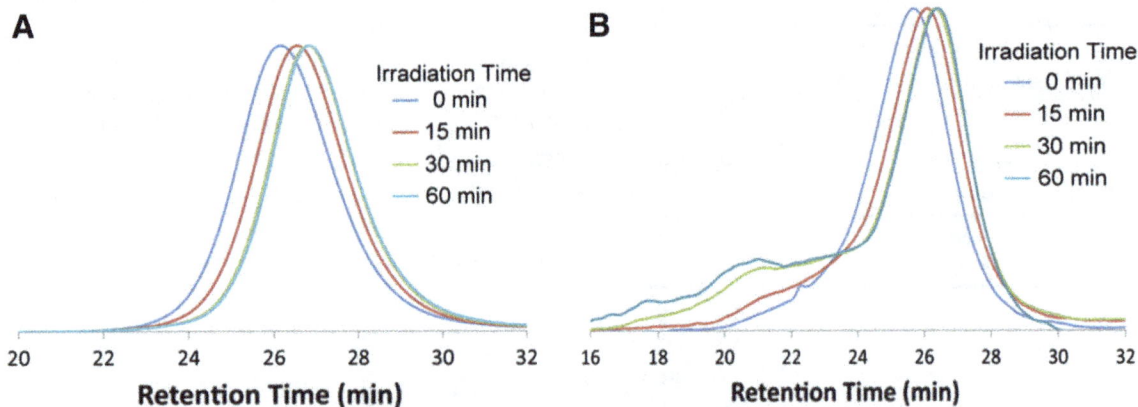

Fig. 6 SEC traces for **2c-46** photodimerization studies showing the appearance of multi-chain aggregates. **a** UV detector traces, **b** MALS detector traces (Frank et al. [49]—Reproduced by permission of John Wiley And Sons Inc)

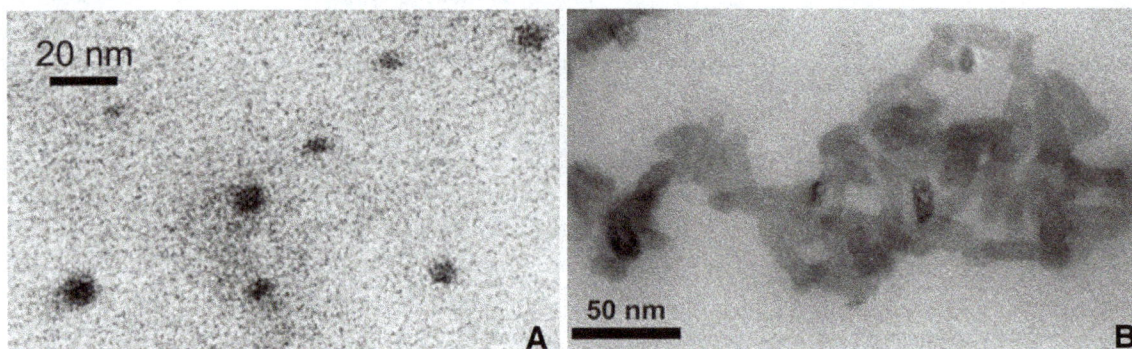

Fig. 7 TEM images of SCNP obtained from **1b** (**a**) and **2a-10** (**b**)

indication that folding is occurring with the formation of anthracene photodimers. Anthracene dimerization was confirmed by the disappearance of the characteristic anthracene absorption peak in the UV–Vis spectra (insets in Fig. 3). The effect of reactive co-monomer incorporation is clearly observed here: samples with 10 % (**2a-10**) anthracene co-monomer showed the least pronounced changes in retention time when compared to the samples with 20 % (**2a-20**) and 46 % (**2a-46**) anthracene co-monomer (Fig. 3).

Conventional SEC reveals the same behavior even when multiple orthogonal intra-chain cross-linking chemistries are used. Figure 4 shows that for polymers in series **3**, the expected shifts in retention time confirm the efficacy of each folding reaction when engaged sequentially. As expected, the SEC trace of the final nanoparticle shifted to the longest retention time indicating a more tightly wrapped structure consistent with theory and experiment [6].

Triple detection SEC

The application of standard SEC as described above provides useful qualitative information about changes in the solution conformation of polymer chains during SCNP fabrication. Still this method is not without ambiguities, especially with regards to actual changes in molecular weight (or lack thereof) when triggering intra-chain cross-linking. Since relative molecular weight and retention time are linked in traditional SEC, it is impossible to formally characterize SCNP molecular weight with this technique. Furthermore, while shifts in retention time certainly signify changes in hydrodynamic volume it is impossible to quantify these changes with standard SEC measurements alone. Dynamic light scattering (DLS) in conjunction with SEC can be a very useful tool for SCNP characterization, but lacks the chromatographic separation provided by SEC.

To alleviate some of these issues our group implemented SEC with multiple modes of detection for SCNP characterization. The use of multi-angle light scattering (MALS) detection with SEC allows the determination of polymer and SCNP molecular weight independent of retention time [54]. The addition of an inline differential viscometer permits characterization of molecular conformation, including hydrodynamic radius, for each thin chromatographic slice as the sample elutes from the SEC column. With this setup, the absolute molecular weight of the parent polymers can be easily measured and used to calculate the expected SCNP molecular weight. Comparing this value with the experimental absolute SCNP molecular weight can provide useful quantitative information about the chemistry involved in this process. Changes in retention time, which are assumed to correlate with changes in

solvated volume, can likewise be readily quantified by this method. Recent work from our lab demonstrating the utility of this technique is highlighted below.

Table 1 summarizes the triple detection data for series **1b.** As expected, the MALS detector trace shows a shift to longer retention time after addition of cross-linker similar to what we observed using standard SEC. Examining the absolute molecular weight reveals that although the SCNP curve shifted to a longer retention time the molecular weight actually increased by the amount predicted with the addition of stoichiometric cross-linker. In addition, the viscometric data confirm and quantify the decrease in hydrodynamic radius. It further reveals a decrease in intrinsic viscosity after SCNP formation consistent with theory and experiment [55].

The SEC data for series **2** showed similar behavior where shifts to longer retention times occur after folding as expected. The viscometric data reveals a decrease in both intrinsic viscosity and hydrodynamic radius. The absolute molecular weight, predicted to remain constant, changes only slightly and is consistent with a primarily unimolecular phenomenon (Table 2). This series also displays the trend, although unsurprising, that the extent of reactive co-monomer incorporation directly dictates how much folding will occur.

SEC MALS data from series **3**, shown in Table 3, again highlight how useful this technique can be as a quantitative tool. For these polymers, the comparison between the predicted and the measured absolute molecular weights demonstrates that the folding chemistry proceeded as designed. In addition, DLS measurements confirmed that retention time shifts witnessed in SEC measurements (Table 3) are indeed due to folding.

Intra-chain versus inter-chain reactions

A crucial bit of qualitative data provided by SEC MALS is the distinction between intra-chain folding and inter-chain coupling. In our experience, MALS is capable of detecting larger nanoaggregates in very minute concentrations while traditional SEC detectors such as UV and RI do not show the characteristic aggregate peak [10, 49]. For series **1**, multi-chain aggregates were detectable via MALS while virtually unseen in RI traces (Fig. 5) [10].

Likewise for series **2**, the UV detector fails to detect inter-chain aggregates that are readily observed by MALS, evident in the emergence of peak shoulders in MALS traces. These data also suggest that at a certain level of reactive co-monomer incorporation inter-chain coupling becomes unavoidable (Fig. 6) [49]. Thus, triple detection SEC provides a robust method for analyzing polymer folding chemistry and confirming single-chain behavior.

Correlating solution measurements with other characterization tools

We highlighted in the previous sections how triple detection SEC can be applied to effectively characterize single-chain folding. Additional secondary characterization techniques can offer invaluable information about SCNP size and shape, including small angle neutron scattering (SANS) [3, 23], thorough DLS measurements [4, 23], atomic force microscopy (AFM) [11, 19, 20, 26, 29, 30, 33, 39, 42, 43, 56, 57], and transmission electron microscopy (TEM) [2, 7, 10, 11, 27, 38, 49]. We found TEM particularly useful as a tool to provide visual evidence for SCNP formation. Figure 7 shows representative images from series **1** (**1b**) and series **2** (**2a-10**). In each case, the images reveal nanostructures with dimensions on the same order of magnitude as numbers obtained from solution measurements. Interestingly, series **2** SCNP adopt an oblong ellipsoid geometry, rather than a spherical conformation. We are still investigating the cause of this behavior, which is consistent with form factor measurements made by others [3, 23].

Summary

In summary, conventionally calibrated SEC is a satisfactory method to characterize SCNP via interpretation of comparative shifts in retention time. Evolving this tool by marrying SEC with MALS and viscometric detection affords a precise and reliable method for the qualitative and quantitative study of SCNP fabrication. This methodology facilitates the detailed investigation of structure property relationships using absolute molecular weight, intrinsic viscosity, and hydrodynamic radius of SCNP systems. During SCNP formation, the decrease in both the intrinsic viscosity and hydrodynamic radius is observed with, or in the absence of, concurrent changes in the absolute molecular weight depending on the folding technique employed. Triple detection SEC in conjunction with additional characterization tools including TEM and DLS provides a variety of useful information for studying the fascinating and rapidly expanding research area of single-chain folding processes and single-chain nanoparticles.

References

1. Willenbacher J, Altintas O, Roesky PW, Barner-Kowollik C (2013) Macromol Rapid Commun 1521–3927

2. He J, Tremblay L, Lacelle S, Zhao Y (2011) Preparation of polymer single chain nanoparticles using intramolecular photo-dimerization of coumarin. Soft Matter 7:2380–2386

3. Sanchez-Sanchez A, Akbari S, Etxeberria A, Arbe A, Gasser U, Moreno AJ, Colmenero J, Pomposo JA (2013) "Michael" nanocarriers mimicking transient-binding disordered proteins. ACS Macro Lett 2:491–495

4. Gillissen MAJ, Voets IK, Meijer EW, Palmans ARA (2012) Single chain polymeric nanoparticles as compartmentalised sensors for metal ions. Polym Chem 3:3166–3174

5. Perez-Baena I, Barroso-Bujans F, Gasser U, Arbe A, Moreno AJ, Colmenero J, Pomposo JA (2013) Endowing single-chain polymer nanoparticles with enzyme-mimetic activity. ACS Macro Lett 2:775–779

6. Hosono N, Gillissen MAJ, Li Y, Sheiko SS, Palmans ARA, Meijer EW (2013) Orthogonal self-assembly in folding block copolymers. J Am Chem Soc 135:501–510

7. Terashima T, Mes T, De Greef TFA, Gillissen MAJ, Besenius P, Palmans ARA, Meijer EW (2011) Single-chain folding of polymers for catalytic systems in water. J Am Chem Soc 133:4742–4745

8. Murray BS, Fulton DA (2011) Dynamic covalent single-chain polymer nanoparticles. Macromolecules 44:7242–7252

9. Wang Y, Dong L, Xiong R, Hu AJ (2013) Practical access to bandgap-like N-doped carbon dots with dual emission unzipped from PAN@PMMA core–shell nanoparticles. Mater Chem C 1:7731–7735

10. Tuten BT, Chao D, Lyon CK, Berda EB (2012) Single-chain polymer nanoparticles via reversible disulfide bridges. Polym Chem 3:3068–3071.

11. Wang P, Pu H, Jin MJ (2011) Single-chain nanoparticles with well-defined structure via intramolecular crosslinking of linear polymers with pendant benzoxazine groups. Polym Sci Part A Polym Chem 49:5133–5141

12. Dirlam PT, Kim HJ, Arrington KJ, Chung WJ, Sahoo R, Hill LJ, Costanzo PJ, Theato P, Char K, Pyun J, Kim J, Chung J (2013) Single chain polymer nanoparticles via sequential ATRP and oxidative polymerization. Polym Chem 4:3765–3773

13. Moad G, Chong YK, Postma A, Rizzardo E, Thang SH (2005) Advances in RAFT polymerization: the synthesis of polymers with defined end-groups. Polymer 46:8458–8468

14. Moad G, Rizzardo E, Thang SH (2005) Living radical polymerization by the RAFT process. Aust J Chem 58:379–410

15. Alfred SF, Lienkamp K, Madkour AE, Tew GN (2008) Water-soluble ROMP polymers from amine-functionalized norbornenes. Polymer 6672–6676

16. Bielawski CW, Grubbs RH (2007) Living ring-opening metathesis polymerization. Prog Polym Sci 32:1–29

17. Matyjaszewski K, Tsarevsky NV (2009) Nanostructured functional materials prepared by atom transfer radical polymerization. Nat Chem 1:276–288

18. Altintas O, Barner-Kowollik C (2012) Single chain folding of synthetic polymers by covalent and non-covalent interactions: current status and future perspectives. Macromol Rapid Commun 33:958–971

19. Ormategui N, García I, Padro D, Cabañero G, Grande HJ, Loinaz I (2012) Synthesis of single chain thermoresponsive polymer nanoparticles. Soft Matter 8:734–740

20. Foster EJ, Berda EB, Meijer EW (2009) Metastable supramolecular polymer nanoparticles via intramolecular collapse of single polymer chains. J Am Chem Soc 131:6964–6966

21. Stals PJM, Gillissen MAJ, Nicolaÿ R, Palmans ARA, Meijer EW (2013) The balance between intramolecular hydrogen bonding, polymer solubility and rigidity in single-chain polymeric nanoparticles. Polym Chem 4:2584–2597

22. Stals PJM, Li Y, Burdyńska J, Nicolaÿ R, Nese A, Palmans ARA, Meijer EW, Matyjaszewski K, Sheiko SS (2013) How far can we push polymer architectures? J Am Chem Soc 135:11421–11424

23. Gillissen MAJ, Terashima T, Meijer EW, Palmans ARA, Voets IK (2013) Sticky supramolecular grafts stretch single polymer chains. Macromolecules 46:4120–4125

24. Altintas O, Willenbacher J, Wuest KNR, Oehlenschlaeger KK, Krolla-Sidenstein P, Gliemann H, Barner-Kowollik C (2013) A mild and efficient approach to functional single-chain polymeric nanoparticles via photoinduced Diels–Alder ligation. Macro-molecules 46:8092–8101

25. Harth E, Van Horn B, Lee VY, Germack DS, Gonzales CP, Miller RD, Hawker CJ (2002) A facile approach to architecturally defined nanoparticles via intramolecular chain collapse. J Am Chem Soc 124:8653–8660

26. Cherian AE, Sun FC, Sheiko SS, Coates GW (2007) Formation of nanoparticles by intramolecular cross-linking: following the reaction progress of single polymer chains by atomic force microscopy. J Am Chem Soc 129:11350–11351

27. De Luzuriaga AR, Ormategui N, Grande HJ, Odriozola I, Pomposo JA, Loinaz I (2008) Intramolecular click cycloaddition: an efficient room-temperature route towards bioconjugable poly-meric nanoparticles. Macromol Rapid Commun 29:1156–1160

28. Schmidt BVKJ, Fechler N, Falkenhagen J, Lutz J-F (2011) Controlled folding of synthetic polymer chains through the for-mation of positionable covalent bridges. Nat Chem 3:234–238

29. Perez-Baena I, Loinaz I, Padro D, García I, Grande HJ, Odriozola I (2010) Single-chain polyacrylic nanoparticles with multiple Gd(iii) centres as potential MRI contrast agents. J Mater Chem 20:6916–6922

30. Oria L, Aguado R, Pomposo JA, Colmenero J (2010) A versatile "click" chemistry precursor of functional polystyrene nanopar-ticles. Adv Mater 22:3038–3041

31. Chao D, Jia X, Tuten B, Wang C, Berda EB (2013) Controlled folding of a novel electroactive polyolefin via multiple sequential orthogonal intra-chain interactions. Chem Commun 49:4178–4180.

32. Ding L, Yang G, Xie M, Gao D, Yu J, Zhang Y (2012) More insight into tandem ROMP and ADMET polymerization for yielding reactive long-chain highly branched polymers and their transformation to functional polymer nanoparticles. Polymer 53:333–341

33. Zhu B, Ma J, Li Z, Hou J, Cheng X, Qian G, Liu P, Hu A (2011) Formation of polymeric nanoparticles via Bergman cyclization mediated intramolecular chain collapse. J Mater Chem 21:2679–2683

34. Zhu B, Sun S, Wang Y, Deng S, Qian G, Wang M, Hu AJ (2013) Preparation of carbon nanodots from single chain polymeric nanoparticles and theoretical investigation of the photolumines-cence mechanism. Mater Chem C 1:580–586

35. Zhu B, Qian G, Xiao Y, Deng S, Wang M, Hu AJ (2011) A convergence of photo-bergman cyclization and intramolecular chain collapse towards polymeric nanoparticles. Polym Sci Part A Polym Chem 49:5330–5338

36. Zheng H, Ye X, Wang H, Yan L, Bai R, Hu W (2011) A facile one-pot strategy for preparation of small polymer nanoparticles by self-crosslinking of amphiphilic block copolymers containing acyl azide groups in aqueous media. Soft Matter 7:3956–3962

37. Sanchez-Sanchez A, Asenjo-Sanz I, Buruaga L, Pomposo JA (2012) Naked and self-clickable propargylic-decorated single-chain nanoparticle precursors via redox-initiated RAFT poly-merization. Macromol Rapid Commun 33:1262–1267

38. Jiang X, Pu H, Wang P (2011) Polymer nanoparticles via intramolecular crosslinking of sulfonyl azide functionalized polymers. Polymer 52:3597–3602

39. Adkins CT, Muchalski H, Harth E (2009) Nanoparticles with indi-vidual site-isolated semiconducting polymers from intramolecular chain collapse processes. Macromolecules 42:5786–5792

40. Njikang G, Liu G, Curda SA (2008) Tadpoles from the intramolecular photo-cross-linking of Diblock copolymers. Macromolecules 41:5697–5702

41. Appel EA, Dyson J, del Barrio J, Walsh Z, Scherman OA (2012) Formation of single-chain polymer nanoparticles in water through host-guest interactions. Angew Chem Int Ed Engl 51:4185–4189

42. Foster EJ, Berda EB, Meijer EWJ (2011) Tuning the size of supramolecular single-chain polymer nanoparticles. Polym Sci Part A Polym Chem 49:118–126

43. Berda EB, Foster EJ, Meijer EW (2010) Toward controlling folding in synthetic polymers: fabricating and characterizing supramolecular single-chain nanoparticles. Macromolecules 43:1430–1437

44. Seo M, Beck BJ, Paulusse JMJ, Hawker CJ, Kim SY (2008) Polymeric nanoparticles via noncovalent cross-linking of linear chains. Macromolecules 41:6413–6418

45. Cheng C, Qi K, Germack DS, Khoshdel E, Wooley KL (2007) Synthesis of core-crosslinked nanoparticles with controlled cylindrical shape and narrowly-dispersed size via core-shell brush block copolymer templates. Adv Mater 19:2830–2835

46. Berda EB, Baughman TW, Wagener KBJ (2006) Precision branching in ethylene copolymers: Synthesis and thermal behavior. Polym Sci Part A Polym Chem 44:4981–4989

47. Odian G (2004) Principles of polymerization, 4th edn. Wiley, New Jersey, p 832

48. Aiertza MK, Odriozola I, Cabañero G, Grande H-J, Loinaz I (2012) Single-chain polymer nanoparticles. Cell Mol Life Sci 69:337–346

49. Frank PG, Tuten BT, Prasher A, Chao D, Berda EB (2014) Intra-chain photodimerization of pendant anthracene units as an effi-cient route to single-chain nanoparticle fabrication. Macromol Rapid Commun 35:249–253.

50. Becker HD (1993) Unimolecular photochemistry of anthracenes. Chem Rev 93:145–172

51. Sun T, Chance RR, Graessley WW, Lohse DJ (2004) A study of the separation principle in size exclusion chromatography. Mac-romolecules 37:4304–4312

52. Grubisic Z, Rempp P, Benoit HJ (1967) A universal calibration for gel permeation chromatography. Polym Sci Part B Polym Lett 5:753–759

53. Moore JCJ (1964) Gel permeation chromatography. I. A new method for molecular weight distribution of high polymers. Polym Sci Part A Gen Pap 2:835–843

54. Wyatt PJ (1993) Light scattering and the absolute characteriza-tion of macromolecules. Anal Chim Acta 272:1–40

55. Beck JB, Killops KL, Kang T, Sivanandan K, Bayles A, Mackay ME, Wooley KL, Hawker CJ (2009) Facile preparation of nanoparticles by intramolecular crosslinking of isocyanate func-tionalized copolymers. Macromolecules 42:5629–5635

56. Kim Y, Pyun J, Fréchet JMJ, Hawker CJ, Frank CW (2005) The dramatic effect of architecture on the self-assembly of block copolymers at interfaces. Langmuir 21:10444–10458

57. Croce TA, Hamilton SK, Chen ML, Muchalski H, Harth E (2007) Alternative o—quinodimethane cross-linking precursors for intramolecular chain collapse nanoparticles. Macromolecules 40:6028–6031

9

Catalysing sustainable fuel and chemical synthesis

Adam F. Lee

Abstract Concerns over the economics of proven fossil fuel reserves, in concert with government and public acceptance of the anthropogenic origin of rising CO_2 emissions and associated climate change from such combustible carbon, are driving academic and commercial research into new sustainable routes to fuel and chemicals. The quest for such sustainable resources to meet the demands of a rapidly rising global population represents one of this century's grand challenges. Here, we discuss catalytic solutions to the clean synthesis of biodiesel, the most readily implemented and low cost, alternative source of transportation fuels, and oxygenated organic molecules for the manufacture of fine and speciality chemicals to meet future societal demands.

Keywords Heterogeneous catalysis · Biofuels · Biodiesel · Selective oxidation · Alcohols

Introduction

Sustainability, in essence the development of methodologies to meet the needs of the present without compromising those of future generations has become a watchword for modern society, with developed and developing nations and multinational corporations promoting international research programmes into sustainable food, energy, materials and even city planning. In the context of energy and materials (specifically synthetic chemicals), despite significant growth in proven and predicted fossil fuel reserves over the next two

A. F. Lee (✉)
European Bioenergy Research Institute, Aston University,
Aston Triangle, Birmingham B4 7ET, UK
e-mail: a.f.lee@warwick.ac.uk; a.f.lee@aston.ac.uk

decades, notably heavy crude oil, tar sands, deepwater wells, and shale oil and gas, there are great uncertainties in the economics of their exploitation via current extraction methodologies, and crucially, an increasing proportion of such carbon resources (estimates vary between 65 and 80 % [1–3]) cannot be burned without breaching the UNFCC targets for a 2 °C increase in mean global temperature relative to the pre-industrial level [4, 5]. There is clearly a tightrope to walk between meeting rising energy demands, predicted to rise 50 % globally by 2040 [6] and the requirement to mitigate current CO_2 emissions and hence climate change. The quest for sustainable resources to meet the demands of a rapidly rising global population represents one of this century's grand challenges [7, 8].

While many alternative sources of renewable energy have the potential to meet future energy demands for stationary power generation, biomass offers the most readily implemented, low cost solution to a drop-in transportation fuel for blending with/replacing conventional diesel [9] via carbohydrate hydrodeoxygenation (HDO) or lipid transesterification illustrated in Scheme 1. First generation bio-based fuels derived from edible plant materials received much criticism over the attendant competition between land usage for fuel crops versus traditional agricultural cultivation [10]. Deforestation practices, notably in Indonesia, wherein vast tracts of rainforest and peat land are being cleared to support palm oil plantations have also provoked controversy [11]. To be considered sustainable, second generation bio-based fuels and chemicals are sought that use biomass sourced from non-edible components of crops, such as stems, leaves and husks or cellulose from agricultural or forestry waste. Alternative non-food crops such as switchgrass or *Jatropha curcas* [12], which require minimal cultivation and do not compete with traditional arable land or drive deforestation, are other potential candidate biofuel

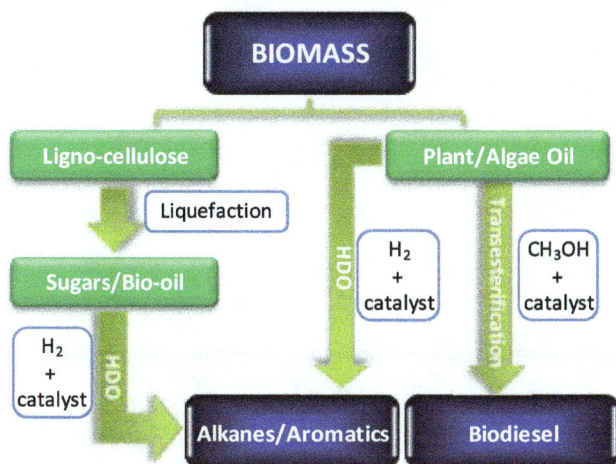

Scheme 1 Chemical conversion routes for the co-production of chemicals and transportation fuels from biomass

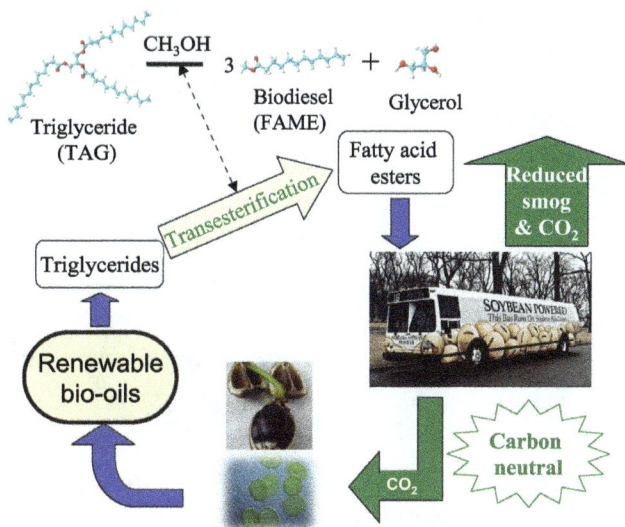

Scheme 2 Carbon cycle for biodiesel production from renewable bio-oils via catalytic transesterification

feedstocks. There is also growing interest in extracting bio-oils from aquatic biomass, which can yield 80–180 times the annual volume of oil per hectare than that obtained from plants [13]. Approximately 9 % of transportation energy needs are predicted to be met via liquid bio-fuels by 2030 [14]. While the abundance of land and aquatic biomass, and particularly of agricultural, forestry and industrial waste, is driving the search for technologies to transform lignocellulose into fuels and chemical, energy and atom-efficient processes to isolate lignin and hemicellulose from the more tractable cellulose component, remain to be identified [15]. Thermal pyrolysis offers one avenue by which to obtain transportation fuels, and wherein catalysis will undoubtedly play a significant role in both pyrolysis of raw biomass and subsequent upgrading of bio-oils via deoxygenation and carbon chain growth. Catalytic depolymerisation of lignin may also unlock opportunities for the production of phenolics and related aromatic compounds for fine chemical and pharmaceutical applications [16].

Biodiesel is a clean burning and biodegradable fuel which, when derived from non-food plant or algal oils or animal fats, is viewed as a viable alternative (or additive) to current petroleum-derived diesel [17]. Commercial biodiesel is currently synthesised via liquid base-catalysed transesterification of C_{14}–C_{20} triacylglyceride (TAG) components of lipids with C_1–C_2 alcohols [18–21] into fatty acid methyl esters (FAMEs) which constitute biodiesel as shown in Scheme 2, alongside glycerol as a potentially valuable by-product [22]. While the use of higher (e.g. C_4) alcohols is also possible [23], and advantageous in respect of producing a less polar and corrosive FAME [24] with reduced cloud and pour points [25], the current high cost of longer chain alcohols, and difficulties associated with separating the heavier FAME product from unreacted alcohol and glycerol, remain problematic.

Unfortunately, homogeneous acid and base catalysts can corrode reactors and engine manifolds, and their removal from the resulting biofuel is particularly problematic and energy intensive, requiring aqueous quench and neutralisation steps which result in the formation of stable emulsions and soaps [9, 26, 27]. Such homogeneous approaches also yield the glycerine by-product, of significant potential value to the pharmaceutical and cosmetic industries, in a dilute aqueous phase contaminated by inorganic salts. Heterogeneous catalysis has a rich history of facilitating energy efficient selective molecular transformations and contributes to 90 % of chemical manufacturing processes and to more than 20 % of all industrial products [28, 29]. While catalysis has long played a pivotal role in petroleum refining and petrochemistry, in a post-petroleum era, it will face new challenges as an enabling technology to overcoming the engineering and scientific barriers to economically feasible routes to bio-fuels. The utility of solid base and acid catalysts for biodiesel production has been extensively reviewed [20, 30–33], wherein they offer improved process efficiency by eliminating the need for quenching steps, allowing continuous operation [34], and enhancing the purity of the glycerol by-product. Technical advances in catalyst and reactor design remain essential to utilise non-food based feedstocks and thereby ensure that biodiesel remains a key player in the renewable energy sector for the 21st century. Select pertinent developments in tailoring the nanostructure of solid acid and base catalysts for TAG transesterification to FAMEs and the related esterification of free fatty acid (FFAs) impurities common in bio-oil feedstocks are therefore discussed herein.

Biomass also offers the only non-fossil fuel route to organic molecules for the manufacture of bulk, fine and

Scheme 3 Cartoon depicting the atom-efficient, chemoselective aerobic selective oxidation of allylic alcohols to aldehydes over a heterogeneous catalyst

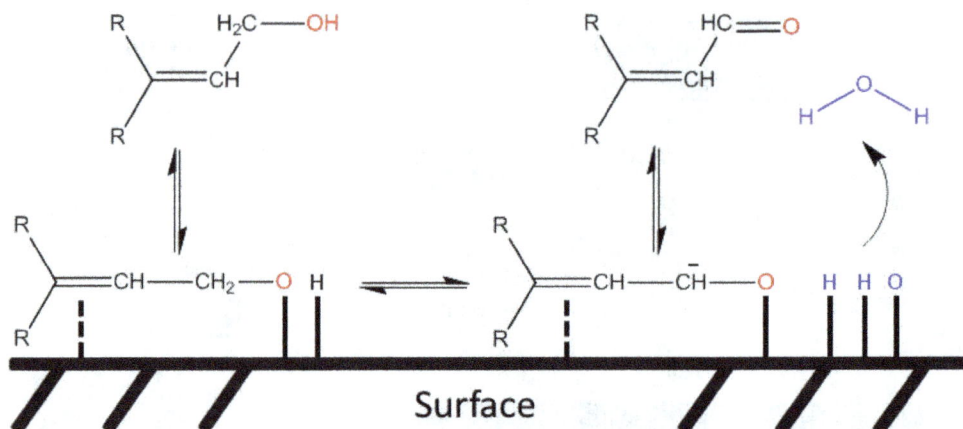

speciality chemicals and polymers [35] required to meet societal demands for advanced materials [8, 36]. The production of such highly functional molecules, whether derived from petroleum feedstocks, requires chemoselective transformations in which e.g. specific heteroatoms or functional groups are incorporated or removed without compromising the underpinning molecular properties. The selective oxidation (selox) of alcohols, carbohydrates and related α,β-unsaturated substrates represent an important class of reactions that underpin the synthesis of valuable chemical intermediates [37, 38]. The scientific, technological and commercial importance of green chemistry presents a significant challenge to traditional selox methods, which previously employed hazardous and toxic stoichiometric oxidants including permanganates, chromates and peroxides, with concomitant poor atom efficiencies and requiring energy-intensive separation steps to obtain the desired carbonyl or acid product. Alternative heterogeneous catalysts utilising oxygen or air as the oxidant offer vastly improved activity, selectivity and overall atom efficiency in alcohol selox (Scheme 3), but are particularly demanding due to the requirement to activate molecular oxygen and C–O bonds in close proximity at a surface in a solid–liquid–gas environment [39–41], and must also be scalable in terms of both catalyst synthesis and implementation. For example, continuous flow microreactors have been implemented in both homogeneous and heterogeneous aerobic selox, providing facile catalyst recovery from feedstreams for the latter [42, 43], but their scale-up/out requires complex manifolding to ensure adequate oxygen dissolution uniform reactant mixing and delivery [44, 45]. Efforts to overcome mass transport and solubility issues inherent to 3-phase catalysed oxidations have centred around the use of supercritical carbon dioxide to facilitate rapid diffusion of substrates to and products from the active catalyst site at modest temperatures [46] affording enhanced turnover frequencies (TOFs), selectivity and on-stream performance versus conventional batch operation in liquid organic solvents [47–51].

The past decade has seen significant progress in understanding the fundamental mode of action of Platinum Group Metal heterogeneous catalysts for aerobic selox and the associated reaction pathways and deactivation processes [41]. This insight has been aided by advances in analytical methodologies, notably the development of in situ or *operando* (under working conditions) spectroscopic [52–54] /microscopic [55–58] tools able to provide quantitative, spatio-temporal information on structure–function relations of solid catalysts in the liquid and vapour phase. Parallel improvements in inorganic synthetic protocols offer finer control over preparative methods to direct the nanostructure (composition, morphology, size, valence and support architecture) of palladium catalysts [59–61] and thereby enhance activity, selectivity and lifetime in an informed manner. Ultimately, heterogeneous catalysts may offer significant advantages over homogeneous analogues in respect of initial catalyst cost, product separation, and metal recovery and recyclability [62]. Catalyst development can thus no longer be considered simply a matter of reaction kinetics, but as a clean technology wherein all aspects of process design, such as solvent selection, batch/flow operation, catalyst recovery and waste production and disposal are balanced [63]. The efficacy of Platinum Group Metals (PGMs) surfaces towards the liquid phase oxidation of alcohols has been known for over 50 years [64], and the development of heterogeneous platinum selox catalysts (and more recently coinage metals such as gold [65, 66]) the subject of recent reviews [39, 67–69] hence only palladium selox catalysis is described herein.

Heterogeneously catalysed routes to biodiesel

Solid acid catalysed biodiesel synthesis

A wide range of inorganic and polymeric solid acids are commercially available, however, their application for the transesterification of oils into biodiesel has only been recently explored, in part reflecting their lower activity compared with base-catalysed routes [27], in turn necessitating higher reaction temperatures to deliver suitable conversions. While their activities are generally low, solid acids have the advantage that they are less sensitive to FFA contaminants than their solid base analogues, and hence can operate with unrefined feedstocks containing 3–6 wt% FFAs [27]. In contrast to solid bases which require feedstock pretreatment to remove fatty acid impurities, solid acids are able to esterify FFAs through to FAME in parallel with transesterification major TAG components without soap formation and thus reduce the number of processing steps to biodiesel [70–72].

Mesoporous silicas from the SBA family [73] have been examined for biodiesel synthesis, and include materials grafted with sulfonic acid groups [74, 75] or SO_4/ZrO_2 surface coatings [76]. Phenyl and propyl sulfonic acid SBA-15 catalysts are particularly attractive materials with activities comparable to Nafion and Amberlyst resins in palmitic acid esterification [77]. Phenylsulfonic acid functionalised silica is reportedly more active than their corresponding propyl analogues, in line with their respective acid strengths but is more difficult to prepare. Unfortunately, conventionally synthesised sulfonic acid functionalised SBA-15 silicas with pore sizes below ∼6 nm possess long, isolated parallel channels and suffer correspondingly slow in-pore diffusion and catalytic turnover in FFA esterification. However, poragens such as trimethylbenzene [78] triethylbenzene or triisopropylbenzene [79] can induce swelling of the Pluronic P123 micelles used to produce SBA-15, enabling ordered

mesoporous silicas with diameters spanning 5–30 nm, and indeed ultra-large-pores with a BJH pore diameter as much as 34 nm [79]. This methodology was recently applied to prepare a range of large pore SBA-15 materials employing trimethylbenzene as the poragen, resulting in the formation of highly ordered periodic mesostructures with pore diameters of ∼6, 8 and 14 nm [80]. These silicas were subsequently functionalised by mercaptopropyl trimethoxysilane (MPTS) and oxidised with H_2O_2 to yield expanded $PrSO_3$-SBA-15 catalysts which were effective in both palmitic acid esterification with methanol and tricaprylin and triolein transesterification with methanol under mild conditions. For both reactions, turnover frequencies dramatically increased with pore diameter, and all sulfonic acid heterogeneous catalysts significantly outperformed a commercial Amberlyst resin (Fig. 1). These rate enhancements are attributed to superior mass transport of the bulky FFA and triglycerides within the expanded $PrSO_3$-SBA-15. Similar observations have been made over Poly(styrene-sulfonic acid)-functionalised ultra-large pore SBA-15 in the esterification of oleic acid with butanol [81].

Improving pore interconnectivity, for example through swapping the *p6 mm* architecture of SBA-15 for the *Ia3d* of KIT-6 was subsequently explored as an alternative means to enhance in-pore active site accessibility (Scheme 1) for FFA esterification [82]. KIT-6 mesoporous materials exhibit improved characteristics for biomolecule immobilisation [83] reflecting superior diffusion within the interconnected cubic structure. A family of pore-expanded propylsulfonic acid KIT-6 analogues were prepared via MPTS grafting and oxidation and screened for FFA

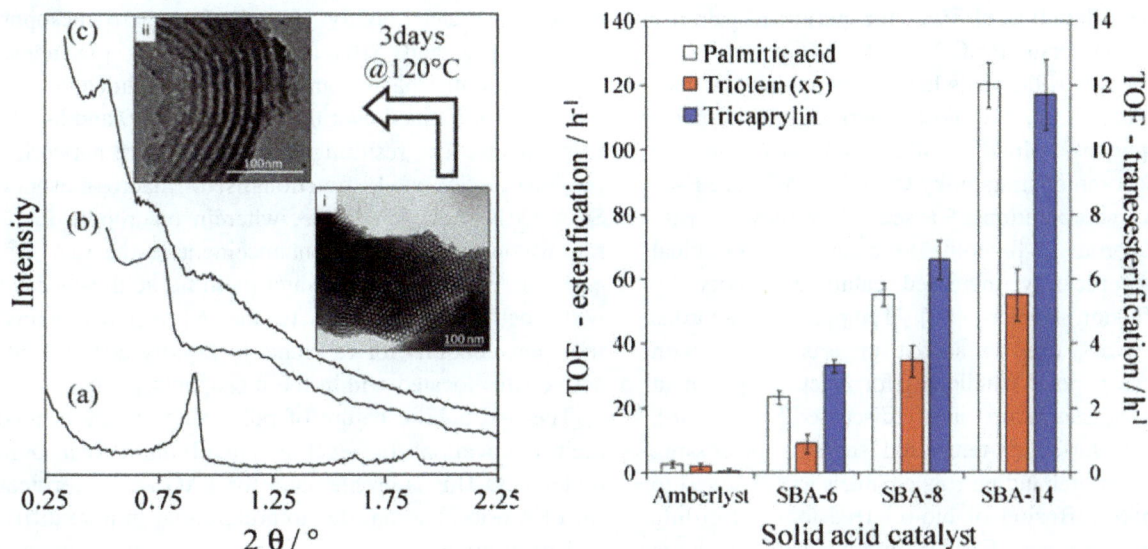

Fig. 1 (*Left*) Low angle powder X-ray diffraction patterns and transmission electron micrographs of propylsulfonic acid functionalised SBA-15 silicas as a function of pore diameter; and (*right*) corresponding catalytic activity in FFA esterification and TAG transesterification compared to a commercial solid acid resin. Adapted from reference [80] with permission from The Royal Society of Chemistry

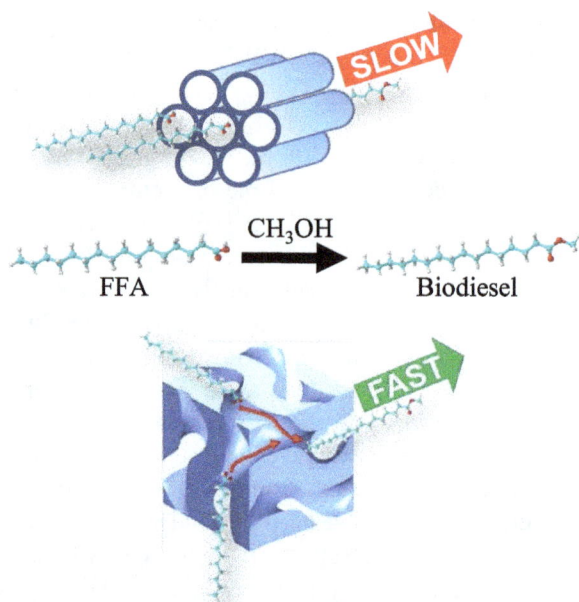

Fig. 2 Superior performance of interconnected, mesoporous propyl-sulfonic acid KIT-6 catalysts for biodiesel synthesis via FFA esterification with methanol versus non-interconnected mesoporous SBA-15 analogue. Adapted from reference [82]. Copyright 2012 American Chemical Society

esterification with methanol as a function of alkyl chain length under mild conditions. As-synthesised PrSO$_3$H-KIT-6 exhibited respective 40 and 70 % TOF enhancements toward propanoic and hexanoic acid esterification compared with a PrSO$_3$H-SBA-15 analogue of comparable (5 nm) pore diameter as a consequence of the improved mesopore interconnectivity. However, pore accessibility remained rate-limiting for esterification of the longer chain lauric and palmitic acids. Hydrothermal aging protocols facilitated expansion of the KIT-6 mesopore up to 7 nm, with consequent doubling of TOFs for lauric and palmitic acid esterification versus PrSO$_3$H-SBA-15 (Fig. 2).

While numerous solid acids have been applied for biodiesel synthesis [27, 32, 84], most materials exhibit micro- and/or mesoporosity which, as illustrated above, are not optimal for accommodating bulky C$_{16}$–C$_{18}$ TAGs of FFAs. For example, incorporation of a secondary mesoporosity into a microporous H-β-zeolite to create a hierarchical solid acid significantly increased catalytic activity by lowering diffusion barriers [85]. Templated mesoporous materials are widely used as catalyst supports [86, 87], with SBA-15 silicas popular candidates for reactions pertinent to biodiesel synthesis as previously discussed [75, 77, 88]. However, such surfactant-templated supports possessing long, isolated parallel and narrow channels are ill-suited to efficient in-pore diffusion of bio-oil feedstocks affording poor catalytic turnover. Further improvements in pore architecture are hence required to optimise mass transport of heavier bulky TAGs and FFAs commonly found in plant and algal oils. Simulations demonstrate that in the Knudsen

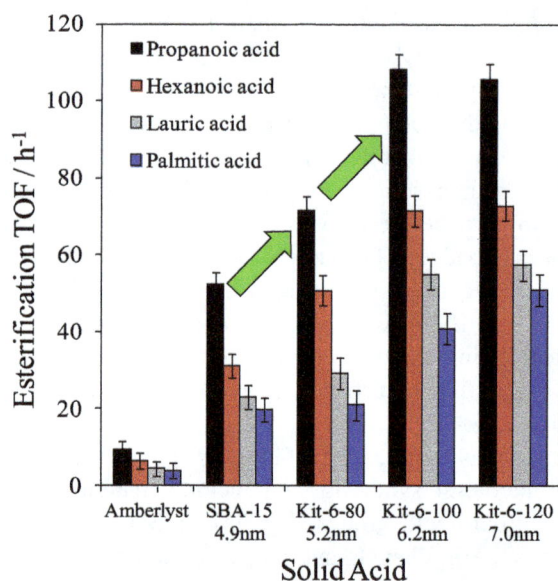

diffusion regime [89], where reactants/products are able to diffuse enter/exit mesopores but experience moderate diffusion limitations, hierarchical pore structures may significantly improve catalyst activity. Materials with interpenetrating, bimodal meso–macropore networks have been prepared using microemulsion [90] or co-surfactant [91] templating routes and are particularly attractive for liquid phase, flow reactors wherein rapid pore diffusion is required. Liquid crystalline (soft) and colloidal polystyrene nanospheres (hard) templating methods have been combined to create highly organised, macro–mesoporous aluminas [92] and 'SBA-15 like' silicas [93] (Scheme 4), in which both macro- and mesopore diameters can be independently tuned over the range 200–500 and 5–20 nm, respectively. The resulting hierarchical pore network of a propylsulfonic acid functionalised macro–mesoporous SBA-15 is shown in Fig. 3, wherein macropore incorporation confers a striking enhancement in the rates of tricaprylin transesterification and palmitic acid esterification with methanol, attributed to the macropores acting as transport conduits for reactants to rapidly access PrSO$_3$H active sites located within the mesopores.

The hydrophilic nature of polar silica surfaces hinders their application for reactions involving apolar organic molecules. This is problematic for TAG transesterification (or FFA esterification) due to preferential in-pore diffusion and adsorption of alcohol versus fatty acid components. Surface hydroxyl groups also favour H$_2$O adsorption, which if formed during FFA esterification can favour the reverse hydrolysis reaction and consequent low FAME

Scheme 4 Liquid crystal and polystyrene nanosphere dual surfactant/physical templating route to hierarchical macroporous–mesoporous silicas

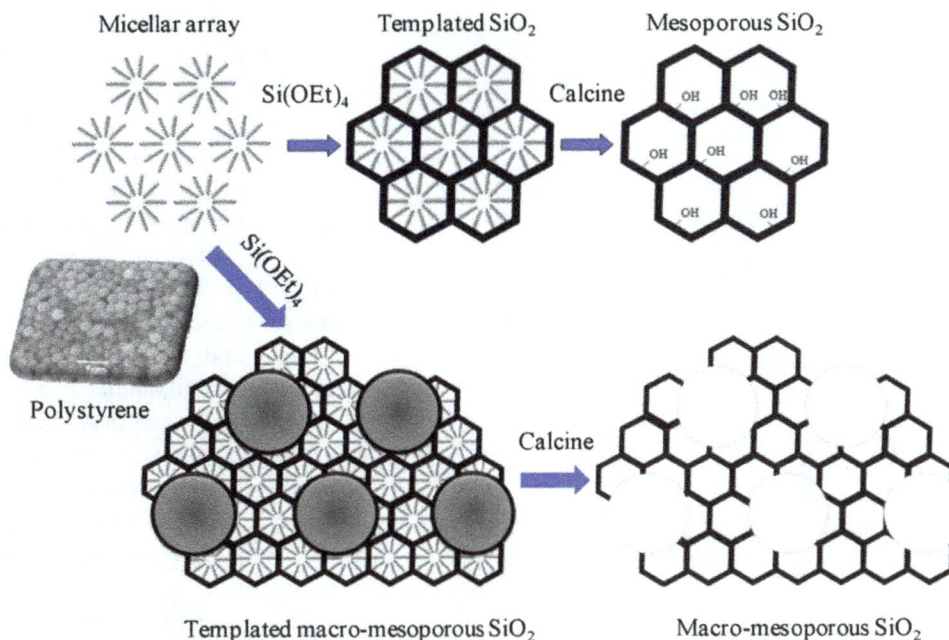

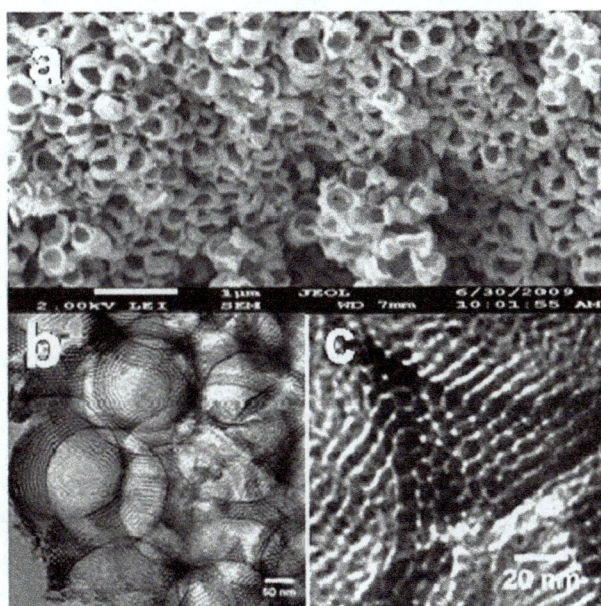

Fig. 3 (*Left*) SEM (**a**) and low and high magnification TEM (**b**, **c**) micrographs of a hierarchical macro–mesoporous Pr-SO₃H-SBA-15; (*right*) corresponding catalytic performance in palmitic acid esterification and tricaprylin transesterification with methanol as a function of macropore density versus a purely mesoporous Pr-SO₃H-SBA-15. Adapted from reference [93] with permission from The Royal Society of Chemistry

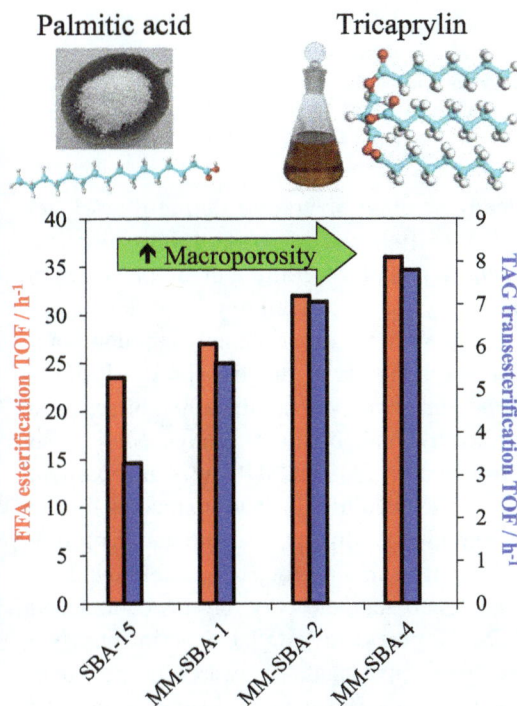

yields. Surface modification via the incorporation of organic functionality into polar oxide surfaces, or dehydroxylation, can lower their polarity and thereby increase initial rates of acid catalysed transformations of liquid phase organic molecules [94]. Surface polarity can also be tuned by incorporating alkyl/aromatic groups directly into the silica framework, for example polysilsesquioxanes can be prepared via the co-condensation of 1,4-bis(triethoxysilyl)benzene (BTEB), or 1,2-bis(trimethoxysilyl)-ethane (BTME), with TEOS and MPTS in the sol–gel process [95, 96] which enhances small molecule esterification [97] and etherification [98]. The incorporation of organic spectator

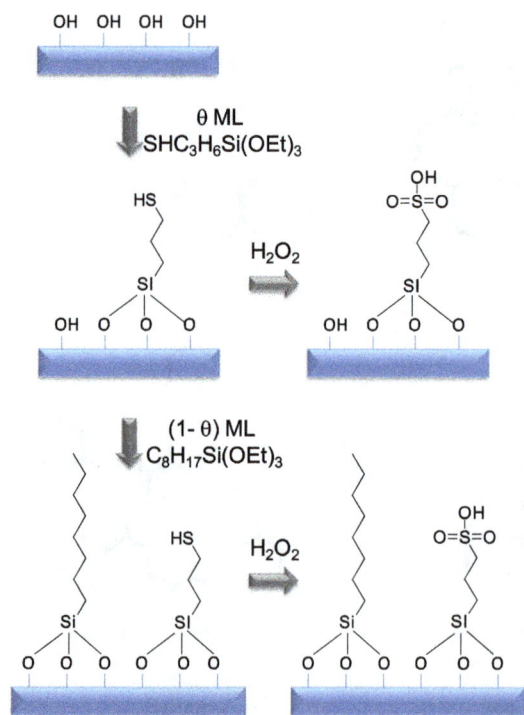

Scheme 5 Protocol for the synthesis of sulfonic acid and octyl co-functionalised sulfonic acid MCM-41catalysts. Adapted from reference [101] with permission from The Royal Society of Chemistry

groups (e.g. phenyl, methyl or propyl) during the sol–gel syntheses of SBA-15 [99] and MCM-41 [100] sulphonic acid silicas is achievable via co-grafting or simple addition of the respective alkyl or aryltrimethoxysilane during co-condensation protocols. An experimental and computational study of sulphonic acid functionalised MCM-41 materials was undertaken to evaluate the effect of acid site density and surface hydrophobicity on catalyst acidity and associated performance [101]. MCM-41 was an excellent candidate due to the availability of accurate models for the pore structure from kinetic Monte Carlo simulations [102], and was modified with surface groups to enable dynamic simulation of sulphonic acid and octyl groups co-attached within the MCM-41 pores. In parallel experiments, two catalyst series were investigated towards acetic acid esterification with butanol (Scheme 5). In one series, the propylsulphonic acid coverage was varied between θ (RSO$_3$H) = 0–100 % ML over the bare silica (MCM–SO$_3$H). For the second octyl co-grafted series, both sulfonic acid and octyl coverages were tuned (MCM–Oc–SO$_3$H). These materials allow the effect of lateral interactions between acid head groups and the role of hydrophobic octyl modifiers upon acid strength and activity to be separately probed.

To avoid diffusion limitations, butanol esterification with acetic acid was selected as a model reaction (Fig. 4). Ammonia calorimetry revealed that the acid strength of polar MCM–SO$_3$H materials increases from 87 to 118 kJ mol^{-1} with sulphonic acid loading. Co-grafted octyl groups dramatically enhance the acid strength of MCM–Oc–SO$_3$H for submonolayer SO$_3$H coverages, with $_-\Delta H_{ads}$(NH$_3$) rising to 103 kJ mol^{-1}. The per site activity of the MCM–SO$_3$H series in butanol esterification with acetic acid mirrors their acidity, increasing with SO$_3$H content. Octyl surface functionalisation promotes esterification for all MCM–Oc–SO$_3$H catalysts, doubling the turnover frequency of the lowest loading SO$_3$H material. Molecular dynamic simulations indicate that the interaction of isolated sulphonic acid moieties with surface silanol groups is the primary cause of the lower acidity and activity of submonolayer samples within the MCM–SO$_3$H series. Lateral interactions with octyl groups help to re-orient sulphonic acid headgroups into the pore interior, thereby enhancing acid strength and associated esterification activity.

In summary, recent developments in tailoring the structure and surface functionality of sulfonic acid silicas have led to a new generation of tunable solid acid catalysts well-suited to the esterification of short and long chain FFAs, and transesterification of diverse TAGs, with methanol under mild reaction conditions. A remaining challenge is to extend the dimensions and types of pore-interconnectivities present within the host silica frameworks, and to find alternative low cost soft and hard templates to facilitate synthetic scale-up of these catalysts for multi-kg production. Surfactant template extraction is typically achieved via energy-intensive solvent reflux, which results in significant volumes of contaminated waste and long processing times, while colloidal templates often require high temperature calcination which prevents template recovery/re-use and releases carbon dioxide. Preliminary steps towards the former have been recently taken, employing room temperature ultrasonication in a small solvent volume to deliver effective extraction of the P123 Pluronic surfactant used in the preparation of SBA-15 in only 5 min, with a 99.9 % energy saving and 90 % solvent reduction over reflux methods, and without compromising textural, acidic or catalytic properties of the resultant Pr-SO$_3$H-SBA-15 in hexanoic acid esterification (Fig. 5) [103].

Solid base-catalysed biodiesel synthesis

Base catalysts are generally more active than acids in transesterification, and hence are particularly suitable for high purity oils with low FFA content. Biodiesel synthesis using a solid base catalyst in continuous flow, packed bed arrangement would facilitate both catalyst separation and co-production of high purity glycerol, thereby reducing production costs and enabling catalyst re-use. Diverse solid

Fig. 4 (*Left*) Molecular dynamics simulations of MCM–SO$_3$H and MCM–Oc–SO$_3$H pore models highlighting the interaction between surface sulfonic acid and hydroxyl groups in the absence of co-grafted octyl chains; (*right*) influence of PrSO$_3$H surface density and co-grafted octyl groups on catalytic performance in acetic acid esterification with butanol. Adapted from reference [101] with permission from The Royal Society of Chemistry

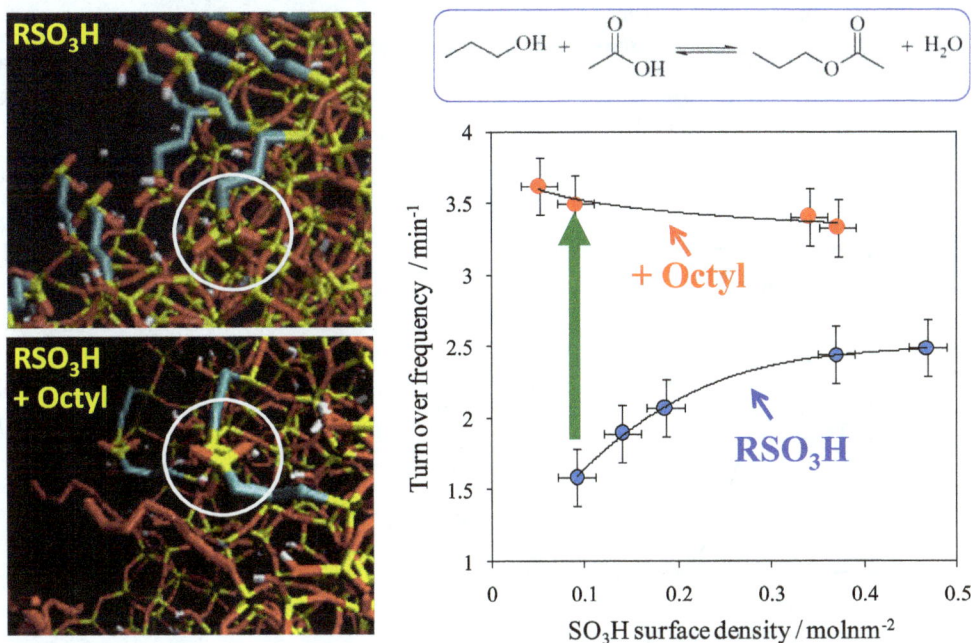

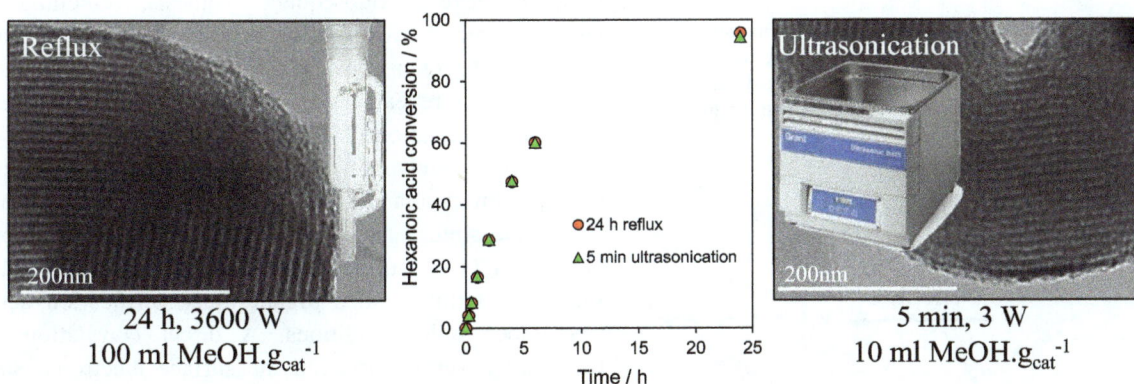

Fig. 5 Surfactant template extraction via energy/atom-efficient ultra-sonication delivers a one-pot PrSO$_3$H-SBA-15 solid acid catalyst with identical structure and reactivity to that obtained by conventional, inefficient reflux. Adapted from reference [103] with permission from The Royal Society of Chemistry

base catalysts are known, notably alkali or alkaline earth oxides, supported alkali metals, basic zeolites and clays such as hydrotalcites and immobilised organic bases [104]. Basicity in alkaline earth oxides is believed to arise from M^{2+}–O^{2-} ion pairs present in different coordination environments [105]. The strongest base sites occur at low coordination defect, corner and edge sites, or on high Miller index surfaces. Such classic heterogeneous base catalysts have been extensively tested for TAG transesterification [106] and there are numerous reports on commercial and microcrystalline CaO applied to rapeseed, sunflower or vegetable oil transesterification with methanol [107, 108]. Promising results have been obtained, with 97 % oil conversion achieved at 75 °C [108], however, concern remains over Ca^{2+} leaching under reaction conditions and associated homogeneous catalytic contributions [109], a common problem encountered in metal catalysed biodiesel production which hampers commercialisation [110].

Alkali-doped CaO and MgO have also been investigated for TAG transesterification [111–113], with their enhanced basicity attributed to the genesis of O^- centres following the replacement of M^+ for M^{2+} and associated charge imbalance and concomitant defect generation. Optimum activity for Li-doped CaO occurs when a saturated Li$^+$ monolayer is formed (Fig. 6) [113], although leaching of the alkali promoter remains problematic [114].

It is widely accepted that the catalytic activity of alkaline earth oxide catalysts is very sensitive to their preparation, and corresponding surface morphology and/or defect density. For example, Parvulescu and Richards demonstrated the impact of the different MgO crystal facets upon the transesterification of sunflower oil by

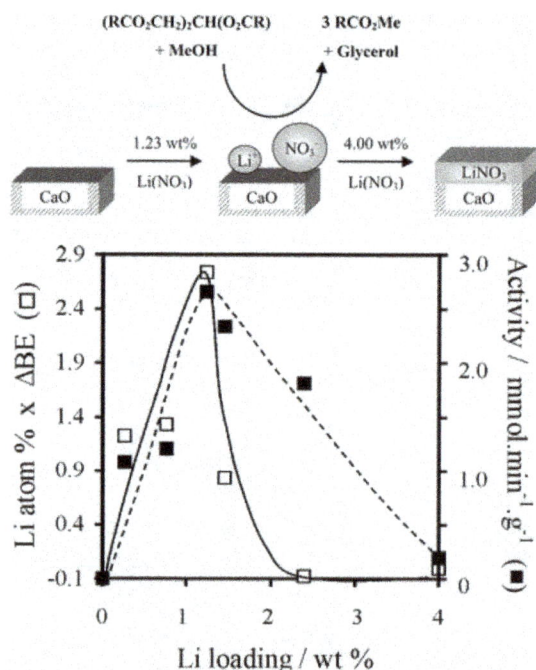

Fig. 6 Correlation between evolving surface composition, density of electronically perturbed Li^+ sites, and corresponding activity in tributyrin transesterification with methanol over Li-doped CaO as a function of Li loading. Adapted from reference [113] with permission from The Royal Society of Chemistry

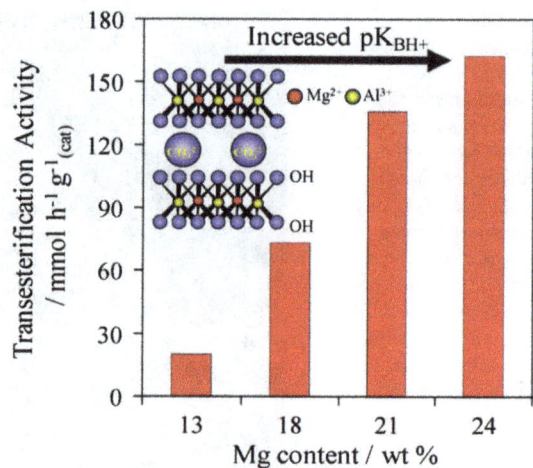

Fig. 8 Impact of Mg:Al hydrotalcite surface basicity on their activity towards tributyrin transesterification. Adapted from reference [118] with permission from Elsevier

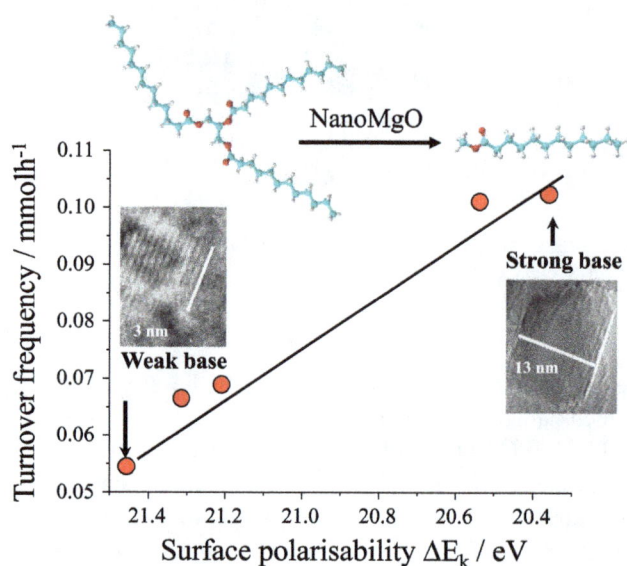

Fig. 7 Relationship between surface polarisability of MgO nanocrystals and their turnover frequency towards tributyrin transesterification. Adapted from reference [117] with permission from The Royal Society of Chemistry

comparing nanoparticles [115] versus (111) terminated nanosheets [116]. Chemical titration reveals that both morphologies possess two types of base sites, with the nanosheets exhibiting well-defined, medium-strong basicity consistent with their uniform exposed facets and

which confer higher FAMe yields during sunflower oil transesterification. Subsequent synthesis, screening and spectroscopic characterisation of a family of size-/shape-controlled MgO nanoparticles prepared via a hydrothermal synthesis revealed small (<8 nm) particles terminate in high coordination (100) facets, and exhibit both weak polarisability and poor activity in tributyrin transesterification with methanol [117]. Calcination drives restructuring and sintering to expose lower coordination stepped (111) and (110) surface planes, which are more polarisable and exhibit much higher transesterification activities under mild conditions. A direct correlation was therefore observed between the surface electronic structure and associated catalytic activity, revealing a pronounced structural preference for (110) and (111) facets (Fig. 7).

Hydrotalcites are another class of solid base catalysts that have attracted recent attention because of their high activity and robustness in the presence of water and FFA [118, 119]. Hydrotalcites ($[M(II)_{1-x}M(III)_x(OH)_2]^{x+}$ $(A^{n-}_{x/n})$ mH_2O) adopt a layered double hydroxide structure with brucite-like $Mg(OH)_2$ hydroxide sheets containing octahedrally coordinated M^{2+} and M^{3+} cations and A^{n-} anions between layers to balance the overall charge [120], and are conventionally synthesised via co-precipitation from their nitrates using alkalis as both pH regulators and a carbonate source. Mg–Al hydrotalcites have been applied for TAG transesterification of poor and high quality oil feeds [121] such as refined and acidic cottonseed oil (9.5 wt% FFA), and animal fat feed (45 wt% water), delivering 99 % conversion within 3 h at 200 °C. It is important to note that many catalytic studies employing hydrotalcites for transesterification are suspect due to their use of Na or K hydroxide/carbonate solutions to precipitate

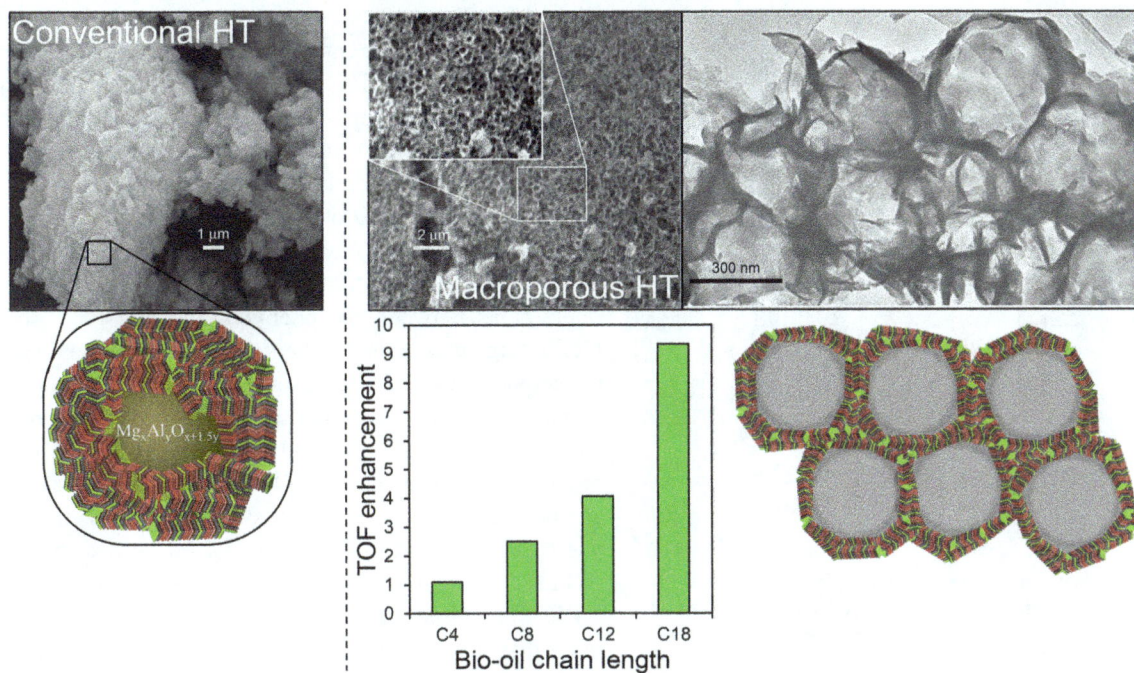

Fig. 9 Superior catalytic performance of a hierarchical macroporous–microporous Mg–Al hydrotalcite solid base catalyst for TAG transesterification to biodiesel versus a conventional microporous analogue. Adapted from reference [124] with permission from The Royal Society of Chemistry

the hydrotalcite phase. Complete removal of alkali residues from the resulting hydrotalcites is inherently difficult, resulting in parallel ill-defined homogeneous contributions to catalysis arising from leached Na or K [122, 123]. This problem has been overcome by the development of alkali-free precipitation routes using NH_3OH and NH_3CO_3, offering well-defined thermally activated and rehydrated Mg–Al hydrotalcites with compositions spanning $x = 0.25 - 0.55$ [118]. Spectroscopic measurements reveal that increasing the Mg:Al ratio enables the surface charge and accompanying base strength to be systematically enhanced, with a concomitant increase in the rate of tributyrin transesterification under mild reaction conditions (Fig. 8).

In spite of their promise for biodiesel production, conventionally prepared hydrotalcites are microporous, and hence poorly suited to application in the transesterification of bulky C_{16}–C_{18} TAG components of bio-oils. This problem was recently tackled by adopting the same hard templating method utilising polystyrene nanospheres described in Scheme 4 to incorporate macroporosity, and thus create a hierarchical macroporous–microporous hydrotalcite solid base catalyst [124]. The introduction of macropores as 'superhighways' to rapidly transport heavy TAG oil components to active base sites present at (high aspect ratio) hydrotalcite nanocrystallites, dramatically enhanced turnover frequencies for triolein transesterification compared with that achievable over an analogous Mg–

Al microporous hydrotalcite (Fig. 9), reflecting superior mass transport through the hierarchical catalyst.

In terms of sustainability, it is important to find low cost routes to the synthesis of solid base catalysts that employ earth abundant elements. Dolomitic rock, comprising alternating $Mg(CO_3)$–$Ca(CO_3)$ layers, is structurally very similar to calcite ($CaCO_3$), with a high natural abundance and low toxicity, and in the UK is sourced from quarries working Permian dolomites in Durham, South Yorkshire and Derbyshire [125]. In addition to uses in agriculture and construction, dolomite finds industrial applications in iron and steel production, glass manufacturing and as fillers in plastics, paints, rubbers, adhesives and sealants. Catalytic applications for powdered, dolomitic rock offer the potential to further valorise this readily available waste mineral, and indeed dolomite has shown promise in biomass gasification [126] as a cheap, disposable and naturally occurring material that significantly reduces the tar content of gaseous products from gasifiers. Dolomite has also been investigated as a solid base catalyst in biodiesel synthesis [127], wherein fresh dolomitic rock comprised approximately 77 % dolomite and 23 % magnesian calcite. High temperature calcination induced Mg surface segregation, resulting in MgO nanocrystals dispersed over $CaO/(OH)_2$ particles, while the attendant loss of CO_2 increases both the surface area and basicity. The resulting calcined dolomite proved an effective catalyst for the transesterification of C_4, C_8 and TAGs with methanol and longer chain C_{16-18}

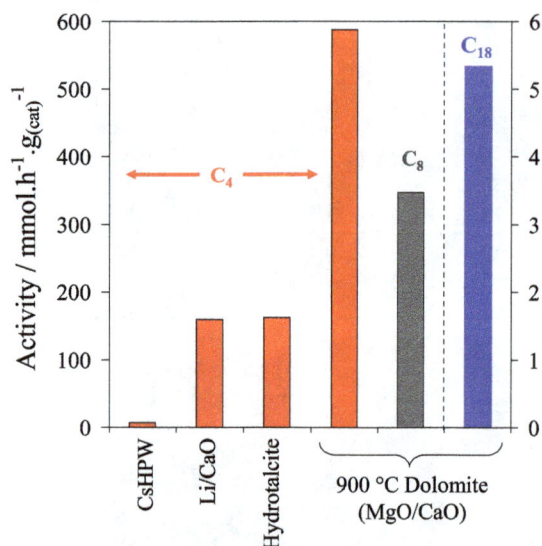

Fig. 10 Catalytic activity of calcined Dolomite for the transesterification of short and long chain TAGs with methanol benchmarked against literature solid acid and base catalysts. Reproduced from reference [127] with permission from The Royal Society of Chemistry

components present within olive oil, with TOFs for tributyrin conversion to methyl butanoate the highest reported for any solid base (Fig. 10). The slower transesterification rates for bulkier TAGs were attributed to diffusion limitations in their access to base sites. Calcined dolomite has also shown promise in the transesterification of canola oil with methanol, achieving 92 % FAME after 3 h reaction with 3 wt% catalyst [128].

In summary, a host of inorganic solid base catalysts have been developed for the low temperature transesterification of triglyceride components of bio-oil feedstocks, offering activities far superior to those achieved via alternative solid acid catalysts to date. However, leaching of alkali and alkaline earth elements and associated catalyst recycling remains a challenge, while improved resilience to water and fatty acid impurities in plant, algal and waste oils feedstocks is required to eliminate additional esterification pre-treatments. To date, only a handful of biodiesel production processes employing heterogeneous catalysts have been commercialised, notably the Esterfip-H process developed by Axens and IFP which utilises a mixture of ZnO and alumina and is operated on a 200 kton per annum scale with parallel production of high quality glycerine [129].

Palladium catalysed aerobic alcohol selox

Particle size effects

Within nanocatalysis, the particle size is a well-documented key parameter influencing both activity and selectivity. This reflects the combination of quantum and geometric effects associated with the respective evolution of electronic properties from atomic like to delocalised bands, and shifting population of low to high coordination surface atoms, with increasing nanoparticle size and dimensionality. Kaneda et al. [130] hypothesised that the unique reactivity of 2060 atom Pd clusters supported on titania towards aromatic alcohol selox arose from a distribution of Pd^0, Pd^+ and Pd^{2+} surfaces sites, with π-bonding interactions between the phenyl group and Pd^{2+} species facilitating subsequent oxidative addition of the O–H bond by neighbouring Pd^0 and eventual β–hydride elimination. Surface hydride was hypothesised to react with oxygen from a neighbouring Pd_2O centre forming H_2O and regenerating the metal site. Optimal activity for cinnamyl alcohol selox to cinnamaldehyde coincided with clusters possessing the maximum fraction of Pd^+ character.

Particle size dependency was also reported for the catalytic transformation of benzyl alcohol over Pd nanoparticles dispersed on alumina, SiO_2 and NaX zeolite supports [131, 132]. For Pd/NaX and Pd/SiO_2-Al_2O_3, benzyl alcohol selox was fastest over particles between 3 and 5 nm, whereas geraniol and 2-octanol were structure-insensitive. Systematic studies of particle size effects in cinnamyl and crotyl alcohol selox over amorphous and mesostructured alumina and silica supports have likewise uncovered pronounced size effects in both initial selox rates and TOFs [133–136], which increase monotonically with shrinking nanoparticle diameters (even down to single atoms) [137]. HAADF–STEM analysis reveals atomically dispersed palladium exhibits maximal rates towards benzyl, cinnamyl and crotyl alcohols, with selectivities to their corresponding aldehydes >70 %. The origin of such size effects is revisited below. The use of colloidal Pd nanoclusters for aqueous phase alcohol selox is limited [138–140], wherein Pd aggregation and Pd black formation hinders catalytic performance. However, the successful stabilisation of 3.6 nm Pd nanoclusters is reported using an amphiphilic nonionic triblock copolymer, Pluronic P123; in the selective oxidation of benzyl alcohol, 100 % aldehyde selectivity and high selox rates are achievable, with high catalytic activity maintained with negligible sintering after 13 recycling reactions [141].

Surface reaction mechanism

The rational design and optimisation of palladium selox catalysts require a microscopic understanding of the active catalytic species responsible for alcohol and oxygen activation, and the associated reaction pathway to the aldehyde/ketone products and any competing processes. A key characteristic of palladium is its ability to perform selox chemistry at temperatures between 60 and 160 °C and with

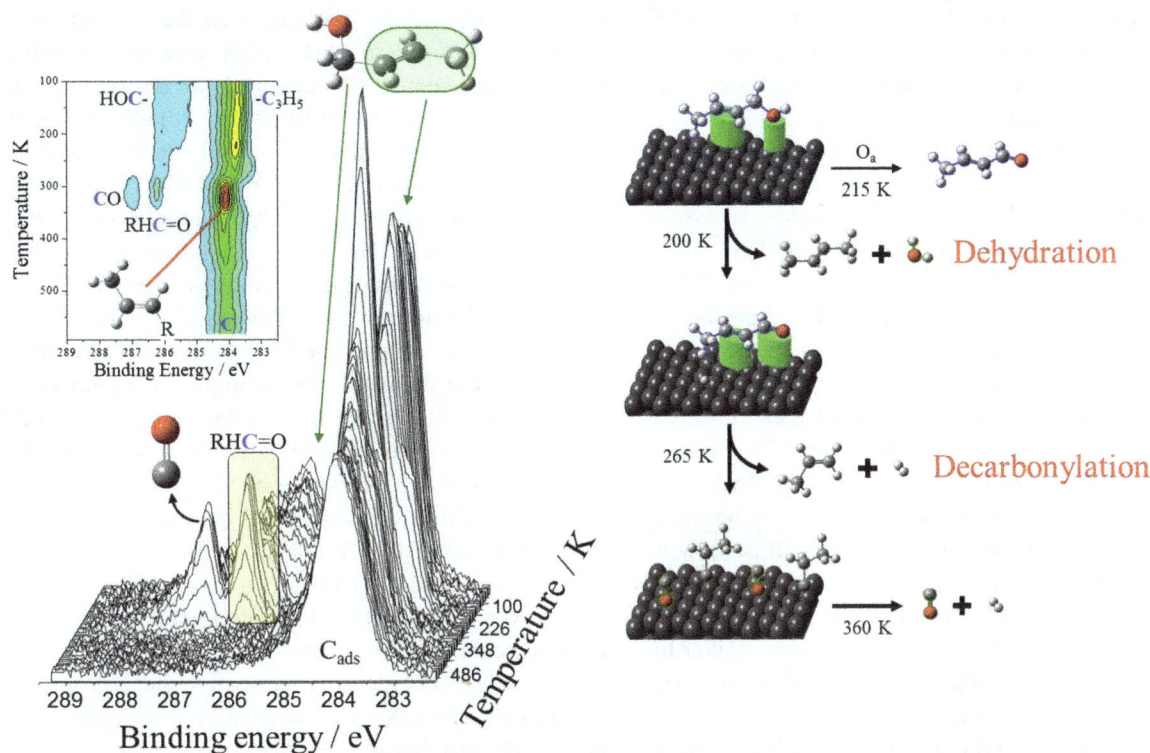

Fig. 11 Temperature-programmed C 1s XP spectra of a reacting crotyl alcohol adlayer over Pd(111) highlighting the primary dehydrogenation pathway and competing decarbonylation pathways. Adapted from reference [143]. Copyright 2007 American Chemical Society

ambient oxygen pressure [39, 142] via the wi dely accepted oxidative dehydrogenation route illustrated in Scheme 3 [39, 67]. Whether O–H or C–H scission of the α-carbon is the first chemical step remains a matter of debate, since the only fundamental studies over well-defined Pd(111) surfaces to date employed temperature-programmed XPS [143] and metastable de-excitation spectroscopy (MDS) [144] with temporal resolutions on the second → minute timescale, over which loss of both hydrogens appears coincident. However, temperature-programmed mass spectrometric [145] and vibrational [146] studies of unsaturated C_1–C_3 alcohols implicate O–H cleavage and attendant alkoxy formation over Pd single crystal surfaces as the first reaction step [142, 147]. It is generally held that the resultant hydrogen adatoms react with dissociatively absorbed oxygen to form water, which immediately desorbs at ambient temperature thereby shifting the equilibrium to carbonyl formation [39, 67]. Temperature-programmed XPS studies of crotyl alcohol adsorbed over clean Pd(111) [143] prove that oxidative dehydrogenation to crotonaldehyde occurs at temperatures as low as −60 °C (Fig. 11), with alcohol dehydration to butane only a minor pathway. These ultra-high vacuum measurements also revealed that reactively formed crotonaldehyde undergoes a competing decarbonylation reaction over metallic palladium above 0 °C yielding strongly bound CO and

propylidene which may act as site-blockers poisoning subsequent catalytic selox cycles, coincident with evolution of propene into the gas phase. Unexpectedly, pre-adsorbed atomic oxygen switched-off undesired decarbonylation chemistry, promoting facile crotonaldehyde desorption.

Nature of the active site

The preceding observation that surface oxygen is not only critical for the removal of hydrogen adatoms but also to suppress decarbonylation of selox products over metallic palladium is in excellent agreement with an in situ ATR-IR study of cinnamyl alcohol selox over Pd/Al2O3 [148]. In related earlier investigations employing aqueous electro-chemical protocols, the same researchers postulated that oxidative dehydrogenation of alcohols requires PGM catalysts in a reduced state, hypothesising that 'over-oxidation' was responsible for deactivation of palladium selox catalysts [69]. A subsequent operando X-ray absorption spectroscopy (XAS) study by Grunwaldt et al. [150], bearing remarkable similarity to an earlier study to the author of this review [149], evidenced in situ reduction of oxidised palladium in an as-prepared Pd/Al2O3 catalyst during cinnamyl alcohol oxidation within a continuous flow fixed-bed reactor. Unfortunately the reaction kinetics were not measured in parallel to explore the impact of palladium reduction, however, a follow-up study of 1-phenylethanol selox employing the same reactor configuration (and oxygen-deficient conditions) evidenced a strong interplay between selox conversion/selectivity and palladium oxidation state [151]. It was concluded that metallic Pd was the catalytically active species, an assertion re-affirmed in subsequent in situ ATR-IR/XAS measurements of benzyl [152–154] and cinnamyl alcohol [155] selox in toluene and under supercritical CO_2, respectively, wherein the C=O stretching intensity was assumed to track alcohol conversion. It is interesting to note that the introduction of oxygen to the reactant feed in these infrared studies dramatically improved alcohol conversion/aldehyde production (Fig. 12), which was attributed to hydrogen abstraction from the catalyst surface [156, 157] rather than to a change in palladium oxidation state. In contrast to their liquid phase experiments, high pressure XANES and EXAFS measurements of Pd/Al_2O_3 catalysed benzyl alcohol selox under supercritical CO_2 led Grunwaldt and Baiker to conclude that maximum activity arose from particles mainly oxidised in the surface/shelfedge [48].

In a parallel research programme, the author's group systematically characterised the physicochemical properties of palladium nanoparticles as a function of size over non-reducible supports to quantify structure–function relations in allylic alcohol selox [133–137, 158, 159]. The combination of XPS and XAS measurements revealed that freshly prepared alumina [134, 137] and silica [135, 158] supported nanoparticles are prone to oxidation as their

diameter falls below ~4 nm, with the fraction of PdO proportional to the support surface area and interconnectivity. Complementary kinetic analyses uncovered a direct correlation between the surface PdO content and activity/TOFs towards cinnamyl and crotyl alcohol selox [134, 137]. Operando liquid phase XAS of Pd/C and Pd/Al_2O_3-SBA-15 catalysts during cinnamyl alcohol selox evidenced in situ reduction of PdO (Fig. 13), however, by virtue of simultaneously measuring the rate of alcohol selox, Lee et al. were able to prove that this oxide → metal structural transition was accompanied by coincident deactivation. Together these findings strongly implicate a (surface) PdO active phase, consistent with surface science predictions that metallic palladium favours aldehyde decarbonylation and consequent self-poisoning by CO and organic residues [143, 160], akin to that reported during fatty acid decarboxylation over Pd/MCF [161].

To conclusively establish whether oxide or metal is responsible for alcohol selox catalysed by dispersed palladium nanoparticles, a multi-dimensional spectroscopic investigation of vapour phase crotyl alcohol selox was undertaken (since XAS is an averaging technique a complete understanding of catalyst operation requires multiple analytical techniques [162–164]). Synchronous, time-resolved DRIFTS/MS/XAS measurements of supported and colloidal palladium were performed in a bespoke environmental cell [165] to simultaneously interrogate adsorbates on the catalyst surface, Pd oxidation state and reactivity under transient conditions in the absence of competitive solvent effects [166, 167]. Under mild reaction

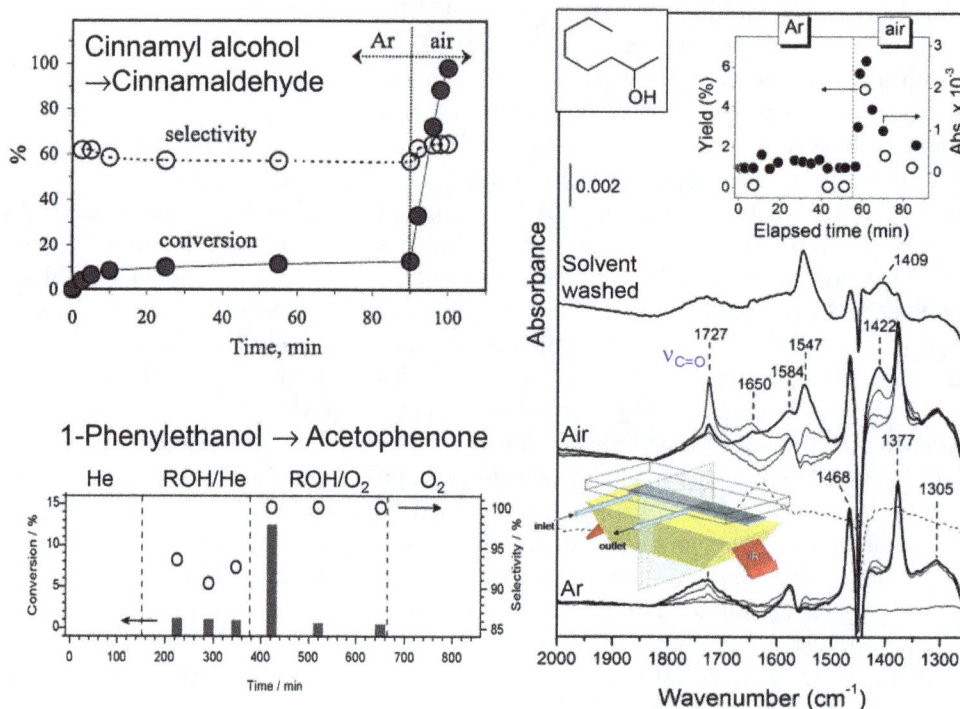

Fig. 12 Impact of oxygen on the selective oxidation of (*top left*) cinnamyl alcohol; (*bottom left*) 1-phenylethanol; and (*right*) 2-octanol. Adapted from references [148, 151, 154] with permission from Elsevier

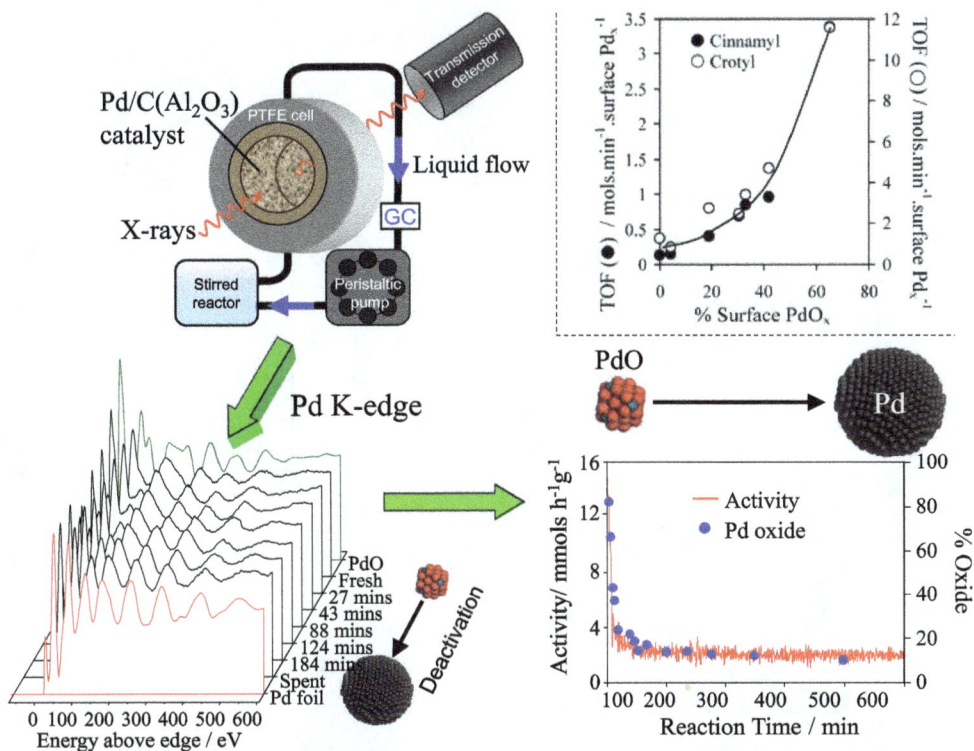

Fig. 13 (*Top right*) Dependence of allylic alcohol selox rate upon surface PdO; (*top left*) schematic of operando liquid phase reactor; (*bottom left*) evolution of Pd K-edge XAS of Pd/Al₂O₃ catalyst during cinnamyl alcohol aerobic selox; (*bottom right*) temporal correspondence between Pd oxidation state and selox activity in cinnamyl alcohol selox. Adapted from references [133, 134] with permission from The Royal Society of Chemistry

temperatures, palladium nanoparticles were partially oxidised, and unperturbed by exposure to sequential alcohol or oxygen pulses (Fig. 14). Crotonaldehyde formed immediately upon contact of crotyl alcohol with the oxide surface, but only desorbed upon oxygen co-adsorption. Higher reaction temperatures induced PdO reduction in response to crotyl alcohol exposure, mirroring that observed during liquid phase selox, however, this reduction could be fully reversed by subsequent oxygen exposure. Such reactant-induced restructuring was exhibited by all palladium nanoparticles, but the magnitude was inversely proportional to particle size [168]. These dynamic measurements decoupled the relative reactivity of palladium oxide from metal revealing that PdO favoured crotyl alcohol selox to crotonaldehyde and crotonic acid, whereas metallic palladium drove secondary decarbonylation to propene and CO in accordance with surface science predictions [143].

Recent ambient pressure XPS investigations of crotyl alcohol/O₂ gas mixtures over metallic and oxidised Pd(111) single crystal surfaces confirmed that only two-dimensional Pd_5O_4 and three-dimensional PdOx surfaces were capable of crotonaldehyde production (Fig. 15) [169]. However, even under oxygen-rich conditions, on-stream reduction of the Pd_5O_4 monolayer oxide occurred >70 °C accompanied by surface poisoning by hydrocarbon residues. In contrast, PdOx multilayers were capable of sustained catalytic turnover of crotyl alcohol to crotonaldehyde, conclusively proving surface palladium oxide as the active phase in allylic alcohol selox.

Establishing support effects

Anchoring Pd nanoparticles onto support structures offers an effective means to tune their physicochemical characteristics and prevent on-stream deactivation e.g. by sintering. Supports employing porous architectures, acid/base character and/or surface redox chemistry e.g. strong metal support interaction (SMSI), afford further opportunities to influence catalyst reactivity [170–173]. Mesoporous silicas are widely used to disperse metal nanoparticles [135, 136, 171, 174, 175]. The transition from low surface area, amorphous silica (200 m²g⁻¹) to two-dimensional non-interconnected pore channels (SBA-15) [73] and three-dimensional interconnected porous frameworks (SBA-16, KIT-6) [73, 176, 177] improved the dispersion of Pd nanoparticles and hence degree of surface oxidation and thus activity in allylic alcohol selox (Fig. 16), but had little impact on the mass transport of small alcohols to/from the active site. [135, 136] The high thermal and chemical stability of such mesoporous silica [178, 179] makes such supports well-suited to commercialisation. Pd nanoparticles confined within such mesoporous silicas demonstrate good selectivity in crotyl and cinnamyl alcohol selox to their respective aldehydes (>70 %), and excellent TOFs of 7,000 and 5,000 h⁻¹ for the respective alcohols. Similar activities are reported for secondary and tertiary allylic alcohols, highlighting the versatility of silica supported Pd nanoparticles [51, 135, 136, 180–182]. Incorporation of macropores into SBA-15 via dual hard/soft templating to

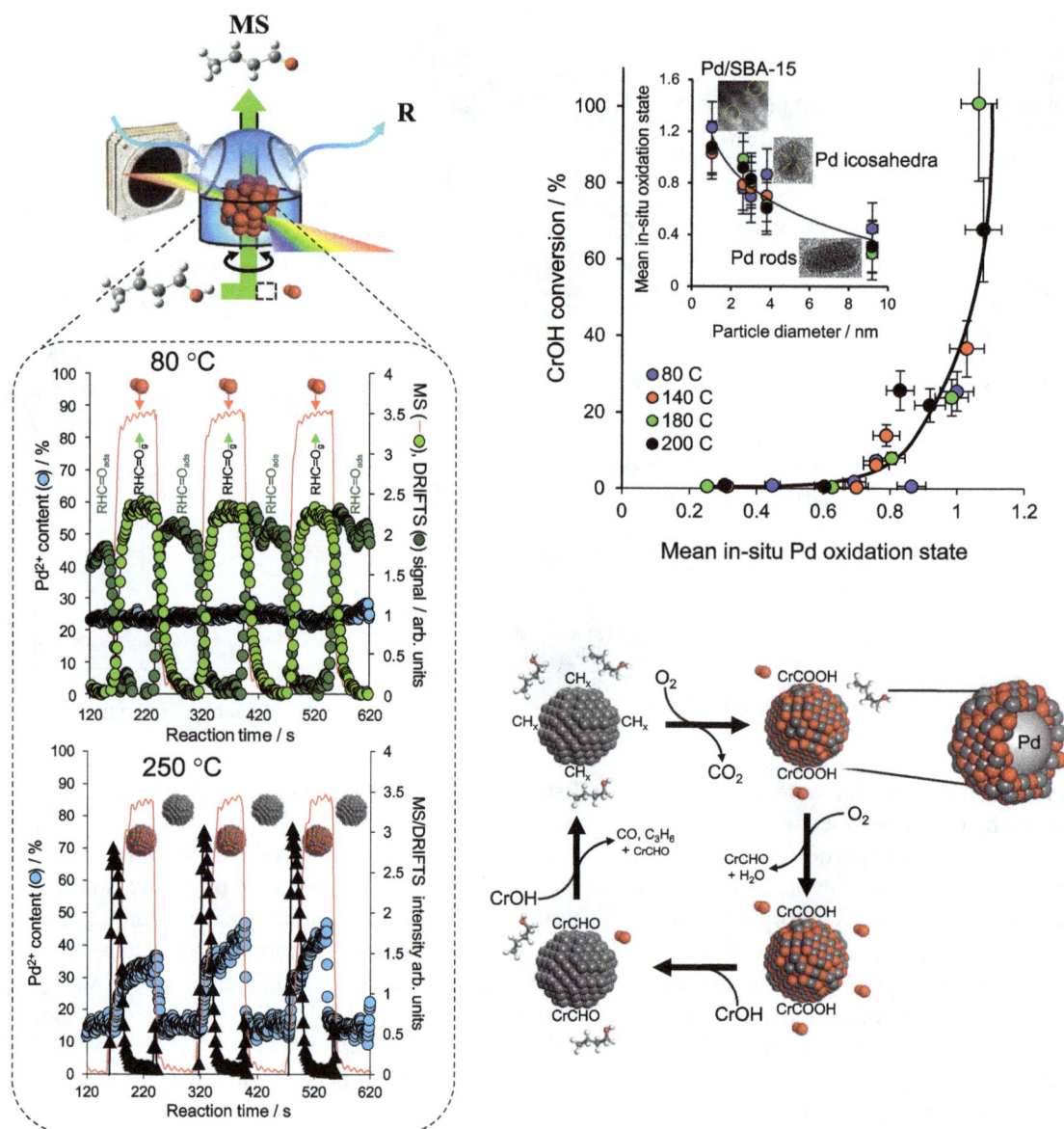

Fig. 14 (*Left*) Cartoon of operando DRIFTS/MS/XAS reaction cell and resulting temperature dependent behaviour of Pd oxidation state and associated reactivity towards crotyl alcohol oxidation over a Pd/meso-Al$_2$O$_3$ catalyst—only selective oxidation over surface PdO occurs at 80 °C, whereas crotonaldehyde decarbonylation and combustion dominate over Pd metal at 250 °C; (*top right*) relationship between Pd oxidation derived in situ and crotyl alcohol conversion; (*bottom right*) summary of reaction-induced redox processes in Pd-catalysed crotyl alcohol selox. Adapted with permission from references [166, 168]. Copyright 2011 and 2012 American Chemical Society

form a hierarchically ordered macroporous–mesoporous Pd/SBA-15 was recently shown to promote the catalytic selox of sterically challenging sesquiterpenoid substrates such as farnesol and phytol via (1) stabilising PdO nanoparticles and (2) dramatically improving in-pore diffusion and access to active sites [158].

The benefits of mesostructured supports are not limited to silica, with ultra-low loadings of palladium impregnated onto a surfactant-templated mesoporous alumina (350

m^2 g^{-1}) generating atomically dispersed Pd^{2+} centres [137]. Such single-site catalysts were 10 times more active in crotonaldehyde and cinnamaldehyde production than comparable materials employing conventional (100 m^2 g^{-1}) γ-alumina, owing to the preferential genesis of higher concentrations of electron-deficient palladium [134, 137], due to either pinning at cation vacancies or metal → support charge transfer [183]. These Pd/meso-Al$_2$O$_3$ catalysts exhibited similar TOFs to their silica counterparts (7,080 and

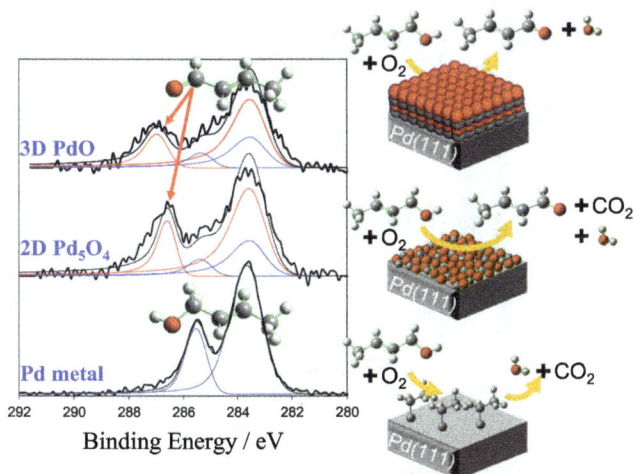

Fig. 15 (*Left*) C 1s XP spectra of crotyl alcohol/O_2 gas mixture over metallic and oxidised Pd(111) surfaces; (*right*) differing reactivity of palladium metal and oxide surfaces. Adapted from reference [169]. Copyright 2012 American Chemical Society

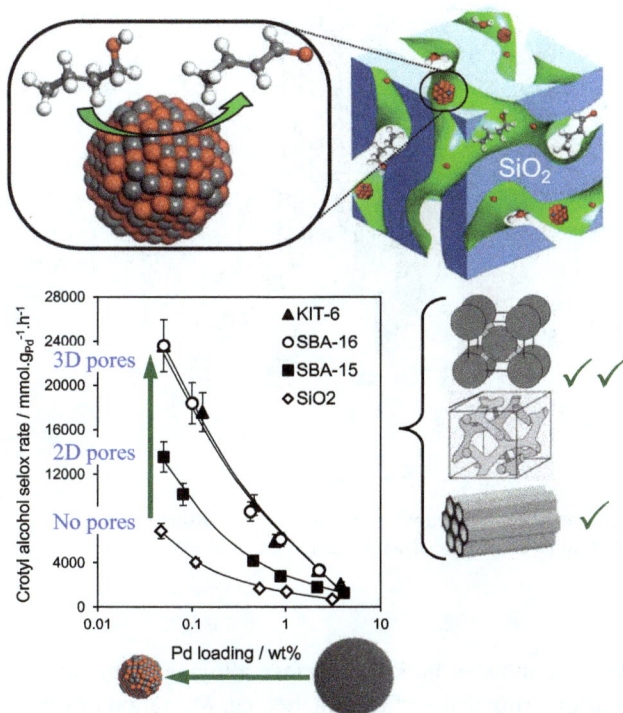

Fig. 16 Comparative activity of Pd nanoparticles dispersed over amorphous, 2D non-interconnected SBA-15 and 3D interconnected SBA-16 and KIT-6 mesoporous silicas in the selective aerobic oxidation of crotyl alcohol. Adapted from reference [135]. Copyright 2011 American Chemical Society

4,400 h^{-1} for crotyl and cinnamyl alcohol selox, respectively) [137], consistent with a common active site and reaction mechanism (Fig. 17).

Mesoporous titania and ceria have also attracted interest as novel catalyst supports. The oxygen storage capacity of ceria-derived materials is of particular interest due to their facile $Ce^{3+} \leftrightarrow Ce^{4+}$ redox chemistry [173, 184–188]. Sacrificial reduction of the ceria supports by reactively formed hydrogen liberated during the oxidative dehydrogenation of alcohols could mitigate in situ reduction of oxidised palladium, and hence maintain selox activity and catalyst lifetime, with Ce^{4+} sites regenerated by dissociatively adsorbed gas phase oxygen [187, 189, 190]. Due to its high density, conventional nanocrystalline cerias possess meagre surface areas (typically ~ 5 m^2g^{-1}), hence Pd/CeO_2 typically exhibit poor selox behaviour due to their resultant low nanoparticle dispersions which favour (self-poisoning) metallic Pd [189, 191, 192].

Bimetallic palladium selox catalysts

Incorporation of a second metal into palladium catalysts can improve both alcohol selox stability and selectivity. Typical promoters such as Ag, Bi, Pb and Sn [157, 193–196], enhance oxidation performance towards challenging substrates such as propylene glycol [197] as well as allylic and benzylic alcohols. Wenkin et al. [194] reported glucose oxidation to gluconates was increased by a factor of 20 over Pd–Bi/C catalysts (Bi/Pd_s = 0.1) versus Pd/C counterparts. In situ XAS and attenuated total reflection infrared spectroscopy (ATR-IR) suggested that Bi residing at the catalyst surface protects palladium from deactivation by either over-oxidation (a hypothesis since disproved [166, 167, 169]) or site-blocking by aromatic solvents [153]. Prati et al. [200] first reported significant rate enhancements and resistance to deactivation phenomena in the liquid phase selox of D-sorbitol to gluconic/gulonic acids upon addition of Au to Pd/C and Pt/C materials [198], subsequently extended to polyol and long chain aliphatic alcohols [199]. A strong synergy between Pd and Au centres was also demonstrated by Hutchings et al., wherein Au–Pd alloy nanoparticles supported on titania exhibited increased reactivity towards a diverse range of primary, allylic and benzylic alkyl alcohols compared to monometallic palladium analogues. The versatility of Au–Pd catalysts has also been shown in selox of saturated hydrocarbons [201], ethylene glycol [202], glycerol [203] and methanol [204], wherein high selectivity and resistance to on-stream deactivation is noted.

The effect of Au–Pd composition has been extensively studied for bimetallic nanoparticles stabilised by PVP surfactants [205]. An optimal Au:Pd composition of 1:3 was identified for 3 nm particles towards the aqueous phase aerobic selox of benzyl alcohol, 1-butanol, 2-butanol, 2-buten-1-ol and 1,4-butanediol; in each case the bimetallic catalysts were superior to palladium alone. Mertens et al. [206] examined similar systems utilising 1.9 nm nanoparticles, wherein an optimal Au content of around 80 % was

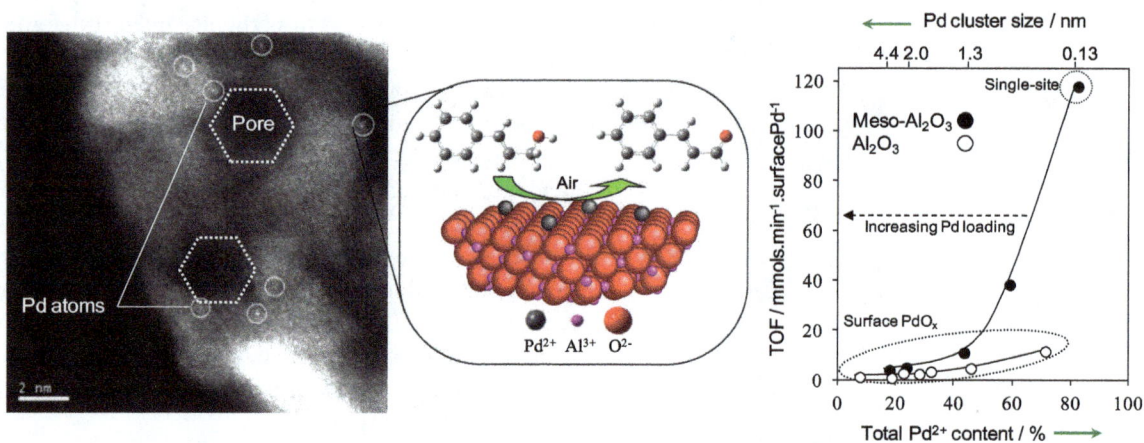

Fig. 17 (*Left*) HAADF–STEM image of atomically dispersed Pd atoms on a mesoporous Al_2O_3 support; and (*right*) associated relationship between Pd^{2+} content/dispersion and activity in crotyl alcohol selox over Pd/alumina catalysts. Adapted with permission from reference [137]. Copyright Wiley–VCH Verlag GmbH & Co. KGaA

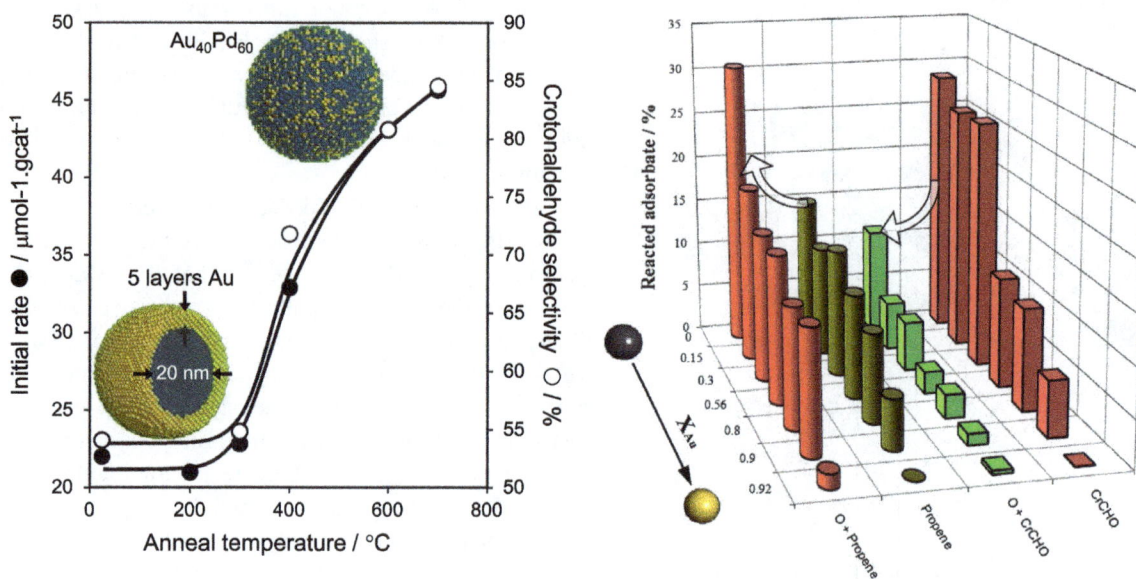

Fig. 18 Impact of thermally induced Au–Pd alloying of (*left*) titania-supported Au shell–Pd core nanoparticles on crotyl alcohol aerobic selox adapted from reference [208], with permission from Elsevier; and (*right*) ultrathin gold overlayers on Pd(111) on crotonaldehyde and propene decomposition with/without co-adsorbed oxygen, adapted from reference [160] with permission from the PCCP Owner Societies

determined for benzyl alcohol selox. The synergic interaction between Au and Pd therefore appears interdependent on nanoparticle size. It is well-known that the catalytic activity of Au nanoparticles increases dramatically <2 nm [207], hence it is interesting to systematically compare phase separated and alloyed catalysts. The author's group prepared titania-supported Au shell (5-layer)-Pd core (20 nm) bimetallic nanoparticles for the liquid phase selox of crotyl alcohol and systematically studied the evolution of their bulk and surface properties as a function of thermal processing by in situ XPS, DRIFTS, EXAFS, XRD and ex-situ HRTEM. Limited Au/Pd alloying occurred below 300 °C in the absence of particle sintering [208]. Higher temperatures induced bulk and surface alloying, with concomitant sintering and surface roughening. Migration of Pd atoms from the core to the surface dramatically enhanced activity and selectivity, with the most active and selective surface alloy containing 40 atom % Au (Fig. 18). This discovery was rationalised in terms of complementary temperature-programmed mass spectometric studies of crotyl alcohol and reactively formed intermediates over Au/Pd(111) model single crystal catalysts which reveal that gold–palladium alloys promote desorption of the desired crotonaldehyde selox product while co-adsorbed oxygen adatoms actually suppress aldehyde combustion. In contrast, the combustion of propene, the undesired secondary product of crotonaldehyde decarbonylation, is enhanced by co-adsorbed oxygen [160].

Scott et al. prepared the inverse Au core-Pd shell nanoparticles and explored the catalytic cycle for alcohol selox to assess their associated stability [205, 209–212]. In situ Pd–K and Pd–L_{III} edge XAS of a Au nanoparticle/Pd(II) salt solution were undertaken to discriminate two possible reaction mechanisms. No evidence was found that crotyl alcohol oxidation was accompanied by Pd^{2+} reduction onto Au nanoparticles, resulting in the formation of a metallic Pd shell (with oxygen subsequently regenerating electron-deficient palladium), and therefore proposed β-H elimination as the favoured pathway. Scott and co-workers proposed that the Au core prevents the re-oxidation of surface Pd^0 atoms; no Pd–O and Pd–Cl contributions were observed by EXAFS.

In summary, the selective oxidation of complex alcohol substrates can be accomplished through Pd-mediated heterogeneous catalysis with high turnover and product selectivity. Application of in situ and operando techniques, such as X-ray and IR spectroscopies, has elucidated the mechanism of alcohol oxidative dehydrogenation and competing aldehyde decarbonylation. Surface PdO has been identified as the active catalytic species, and deactivation the result of reduction to metallic palladium and concomitant self-poisoning by strongly bound CO and carbonaceous residues. Breakthroughs in analytical tools and synthetic approaches to engineering nanoporous supports and shape/size controlled nanoparticles have delivered significant progress towards improved atom and energy efficiency and catalyst stability, however, next generation palladium selox catalysts necessitate improved synthetic protocols to create higher densities of ultra-dispersed Pd^{2+} centres with superior resistance to on-stream reduction under atmospheric oxygen.

Acknowledgments A.F.L. thanks the EPSRC (EP/G007594/3) for financial support and a Leadership Fellowship, and acknowledges the invaluable contributions of Prof Karen Wilson (European Bioenergy Research Institute, Aston University).

References

1. S. Kretzmann, http://priceofoil.org/2013/11/26/new-analysis-shows-growing-fossil-reserves-shrinking-carbon-budget/
2. C.C. Authority, Reducing Australia's greenhouse gas emissions—targets and progress review draft report, commonwealth of Australia, 2013
3. C.C. Secretariat, The critical decade 2013 climate change science, risks and responses, commonwealth of Australia, 2013
4. I.E. Agency, prospect of limiting the global increase in temperature to 2 °C is getting bleaker, http://www.iea.org/newsroomandevents/news/2011/may/name,19839,en.html
5. I.E. Agency, Redrawing the energy climate map, 2013
6. U.S.E.I. Administration, international energy outlook 2013, 2013
7. Armaroli N, Balzani V (2007) Angew Chem Int Ed 46:52–66
8. Azadi P, Inderwildi OR, Farnood R, King DA (2013) Renew Sustain Energy Rev 21:506–523
9. Demirbas A (2007) Energy Policy 35:4661–4670
10. McLaughlin DW (2011) Conserv Biol 25:1117–1120
11. Danielsen F, Beukema H, Burgess ND, Parish F, BrÜHl CA, Donald PF, Murdiyarso D, Phalan BEN, Reijnders L, Struebig M, Fitzherbert EB (2009) Conserv Biol 23:348–358
12. Achten WMJ, Verchot L, Franken YJ, Mathijs E, Singh VP, Aerts R, Muys B (2008) Biomass Bioenergy 32:1063–1084
13. Mata TM, Martins AA, Caetano NS (2010) Renew Sustain Energy Rev 14:217–232
14. BP, BP energy outlook 2030, 2011
15. Sheldon RA (2014) Green Chem 16:950–963
16. Pandey MP, Kim CS (2011) Chem Eng Technol 34:29–41
17. Knothe G (2010) Top Catal 53:714–720
18. Climent MJ, Corma A, Iborra S, Velty A (2004) J Catal 221:474–482
19. Constantino U, Marmottini F, Nocchetti M, Vivani R (1998) Eur J Inorg Chem 10:1439–1446
20. Narasimharao K, Lee A, Wilson K (2007) J Biobased Mater Bioenergy 1:19–30
21. Othman MR, Helwani Z, Martunus, Fernando WJN (2009) Appl Organomet Chem 23:335–346
22. Liu Y, Lotero E, Goodwin JG, Mo X (2007) Appl Catal A 33:138–148
23. Geuens J, Kremsner JM, Nebel BA, Schober S, Dommisse RA, Mittelbach M, Tavernier S, Kappe CO, Maes BUW (2007) Energy Fuels 22:643–645
24. Hu, Du, Tang Z, Min (2004) Ind Eng Chem Res 43:7928–7931
25. Knothe G (2005) Fuel Process Technol 86:1059–1070
26. Ma F, Hanna MA (1999) Bioresour Technol 70:1–15
27. Lotero E, Liu Y, Lopez DE, Suwannakarn K, Bruce DA, Goodwin JG (2005) Ind Eng Chem Res 44:5353–5363
28. Thomas JM (2012) Proc R Soc A Math Phys Eng Sci 468:1884–1903
29. Somorjai GA, Frei H, Park JY (2009) J Am Chem Soc 131:16589–16605
30. Luque R, Herrero-Davila L, Campelo JM, Clark JH, Hidalgo JM, Luna D, Marinas JM, Romero AA (2008) Energy Environ Sci 1:542–564
31. Luque R, Lovett JC, Datta B, Clancy J, Campelo JM, Romero AA (2010) Energy Environ Sci 3:1706–1721
32. Dacquin J-P, Lee AF, Wilson K (2010) Thermochemical conversion of biomass to liquid fuels and chemicals. The Royal Society of Chemistry, UK, pp 416–434
33. Wilson K, Lee AF (2012) Catal Sci Technol 2:884–897
34. Eze VC, Phan AN, Pirez C, Harvey AP, Lee AF, Wilson K (2013) Catal Sci Technol 3:2373–2379
35. Chen G-Q, Patel MK (2011) Chem Rev 112:2082–2099
36. Bozell JJ, Petersen GR (2010) Green Chem 12:539–554
37. Sheldon RA (2005) Green Chem 7:267–278
38. Sheldon RA, Arends I, Hanefeld U (2007) Green chemistry and catalysis. Wiley-VCH Verlag GmbH & Co. KGaA, Weinheim
39. Mallat T, Baiker A (2004) Chem Rev 104:3037–3058
40. Lee AF (2014). In: Wilson K, Lee AF (eds) Heterogeneous catalysts for clean technology: spectroscopy, design, and

monitoring chapt. 2. Wiley-VCH Verlag GmbH & Co. KGaA, Weinheim, pp 11–33

41. Vinod CP, Wilson K, Lee AF (2011) J Chem Technol Biotechnol 86:161–171
42. Kobayashi S, Miyamura H, Akiyama R, Ishida T (2005) J Am Chem Soc 127:9251–9254
43. Kaizuka K, Lee KY, Miyamura H, Kobayashi S (2012) J Flow Chem 2:1
44. Ye XA, Johnson MD, Diao TN, Yates MH, Stahl SS (2010) Green Chem 12:1180–1186
45. Ayude A, Cechini J, Cassanello M, Martinez O, Haure P (2008) Chem Eng Sci 63:4969–4973
46. Tschan R, Wandeler R, Schneider MS, Schubert MM, Baiker A (2001) J Catal 204:219–229
47. Caravati M, Grunwaldt JD, Baiker A (2004) Catal Today 91–2:1–5
48. Grunwaldt JD, Caravati M, Baiker A (2006) J Phys Chem B 110:9916–9922
49. Burgener M, Tyszewski T, Ferri D, Mallat T, Baiker A (2006) Appl Catal A Gen 299:66–72
50. Hou ZS, Theyssen N, Leitner W (2007) Green Chem 9:127–132
51. Hou Z, Theyssen N, Brinkmann A, Klementiev KV, Grünert W, Bühl M, Schmidt W, Spliethoff B, Tesche B, Weidenthaler C (2008) J Catal 258:315–323
52. Beale AM, Jacques SDM, Weckhuysen BM (2010) Chem Soc Rev 39:4656–4672
53. Grunwaldt JD, Schroer CG (2010) Chem Soc Rev 39:4741–4753
54. Lee AF (2012) Aust J Chem 65:615–623
55. Gai PL, Sharma R, Ross FM (2008) MRS Bull 33:107–114
56. Jungjohann KL, Evans JE, Aguiar JA, Arslan I, Browning ND (2012) Microsc Microanal 18:621–627
57. Browning ND, Bonds MA, Campbell GH, Evans JE, Lagrange T, Jungjohann KL, Masiel DJ, Mckeown J, Mehraeen S, Reed BW, Santala M (2012) Curr Opin Solid State Mat Sci 16:23–30
58. Boyes ED, Ward MR, Lari L, Gai PL (2013) Ann Phys Berlin 525:423–429
59. Xiong Y, Cai H, Wiley BJ, Wang J, Kim MJ, Xia Y (2007) J Am Chem Soc 129:3665–3675
60. Zhang H, Jin M, Xiong Y, Lim B, Xia Y (2012) Acc Chem Res 46:1783–1794
61. Xia X, Choi S-I, Herron JA, Lu N, Scaranto J, Peng H-C, Wang J, Mavrikakis M, Kim MJ, Xia Y (2013) J Am Chem Soc 135:15706–15709
62. Astruc D, Lu F, Aranzaes JR (2005) Angew Chem Int Ed 44:7852–7872
63. Hermans I, Spier ES, Neuenschwander U, Turrà N, Baiker A (2009) Top Catal 52:1162–1174
64. Heyns K, Paulsen H (1957) Angew Chem 69:600–608
65. Liu XY, Madix RJ, Friend CM (2008) Chem Soc Rev 37:2243–2261
66. Dimitratos N, Lopez-Sanchez JA, Hutchings GJ (2012) Chem Sci 3:20–44
67. Besson M, Gallezot P (2000) Catal Today 57:127–141
68. Kluytmans J, Markusse A, Kuster B, Marin G, Schouten J (2000) Catal Today 57:143–155
69. Mallat T, Baiker A (1994) Catal Today 19:247–283
70. Narasimharao K, Brown DR, Lee AF, Newman AD, Siril PF, Tavener SJ, Wilson K (2007) J Catal 248:226–234
71. Suwannakarn K, Lotero E, Ngaosuwan K, Goodwin JG (2009) Ind Eng Chem Res 48:2810–2818
72. Kouzu M, Nakagaito A, Hidaka J-s (2011) Appl Catal A 405:36–44
73. Zhao D, Huo Q, Feng J, Chmelka BF, Stucky GD (1998) J Am Chem Soc 120:6024–6036
74. Mbaraka IK, Shanks BH (2005) J Catal 229:365–373
75. Melero JA, Bautista LF, Morales G, Iglesias J, Briones D (2008) Energy Fuels 23:539–547
76. Chen X-R, Ju Y-H, Mou C-Y (2007) J Phys Chem C 111:18731–18737
77. Mbaraka IK, Radu DR, Lin VSY, Shanks BH (2003) J Catal 219:329–336
78. Chen D, Li Z, Wan Y, Tu X, Shi Y, Chen Z, Shen W, Yu C, Tu B, Zhao D (2006) J Mater Chem 16:1511–1519
79. Cao L, Man T, Kruk M (2009) Chem Mater 21:1144–1153
80. Dacquin JP, Lee AF, Pirez C, Wilson K (2012) Chem Commun 48:212–214
81. Martin A, Morales G, Martinez F, van Grieken R, Cao L, Kruk M (2010) J Mater Chem 20:8026–8035
82. Pirez C, Caderon J-M, Dacquin J-P, Lee AF, Wilson K (2012) ACS Catal 2:1607–1614
83. Vinu A, Gokulakrishnan N, Balasubramanian VV, Alam S, Kapoor MP, Ariga K, Mori T (2008) Chem A Eur J 14:11529–11538
84. Melero JA, Iglesias J, Morales G (2009) Green Chem 11:1285–1308
85. Carrero A, Vicente G, Rodríguez R, Linares M, del Peso GL (2011) Catal Today 167:148–153
86. Ying JY, Mehnert CP, Wong MS (1999) Angew Chem Int Ed 38:56–77
87. Lu Y (2006) Angew Chem Int Ed 45:7664–7667
88. Garg S, Soni K, Kumaran GM, Bal R, Gora-Marek K, Gupta JK, Sharma LD, Dhar GM (2009) Catal Today 141:125–129
89. Gheorghiu S, Coppens M-O (2004) AIChE J 50:812–820
90. Zhang X, Zhang F, Chan K-Y (2004) Mater Lett 58:2872–2877
91. Sun J-H, Shan Z, Maschmeyer T, Coppens M-O (2003) Langmuir 19:8395–8402
92. Dacquin J-P, Dhainaut JRM, Duprez D, Royer SB, Lee AF, Wilson K (2009) J Am Chem Soc 131:12896–12897
93. Dhainaut J, Dacquin J-P, Lee AF, Wilson K (2010) Green Chem 12:296–303
94. Wilson K, Rénson A, Clark JH (1999) Catal Lett 61:51–55
95. Rác B, Hegyes P, Forgo P, Molnár Á (2006) Appl Catal A 299:193–201
96. Yang Q, Liu J, Yang J, Kapoor MP, Inagaki S, Li C (2004) J Catal 228:265–272
97. Yang Q, Kapoor MP, Shirokura N, Ohashi M, Inagaki S, Kondo JN, Domen K (2005) J Mater Chem 15:666–673
98. Morales G, Athens G, Chmelka BF, van Grieken R, Melero JA (2008) J Catal 254:205–217
99. Margolese D, Melero JA, Christiansen SC, Chmelka BF, Stucky GD (2000) Chem Mater 12:2448–2459
100. Díaz I, Márquez-Alvarez C, Mohino F, Pérez-Pariente JN, Sastre E (2000) J Catal 193:283–294
101. Dacquin J-P, Cross HE, Brown DR, Duren T, Williams JJ, Lee AF, Wilson K (2010) Green Chem 12:1383–1391
102. Schumacher C, Gonzalez J, Wright PA, Seaton NA (2005) J Phys Chem B 110:319–333
103. Pirez C, Wilson K, Lee AF (2014) Green Chem 16:197–202
104. Ono Y, Baba T (1997) Catal Today 38:321–337
105. Hattori H (1995) Chem Rev 95:537–558
106. Albuquerque MCG, Azevedo DCS, Cavalcante CL Jr, Santamaría-González J, Mérida-Robles JM, Moreno-Tost R, Rodríguez-Castellón E, Jiménez-López A, Maireles-Torres P (2009) J Mol Catal A: Chem 300:19–24
107. Peterson GR, Scarrah WP (1984) J Am Oil Chem Soc 61:1593–1597
108. Verziu M, Coman SM, Richards R, Parvulescu VI (2011) Catal Today 167:64–70
109. Granados ML, Alonso DM, Alba-Rubio AC, Mariscal R, Ojeda M, Brettes P (2009) Energy Fuels 23:2259–2263

110. Di Serio M, Tesser R, Casale L, D'Angelo A, Trifuoggi M, Santacesaria E (2010) Top Catal 53:811–819
111. MacLeod CS, Harvey AP, Lee AF, Wilson K (2008) Chem Eng J 135:63–70
112. Montero J, Wilson K, Lee A (2010) Top Catal 53:737–745
113. Watkins RS, Lee AF, Wilson K (2004) Green Chem 6:335–340
114. Alonso DM, Mariscal R, Granados ML, Maireles-Torres P (2009) Catal Today 143:167–171
115. Verziu M, Cojocaru B, Hu J, Richards R, Ciuculescu C, Filip P, Parvulescu VI (2008) Green Chem 10:373–381
116. Zhu K, Hu J, Kübel C, Richards R (2006) Angew Chem Int Ed 45:7277–7281
117. Montero JM, Gai P, Wilson K, Lee AF (2009) Green Chem 11:265–268
118. Cantrell DG, Gillie LJ, Lee AF, Wilson K (2005) Appl Catal A 287:183–190
119. Di Serio M, Ledda M, Cozzolino M, Minutillo G, Tesser R, Santacesaria E (2006) Ind Eng Chem Res 45:3009–3014
120. Cavani F, Trifirò F, Vaccari A (1991) Catal Today 11:173–301
121. Barakos N, Pasias S, Papayannakos N (2008) Bioresour Technol 99:5037–5042
122. Fraile JM, García N, Mayoral JA, Pires E, Roldán L (2009) Appl Catal A 364:87–94
123. Cross HE, Brown DR (2010) Catal Commun 12:243–245
124. Woodford JJ, Dacquin J-P, Wilson K, Lee AF (2012) Energy Environ Sci 5:6145–6150
125. D Highley, A Bloodworth, R Bate Dolomite-mineral planning factsheet, British Geological Survey, 2006
126. Sutton D, Kelleher B, Ross JRH (2001) Fuel Process Technol 73:155–173
127. Wilson K, Hardacre C, Lee AF, Montero JM, Shellard L (2008) Green Chem 10:654–659
128. Ilgen O (2011) Fuel Process Technol 92:452–455
129. Scharff Y, Asteris D, Fédou S (2013) OCL 20:D502
130. Choi K-M, Akita T, Mizugaki T, Ebitani K, Kaneda K (2003) New J Chem 27:324–328
131. Chen J, Zhang Q, Wang Y, Wan H (2008) Adv Synth Catal 350:453–464
132. Li F, Zhang Q, Wang Y (2008) Appl Catal A 334:217–226
133. Lee AF, Wilson K (2004) Green Chem 6:37
134. Lee AF, Hackett SF, Hargreaves JS, Wilson K (2006) Green Chem 8:549–555
135. Parlett CMA, Bruce DW, Hondow NS, Lee AF, Wilson K (2011) ACS Catal 1:636–640
136. Parlett CMA, Bruce DW, Hondow NS, Newton MA, Lee AF, Wilson K (2013) Chem Cat Chem 5:939–950
137. Hackett SF, Brydson RM, Gass MH, Harvey I, Newman AD, Wilson K, Lee AF (2007) Angew Chem 119:8747–8750
138. Baeza J, Calvo L, Gilarranz M, Mohedano A, Casas J, Rodriguez J (2012) J Catal 293:85–93
139. Wang X, Yang H, Feng B, Hou Z, Hu Y, Qiao Y, Li H, Zhao X (2009) Catal Lett 132:34–40
140. Yudha S, Dhital RN, Sakurai H (2011) Tetrahedron Lett 52:2633–2637
141. Dun R, Wang X, Tan M, Huang Z, Huang X, Ding W, Lu X (2013) ACS Catal 3:3063–3066
142. Weldon MK, Friend CM (1996) Chem Rev 96:1391–1412
143. Lee AF, Chang Z, Ellis P, Hackett SF, Wilson K (2007) J Phys Chem C 111:18844–18847
144. Naughton J, Pratt A, Woffinden CW, Eames C, Tear SP, Thompson SM, Lee AF, Wilson K (2011) J Phys Chem C 115:25290–25297
145. Davis JL, Barteau MA (1987) Surf Sci 187:387–406
146. Davis JL, Barteau MA (1990) Surf Sci 235:235–248
147. Zaera F (2003) Catal Lett 91:1–10
148. Keresszegi C, Burgi T, Mallat T, Baiker A (2002) J Catal 211:244–251
149. Lee AF (2001) Abstr Pap Am Chem Soc 221:U335–U336
150. Grunwaldt J-D, Keresszegi C, Mallat T, Baiker A (2003) J Catal 213:291–295
151. Keresszegi C, Grunwaldt J-D, Mallat T, Baiker A (2004) J Catal 222:268–280
152. Grunwaldt J-D, Caravati M, Baiker A (2006) J Phys Chem B 110:25586–25589
153. Mondelli C, Ferri D, Grunwaldt J-D, Krumeich F, Mangold S, Psaro R, Baiker A (2007) J Catal 252:77–87
154. Keresszegi C, Ferri D, Mallat T, Baiker A (2005) J Catal 234:64–75
155. Caravati M, Meier DM, Grunwaldt J-D, Baiker A (2006) J Catal 240:126–136
156. Mallat T, Baiker A (1995) Catal Today 24:143–150
157. Mallat T, Bodnar Z, Hug P, Baiker A (1995) J Catal 153:131–143
158. Parlett CMA, Keshwalla P, Wainwright SG, Bruce DW, Hondow NS, Wilson K, Lee AF (2013) ACS Catal 3:2122–2129
159. Parlett CMA, Durndell LJ, Wilson K, Bruce DW, Hondow NS, Lee AF (2014) Catal Commun 44:40–45
160. Naughton J, Lee AF, Thompson S, Vinod CP, Wilson K (2010) Phys Chem Chem Phys 12:2670–2678
161. Ping EW, Pierson J, Wallace R, Miller JT, Fuller TF, Jones CW (2011) Appl Catal A 396:85–90
162. Newton MA, Jyoti B, Dent AJ, Fiddy SG, Evans J (2004) Chem Commun 2004:2382–2383
163. Ferri D, Baiker A (2009) Top Catal 52:1323–1333
164. Newton MA (2009) Top Catal 52:1410–1424
165. Newton MA, Dent AJ, Fiddy SG, Jyoti B, Evans J (2007) Catal Today 126:64–72
166. Lee AF, Ellis CV, Naughton JN, Newton MA, Parlett CM, Wilson K (2011) J Am Chem Soc 133:5724–5727
167. Parlett CMA, Gaskell CV, Naughton JN, Newton MA, Wilson K, Lee AF (2013) Catal Today 205:76–85
168. Gaskell CV, Parlett CMA, Newton MA, Wilson K, Lee AF (2012) ACS Catal 2:2242–2246
169. Lee AF, Naughton JN, Liu Z, Wilson K (2012) ACS Catal 2:2235–2241
170. Tauster S (1987) Acc Chem Res 20:389–394
171. Ghedini E, Menegazzo F, Signoretto M, Manzoli M, Pinna F, Strukul G (2010) J Catal 273:266–273
172. Hicks RF, Qi H, Young ML, Lee RG (1990) J Catal 122:295–306
173. Beckers J, Rothenberg G (2010) Green Chem 12:939–948
174. Inumaru K, Nakamura K, Ooyachi K, Mizutani K, Akihara S, Yamanaka S (2010) J Porous Mater 18:455–463
175. Liu Y-M, Cao Y, Yi N, Feng W-L, Dai W-L, Yan S-R, He H-Y, Fan K-N (2004) J Catal 224:417–428
176. Kim T-W, Kleitz F, Paul B, Ryoo R (2005) J Am Chem Soc 127:7601–7610
177. Zhao D, Feng J, Huo Q, Melosh N, Fredrickson GH, Chmelka BF, Stucky GD (1998) Science 279:548–552
178. Sneh O, George SM (1995) J Phys Chem 99:4639–4647
179. Kim JM, Kwak JH, Jun S, Ryoo R (1995) J Phys Chem 99:16742–16747
180. Kaneda K, Fujii M, Morioka K (1996) J Org Chem 61:4502–4503
181. Polshettiwar V, Varma RS (2009) Org Biomol Chem 7:37–40
182. Peterson KP, Larock RC (1998) J Org Chem 63:3185–3189
183. Behafarid F, Ono LK, Mostafa S, Croy JR, Shafai G, Hong S, Rahman TS, Bare SR, Cuenya BR (2012) Phys Chem Chem Phys 14:11766–11779
184. Yuan Q, Duan HH, Li LL, Sun LD, Zhang YW, Yan CH (2009) J Colloid Interface Sci 335:151–167

185. Lykhach Y, Staudt T, Lorenz MPA, Streber R, Bayer A, Steinrück H-P, Libuda J (2010) Chem Phys Chem 11:1496–1504
186. Tana, Zhang M, Li J, Li H, Li Y, Shen W (2009) Catal Today 148:179–183
187. Bensalem A, Bozon-Verduraz F, Perrichon V (1995) J Chem Soc Faraday Trans 91:2185–2189
188. Harrison B, Diwell A, Hallett C (1988) Plat Met Rev 32:73–83
189. Zhu Y, Zhang S, Shan J-j, Nguyen L, Zhan S, Gu X, Tao F (2013) ACS Catal 3:2627–2639
190. Maillet T, Madier Y, Taha R, Barbier J Jr, Duprez D (1997) Stud Surf Sci Catal 112:267–275
191. Oh S-H, Hoflund GB (2006) J Phys Chem A 110:7609–7613
192. Badri A, Binet C, Lavalley J-C (1996) J Chem Soc Faraday Trans 92:1603–1608
193. Lee AF, Gee JJ, Theyers HJ (2000) Green Chem 2:279–282
194. Wenkin M, Ruiz P, Delmon B, Devillers M (2002) J Mol Catal A: Chem 180:141–159
195. Alardin F, Delmon B, Ruiz P, Devillers M (2000) Catal Today 61:255–262
196. Keresszegi C (2004) J Catal 225:138–146
197. Pinxt H, Kuster B, Marin G (2000) Appl Catal A 191:45–54
198. Dimitratos N, Porta F, Prati L, Villa A (2005) Catal Lett 99:181–185
199. Prati L, Villa A, Campione C, Spontoni P (2007) Top Catal 44:319–324
200. Enache DI, Edwards JK, Landon P, Solsona-Espriu B, Carley AF, Herzing AA, Watanabe M, Kiely CJ, Knight DW, Hutchings GJ (2006) Science 311:362–365
201. Long J, Liu H, Wu S, Liao S, Li Y (2013) ACS Catal 3:647–654
202. Griffin MB, Rodriguez AA, Montemore MM, Monnier JR, Williams CT, Medlin JW (2013) J Catal 307:111–120
203. Liu Z, Xu J, Zhang H, Zhao Y, Yu B, Chen S, Li Y, Hao L, Green Chem 2013
204. Wang R, Wu Z, Chen C, Qin Z, Zhu H, Wang G, Wang H, Wu C, Dong W, Fan W (2013) Chem Commun 49:8250–8252
205. Hou W, Dehm NA, Scott RW (2008) J Catal 253:22–27
206. Mertens P, Corthals S, Ye X, Poelman H, Jacobs P, Sels B, Vankelecom I, De Vos D (2009) J Mol Catal A: Chem 313:14–21
207. Hutchings GJ (2005) Catal Today 100:55–61
208. Lee AF, Ellis CV, Wilson K, Hondow NS (2010) Catal Today 157:243–249
209. Scott RW, Wilson OM, Oh S-K, Kenik EA, Crooks RM (2004) J Am Chem Soc 126:15583–15591
210. Scott RW, Sivadinarayana C, Wilson OM, Yan Z, Goodman DW, Crooks RM (2005) J Am Chem Soc 127:1380–1381
211. Dash P, Bond T, Fowler C, Hou W, Coombs N, Scott RWJ (2009) J Phys Chem C 113:12719–12730
212. Maclennan A, Banerjee A, Hu Y, Miller JT, Scott RWJ (2013) ACS Catal 3:1411–1419

10

Ethanol photoreactions over Au–Pd/TiO$_2$

A. K. Wahab · S. Bashir · Y. Al-Salik ·
H. Idriss

Abstract A prototype Au–Pd/TiO$_2$ catalyst was prepared, characterized and tested for the photoreaction of ethanol. XPS Au4f and Pd3d indicated that the as-prepared material is composed of metallic Au, metallic Pd as well oxidized Pd (Pd^{2+}). Ar ion sputtering (5 min) of the catalyst surface resulted in almost total reduction of Pd^{2+} to metallic Pd in addition to considerable reduction of surface Ti cations to Ti^{3+} and Ti^{2+} cations; XPS Au4f lines were not affected. Transmission electron microscopic studies indicated that Au particles have a mean particle size of about 3.5 nm while Pd particles are smaller 1–1.5 nm in size. UV excitation of the catalyst in ultrahigh vacuum (UHV) conditions resulted in the formation of acetaldehyde and hydrogen in addition to photodesorption of the reactant ethanol. Hydrogen production, representing ca. 30 % of the desorbing products, was delayed compared to acetaldehyde desorption. This was interpreted as due to kinetic effect whereby initially most electrons transferred to the conduction band are trapped by surface hydroxyls as well inevitable presence of oxygen in the powder accelerating acetaldehyde formation (dehydrogenation). Only once most oxygen-containing species have reacted and molecular hydrogen was formed. To our knowledge, this is the first UHV in situ study of hydrogen production from ethanol photocatalytically over M/TiO$_2$ system.

Keywords Hydrogen production · Ethanol photoreaction · Gold plasmon · Synergism · Ethanol dehydrogenation

Introduction

Light from the sun, the most abundant source of energy on Earth, contributes by <0.05 % of the total power (15,000 GW annual) used by humans (excluding solar heating which contributes around 0.6 %). The estimated practical and convertible power[1] that the Earth surface receives is equivalent to that provided by 600,000 nuclear reactors (one nuclear power plant generates on average 1 GW power). One mode of solar energy utilization is the use of sunlight to generate energy carriers such as hydrogen from renewable sources (e.g., ethanol and water) using semiconductor photocatalysts.

The photo-assisted splitting of water into hydrogen and oxygen was first achieved by Fujishima and Honda [1], who showed that hydrogen and oxygen could be generated in an electrochemical cell containing a titania photoelectrode, provided an external bias was applied. Since that time, numerous researchers have explored ways of achieving direct water dissociation without the need for an external bias. Much work has been conducted since, a large fraction of which is discussed in recent reviews [2–6]. Among the many issues affecting direct water splitting is the need to separate hydrogen from oxygen and the relatively low hydrogen evolution rates so far achieved. These in addition to the need of using UV light (>3 eV) to excite TiO$_2$ and other related materials has been one of the main obstacles for practical applications. Many authors have sought modified photocatalysts which, unlike pure TiO$_2$, respond to visible (sunlight) excitation, with limited success to date; see some of these materials in ref. [3].

A. K. Wahab · S. Bashir · Y. Al-Salik · H. Idriss (✉)
SABIC-Corporate Research and Innovation (CRI) at KAUST,
Thuwal, Saudi Arabia
e-mail: idrissh@sabic.com

---

[1] The total amount of sun light reaching the earth surface is orders of magnitude higher than the quoted figure.

Fig. 1 Schematic
representation of the main steps
during a photocatalytic reaction
involving a metal (*small black
circle*) deposited on a metal
oxide semiconductor (*large
circle*). Also shown are the
valence and conduction bands
energy with respect to the
normal hydrogen electrode
(NHE) scale. Adsorbed species
are represented by A (electron
acceptor) and D (electron
donor)

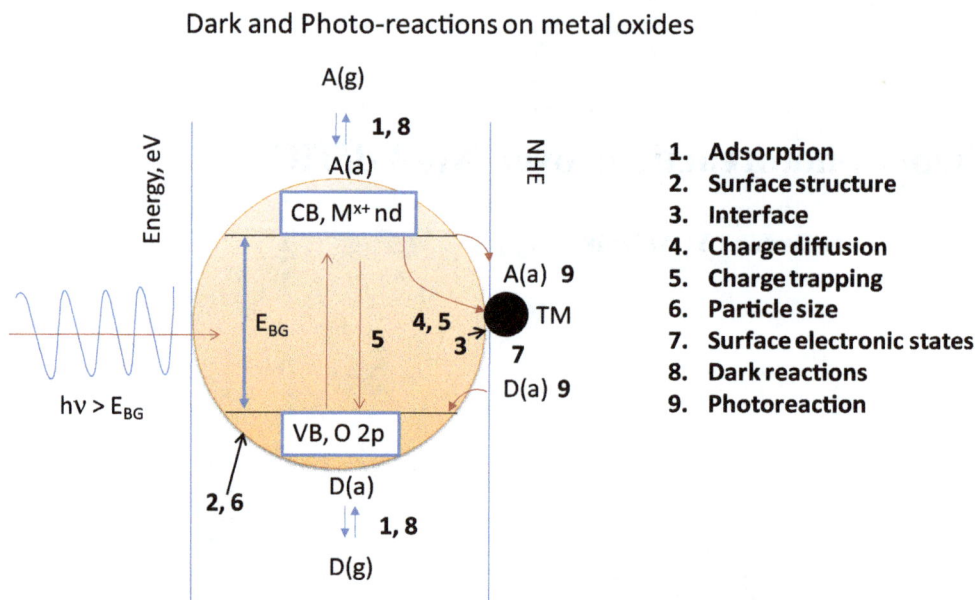

Dark and Photo-reactions on metal oxides

1. **Adsorption**
2. **Surface structure**
3. **Interface**
4. **Charge diffusion**
5. **Charge trapping**
6. **Particle size**
7. **Surface electronic states**
8. **Dark reactions**
9. **Photoreaction**

Many approaches have been conducted to design a photocatalyst that can work under direct sunlight in stable conditions. To achieve this, a few key factors need to be overcome chiefs of them are light absorption efficiency, charge carrier life time and materials stability. To enhance light absorption, a large number of photocatalysts were designed based on visible light-range band gap either by solid solutions, hybrid materials or doping of wide-band gap semiconductors. To decrease the charge carriers' life time and also hydride semiconductors, metal nanoparticles are added and sacrificial agents are used [2–6]. The scheme in Fig. 1 presents several steps and concepts involved in photoreactions composed of a semiconductor on which a metal particle is deposited. As can be seen, many factors can affect the reaction in addition to basic thermodynamic requirements and these include charge carrier (electrons and holes) diffusion (step 4) from the bulk to the surface [and interface (step 3)] both affected by bulk and surface structures (step 2) which are in turn related to particle sizes (step 6) and oxidation states (step 7). Kinetics is also important to consider in particular when sacrificial organic compounds are used, therefore, the effect of concentrations and temperatures on the surface populations (step 8) and ultimately the photoreaction rate (step 9) is not negligible [7].

In this work, we present a study on Au–Pd/TiO$_2$ that is found active for hydrogen production [8]. The Au/TiO$_2$ system has been studied by us and others in numerous works previously [9–13]. Au metal has at least two important roles in photoreaction, first it acts as sink for transferred electrons from the conduction band and second it contributes by its plasmon response (excited by visible light) in the electron transfer reaction [14–16] albeit is not

a fully understood mechanism. To act as an efficient plasmon, Au particles need to be relatively large. Due to this requirement they are less dispersed on the surface of TiO$_2$. Pd is used to decorate the surface around Au so in essence Pd acts as efficient electron trap center, while Au induces the electric field due to it is electronic oscillation; Pd particles are easily prepared in 1 nm size on top of TiO$_2$. The TiO$_2$ used in this work is composed of anatase and rutile phases in ca. 80–20 ratio. This ratio is thought to be in the range where considerable synergism occurs between the two phases further increasing the reaction rate [17]. The catalyst has therefore a combination of properties that are important in photocatalysis: surface plasmons (Au), synergism (anatase + rutile), metal–semiconductor interface (Au/TiO$_2$ and Pd/TiO$_2$). In this work, we focus on its properties as well as a mechanistic study of ethanol reaction under UV excitation in which the channels for hydrogen desorption and acetaldehyde desorption seem to be decoupled on the powder material.

Experimental

X-ray photoelectron spectroscopy was conducted using a Thermo scientific ESCALAB 250 Xi. The base pressure of the chamber was typically in the low 10^{-10} to high 10^{-11} mbar range. Charge neutralization was used for all samples. Spectra were calibrated with respect to C1s at 285.0 eV. Au4f, Pd 3d, O1s, Ti2p, C1s and valence band energy regions were scanned for all materials. Typical acquisition conditions were as follows: pass energy = 30 eV and scan rate = 0.1 eV per 200 ms. Ar ion bombardment was performed with an EX06 ion gun at

1 kV beam energy and 10 mA emission current; sample current was typically 0.9–1.0 nA. Self-supported oxide disks of approximately 0.5 cm diameter were loaded into the chamber for analysis. UV–Vis absorbance spectra of the powdered catalysts were collected over the wavelength range of 250–900 nm on a Thermo Fisher Scientific UV–Vis spectrophotometer equipped with praying mantis diffuse reflectance accessory. Absorbance (A) and reflectance (% R) of the samples were measured. The reflectance (% R) data were used to calculate the band gap of the samples using the Tauc plot (Kubelka–Munk function). The TiO$_2$ exhibited the typical two band gaps of anatase and rutile at 3.2 and 3.0 eV, respectively. BET surface area was measured using Quantachrome Autosorb analyzer by N$_2$ adsorption. The surface area was found to be 55 m^2/g$_{Catal}$. Transmission electron microscopy studies were performed at 200 kV with a JEOL JEM 2010F instrument equipped with a field emission source. The microscope was operated in HAADF-STEM mode (Z-contrast). Samples were dispersed in alcohol in an ultrasonic bath and a drop of supernatant suspension was poured onto a holey carbon-coated grid. In situ study was conducted in an ultrahigh vacuum chamber equipped with an online quadrupole mass spectrometer (RGA Hiden; 300 AMU), a sputter ion gun (PHI), a dosing line for ethanol exposure. The chamber is pumped down with an ion pump (PHI), turbo pump (Pfeiffer) and a titanium sublimation pump (PHI). Base pressure was in the low to mid 10^{-9} torr. Catalyst (as a pressed pellet of about 0.5 cm in diameter; total amount about 50 mg) was loaded into an x, y, z R sample mount on top of a Mo sample holder. Prior to reaction the catalyst was cleaned using UV light in presence of 10^{-6} torr of O$_2$ for 60 min (twice). Ethanol was subjected to freeze–thaw pump to remove residual water and CO$_2$ and its purity is checked by the online mass spectrometer until a typical ethanol fragment is obtained [18]. Six experiments with different dosing of ethanol were conducted from 90 s to 10 min at a chamber of pressure close to 10^{-7} torr. After dosing, the chamber was pumped down to the 10^{-9} torr range. Prior to UV light form a 100 W ultraviolet lamp (H-144GC-100, Sylvania par 38) with a flux of ca. 10 mW/cm^2 at a distance of 25 cm the sample was rotated away from the mass spectrometer which was turned on to measure the back ground of the chamber. A negligible signal of m/e 31, the parent ion of the ethanol compared to that of residual N$_2$/CO (m/e 28) was indicative of the removal of gas-phase ethanol. The following masses were scanned at a dwell time of 50 ms for each (m/e 2, 12, 14, 15, 16, 18, 28, 29, 31 and 44) for each run. The catalyst pellet was then turned toward the mass spectrometer which is enveloped with a shroud having an orifice smaller than that of the pellet to minimize stray desorption, at the same time the UV shutter was removed to start the excitation. Au–Pd/

TiO$_2$ (85 % anatase/15 % rutile) catalysts were synthesized by co-impregnation method (0.13 wt% of gold and 0.2 wt% of Pd; 1:3 molar ratio). The precursor of gold and palladium were AuCl$_4$ (dissolved in aqua regia) and PdCl$_2$ in 1 normal HCl. TiO$_2$ semiconductor, ca. 85 % anatase and 15 % rutile, was used as a support martial. First, TiO$_2$ was placed into Pyrex beaker. Then, the aqua regia solution of Au and Pd in 1 normal HCl were, respectively, poured into a certain amount of TiO$_2$ under magnetic stirring (170 rpm) at 80 °C for 12–24 h. The precipitate formed was dried for >4 h, at 120 °C. Finally, the material was calcined at 300 °C for 5 h; afterward it was crushed using a mortar to fine powder.

Results and discussion

Figure 2 presents XPS of the as-prepared Au–Pd/TiO$_2$ catalyst. Ti2p peaks at 459.2 and 464.6 eV are attributed to the Ti2p$_{3/2}$ and Ti2p$_{1/2}$, respectively. Both the narrow FWHM (1.1 eV) of the Ti2p$_{3/2}$ and the splitting of 5.4 eV are consistent with the presence of Ti^{4+} cations only. O1s peak at 530.3 eV is due to lattice oxygen while the large peak at 532.6 eV is due to irreversibly adsorbed water on the surface; surface hydroxyls at ca. 531.5 eV are masked by the large water contribution. Gold presence can be seen by their XPS Au4f at 83.8 and 87.4 eV attributed to Au4f$_{7/2}$ and Au4f$_{5/2}$, respectively. Both the peak positions and spin splitting of 3.6 eV are characteristic of metallic gold [the 0.2 eV deviation from pure metallic gold (binding energy at 84.0 eV) is slightly larger than the resolution limit and most likely not due to calibration issues]. It has been previously indicated that a charge transfer from TiO$_2$ to Au results in lowering the binding energy of small metallic particles of Au [19] and this might be the cause. It has also been reported that alloy of Au with Pd results in shifting the Au4f by up to 0.5 eV and at the same time increase in the binding energy of Pd3d by up to 0.8 eV [20]. Also seen in the figure is the contribution of Pd4s lines [because Pd loading is larger (3–1 molar ratio) and their particles are smaller therefore more dispersed; see TEM pictures below]. XPS Pd3d region indicates that Pd exists in two oxidation states, Pd^0 and Pd^{2+}. It is also worth noting that almost at the same binding energy position of Pd3d, Au4d lines are overlapping [21]. However, the signal of Au4d lines is about 5 times weaker than that of Au4f [22]. Given the weak signal of Au4f it is safe to neglect its contribution into the spectrum. The splitting between the XPS 3d$_{5/2}$ and 3d$_{3/2}$ of both oxidation states of 5.2–5.3 eV is consistent with many other works on both oxidation states [26]. The binding energy of Pd^0 is, however, slightly shifted upward (by 0.3 eV) when compared to Pd metal [20], which may also indicate the presence of some alloy (Au–Pd) (Fig. 3).

Fig. 2 XPS Au4f, Pd3d, Ti2p and O1s of as-prepared Au–Pd/ TiO₂

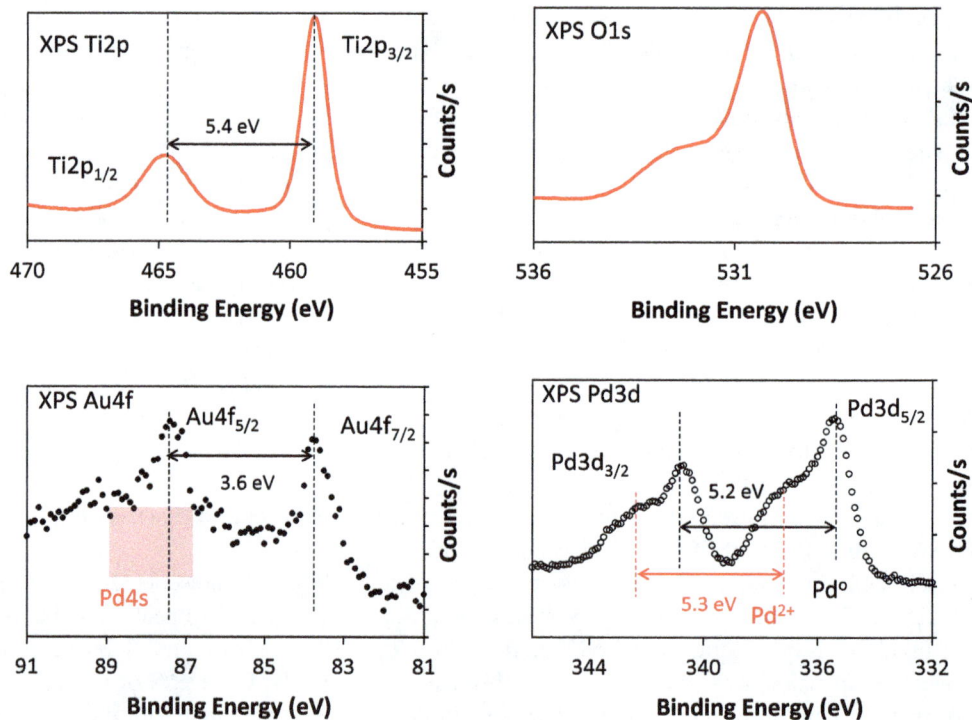

Fig. 3 XPS Au4f, Pd3d, Ti2p and O1s of Ar⁺ sputtered sample (5 min) Au–Pd/TiO₂

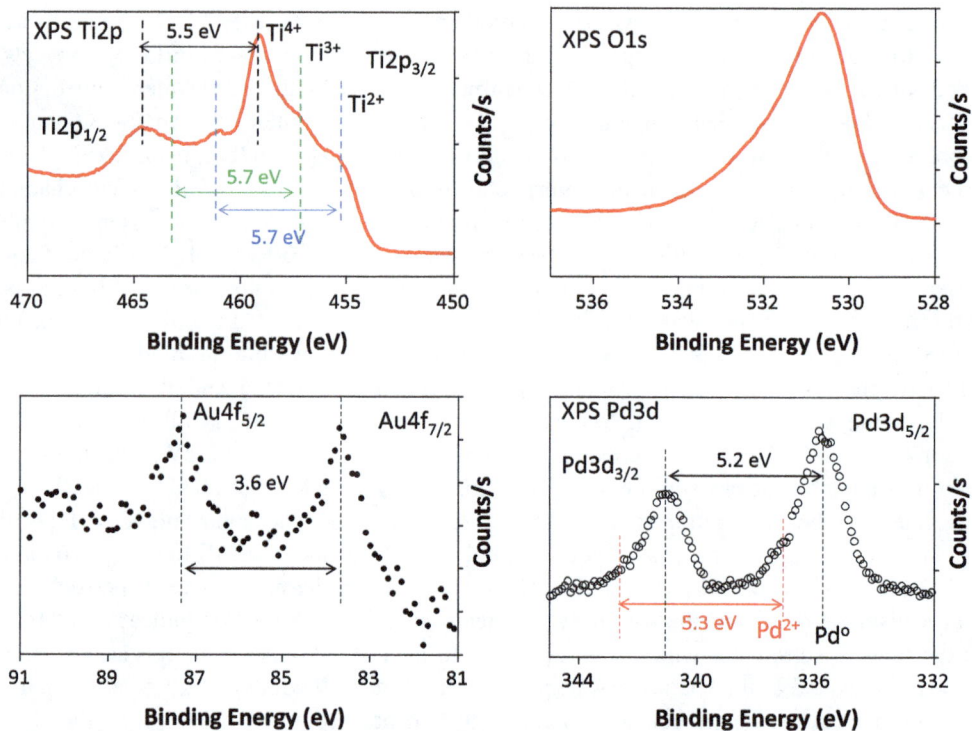

Figure 2 presents the same lines of Fig. 1 but after Ar⁺ sputtering for 5 min to reduce the materials and see for the effect of these on possible electronic shifts. The Ti2p region is that of typical reduced Ti cations (Ti^{4+}, Ti^{3+} and Ti^{2+} and possibly less reduced states with smaller contributions). The XPS $Ti2p_{3/2}$ lines attributed to Ti^{3+} and Ti^{2+}

are at 457.3 and 455.3 eV, respectively [23, 24] in agreement with other works. The $Ti2p_{1/2}$ of Ti^{2+} is clear while that of Ti^{3+} is masked by the $Ti2p_{1/2}$ of Ti^{4+} (although one can see a considerable broadening of the peak at the lower binding energy side due to multiple electronic states). The splitting of the three states is very similar (5.5–5.7 eV).

Table 1 XPS surface composition of As prepared and Ar ion reduced Au-Pd/TiO$_2$

	Area CPS × eV	Sensitivity factor	Corrected area	Atomic (%)
As prepared				
Au4f	222.43	5.24	42.45	0.02
Pd3d	2,423.01	4.64	521.98	0.25
Ti2p	63,711.57	1.80	35,434.69	17.25
O1s	85,127.25	0.71	119,728.90	58.27
C1s	14,724.15	0.30	49,743.75	24.21
Ar$^+$ sputtered				
Au4f	178.04	5.24	33.98	0.02
Pd3d	1,195.65	4.64	257.57	0.14
Ti2p	76,212.77	1.80	42,387.53	22.44
O1s	64,737.00	0.71	91,050.63	48.20
C1s	16,327.96	0.30	55,162.03	29.20

Table 2 Photoreaction for hydrogen production from water in presence of 5 vol% of organic sacrificial agent in water (ethylene glycol in the case of Pd/TiO$_2$ and Au–Pd/TiO$_2$ and ethanol in the case of Au/TiO$_2$—both sacrificial agents are very efficient for the hole capturing with slight variations (ca. 10 %)

Catalyst	Rate (mol/g$_{Catal}$ min)
0.15 at.% Pd/TiO$_2$	2.3×10^{-5}
0.20 at.% Au/TiO$_2$	1.7×10^{-4}
0.05 at.% Au–0.15 at.% Pd/TiO$_2$	2.8×10^{-4}

Amount of catalyst = 25 mg; reactor volume 100 mL, light Hg mercury lamp with 360 nm cutoff filter. Flux ca. 4–5mW/cm^2 (comparable to that provided from the sun in summer midday at a longitude = 25°N)

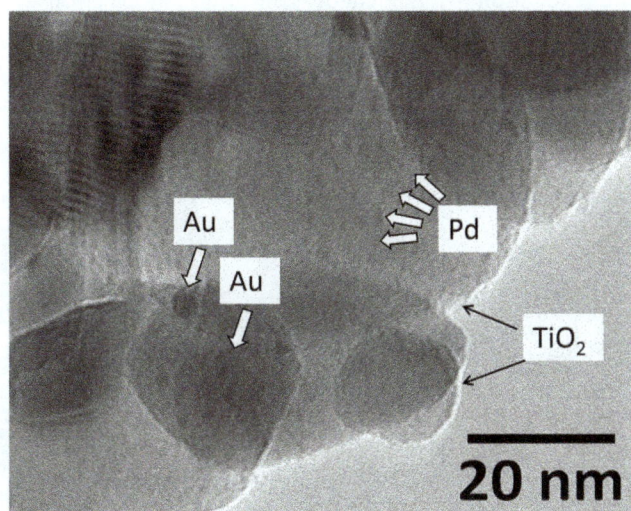

Fig. 4 Transmission electron microscopy of Au/Pd TiO$_2$ (anatase + rutile)

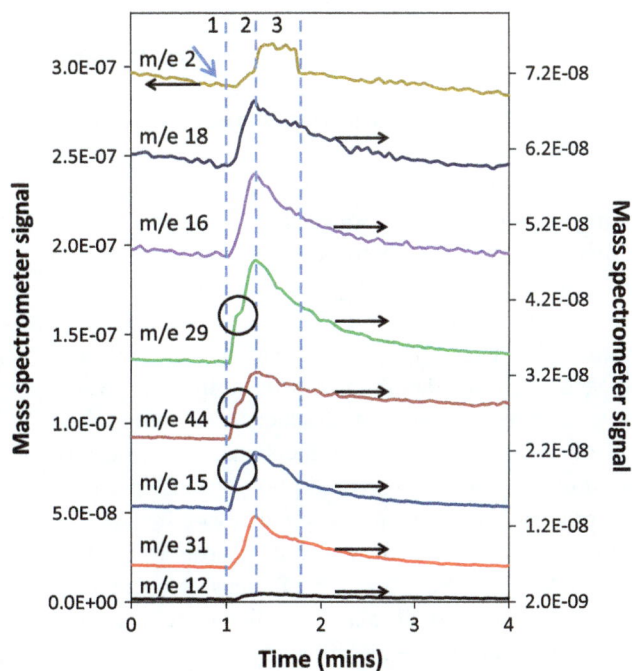

Fig. 5 Mass spectrometer fragmentation pattern of products collected upon ethanol photoreaction under UV excitation for a previously exposed Au–Pd/TiO$_2$ to ethanol for 4 min at 10^{-7} torr (this exposure was found to ensure surface saturation based on monitoring of desorption products)

The O1s region also has the typical reduced shape as seen by Doniac broadening at the high binding energy side [25]. The XPS Au4f region has changed slightly also the background appeared to be less prominent probably because of a more homogenous electronic distribution of Pd particles (contributing by their 4s into the spectrum). The XPS Pd3d indicates that most of Pd particles are now in their reduced state. Both XPS Au4f and 3d main peak positions did not change considerably upon Ar ion sputtering. Table 1 presents the atomic % of the Au–Pd/TiO$_2$ before and after Ar ions sputtering.

In Fig. 4 a representative transmission electron microscopy image of the Au–Pd/TiO$_2$ catalyst. A large number of images were collected. The mean particle size of Au is about 3.5 nm while that of Pd is 1.5 nm (although the Pd particle size distribution was asymmetric at the low size

indicating a larger contribution of small particles <1 nm); there was negligible number of Pd particles above 2 nm. The TEM images of the anatase and rutile phases of this TiO$_2$ have been presented elsewhere; the difference between the two phases can be made based on their local diffraction lines as well as the lattice spacing from high-resolution images.

Table 2 presents the rate of reaction for hydrogen production for Pd/TiO$_2$, Au/TiO$_2$ and Pd–Au/TiO$_2$. The support (85 % anatase/15 % rutile) was the same for the three catalysts. It is to be noted that variations from preparations to the other may contribute by up to 25 % of changes in the

Table 3 CF for mass spectrometer correction factor

m/e	Peak area	Product	CF	Peak area	Selectivity (%)
2	6.9×10^{-7}	Hydrogen	0.5	3.5×10^{-7}	29.4
12	1.1×10^{-8}				
15	5.9×10^{-8}				
16	1.9×10^{-7}				
18	1.7×10^{-7}	Water	1.1	1.9×10^{-7}	15.9
29	2.6×10^{-7}	Acetaldehyde	2.1	5.5×10^{-7}	46.5
31	5.4×10^{-8}	Ethanol	1.8	9.7×10^{-8}	8.3
44	1.2×10^{-7}		1.2		
Total	1.6×10^{-6}			1.2×10^{-6}	

reaction rate and that the rate is dependent on the % of the metal in a non-correlated way as often observed (see for example ref. [12]). Still it is clear that the catalyst composed of Pd–Au/TiO$_2$ is more active than that from Au/TiO$_2$ and Pd/TiO$_2$. A similar observation was reported on Pt–Au/TiO$_2$ previously [26].

To analyze the reaction products and their evolution with time upon radiation we have opted to conduct experiment in which the photocatalyst is illuminated with online mass spectrometer. Figure 5 presents the UV-excited photoreaction over Au–Pd/TiO$_2$ that was initially saturated with ethanol (prior to excitation). Upon illumination (indicated by the thick arrow at the top of the figure) fragments of reaction products are observed to desorb followed by a decrease in the signal due to surface depletion of the reactant. The desorption of these products can be grouped in three modes: photodesorption, photoreduction and photooxidation. Ethanol adsorption on TiO$_2$, as well as M/TiO$_2$ (M = Au, Pd or both) at room temperature gives both molecular and dissociated species. This has been seen both experimentally [27] and by DFT computation methods [28]. As indicated in Fig. 2, the surface contains (in addition to ethanol) non-negligible amounts of irreversibly adsorbed water and these ultimately give hydroxyl radicals upon photoexcitation. There are three regions in the figure denoted by lines 1, 2 and 3. In region 1–2, desorption of all products occurs (except hydrogen). In addition, this desorption is different for the set of products monitored. Ethanol (m/e 31), water (m/e 18) and oxygen (m/e 16; m/e 32 has the same desorption profile) all have similar shape and these may be considered as due to photodesorption upon UV excitation. In addition, fragments related to acetaldehyde are clearly noticed (m/e 29 and m/e 44). m/e 29 is largely due to acetaldehyde (CHO) fragment as ethanol contribution into this fragment is small (typically <30 % of m/e 31). m/e 44 (parent acetaldehyde molecule) has contribution from CO$_2$ in addition. The ratio m/e 29 to m/e 44 of about 5–2 indicates that in this region most of desorption is due to

acetaldehyde [29]. m/e 15 (CH$_3$) has contribution from both acetaldehyde and ethanol. Seen in this region is a clear shoulder (indicated by a circle) composed of these three masses only. This is a clear indication of acetaldehyde desorption due to ethanol dehydrogenation. The formation of acetaldehyde from ethanol over TiO$_2$ upon photoreaction in presence of oxygen on TiO$_2$ single crystal has been reported before [27]. The reaction that involves two-electron injections into the valence band can be explained as follow.

Dissociative adsorption of ethanol and water occurs on the surface of TiO$_2$ in the presence or absence of light.

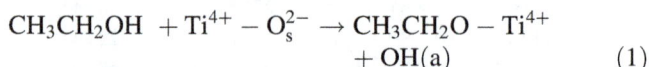

$$CH_3CH_2OH + Ti^{4+} - O_s^{2-} \rightarrow CH_3CH_2O - Ti^{4+} + OH(a) \tag{1}$$

$$H_2O + Ti^{4+} - O_s^{2-} \rightarrow HO - Ti^{4+} + OH(a) \tag{2}$$

s for surface, (a) for adsorbed.

Light excitation of the catalyst resulting in electron (e$^-$)–hole (h$^+$) pair formation.

$$TiO_2 + nUV \rightarrow xe^- + xh^+ \tag{3}$$

where $n > x$ due to scattering and absorption cross section.

Hole scavenging (two electrons injected per ethoxide into the VB of TiO$_2$) followed by acetaldehyde formation.

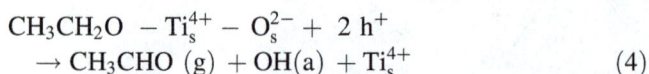

$$CH_3CH_2O - Ti_s^{4+} - O_s^{2-} + 2 h^+ \rightarrow CH_3CHO (g) + OH(a) + Ti_s^{4+} \tag{4}$$

On average, acetaldehyde maximum separate desorption (in region 1–2) occurred after 3 s from UV excitation. The maximum desorption is, however, seen together with ethanol and water desorption at about 20 s of light excitation. In region 2–3, all products start to decrease but hydrogen desorption increased reached a short plateau before a sharp decrease. The observed profile can be rationalized kinetically and due to diffusion. Initially the surface contains large amounts of adsorbed ethanol but also adsorbed water and irreversibly adsorbed O$_2$ molecules (in the sample cleaning process). Upon illumination, the initial reaction is photooxidation producing acetaldehyde. The formation of acetaldehyde can, if coupled to oxygen radical species, prevent the reduction of H ions because O$_2$ is a faster electron scavenger. Once a large fraction of oxygen atoms and molecules have been consumed the remaining ethanol and water could farther react with the generated holes. Electrons transferred to the conduction band can then be transferred to Au and Pd particles that in turn reduce hydrogen ions (of surface hydroxyls) to molecular hydrogen; as follow [considering Eqs. (1)–(4) above].

Electron transfer from the CB of TiO$_2$ to H$^+$ (via M nanoparticles) resulting in molecular hydrogen formation, while the hole transfer occurs from one OH species of water in addition.

$$4 \text{ OH (a)} + 4 \text{ e}^- + 2 \text{ h}^+ \rightarrow 3O_s^{2-} + \frac{1}{2} O_2 + 2 H_2 \tag{5}$$

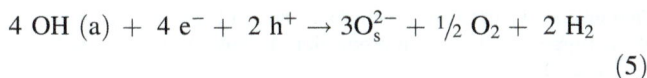

Because of the small size of hydrogen ions, intra and inter diffusion is much faster than for ethanol, acetaldehyde and water, is a plausible explanation of the abrupt decrease of the signal. The signal of all, but m/e 44 products return to the base line. The remaining m/e 44 signal is due to total oxidation of traces of ethanol to CO_2 (as it is not mirrored by m/e 29 and m/e 15; fingerprint of acetaldehyde); it is worth indicating that photooxidation of ethanol is a set of consecutive reactions studied in some details previously on TiO_2 [30] as well as on Ru/TiO_2 [31].

Table 3 presents the computed peak areas of the mass spectrometer signals of the main products. In addition, the corrected peak areas, to the mass spectrometer sensitivity factors for hydrogen, water, ethanol and acetaldehyde are given. While the results are preliminary it is, however, clear that hydrogen desorption contributes by a non-negligible fraction of the overall desorption and that acetaldehyde represents the largest fraction. In the absence of molecular oxygen one would expect that the amount of hydrogen would be at least equal to that of acetaldehyde if all hydrogen is made from ethanol and would be larger if additional hydrogen is made from adsorbed water.

Conclusions

Au–Pd/TiO$_2$ catalysts combining both plasmonic and synergism behavior are prepared, characterized and tested for the photoreaction of ethanol under ultrahigh vacuum condition upon excitation with UV light. The catalyst is composed of metallic Au and Pd in their reduced state. The photoreaction of ethanol results in the production of acetaldehyde and hydrogen. Hydrogen production is retarded due to the initial reaction of surface hydroxyls and adsorbed oxygen atoms and or molecules. Only once most of these species were consumed and molecular hydrogen was formed.

References

1. Fujishima A, Honda K (1972) Electrochemical photolysis of water at a semiconductor electrode. Nature 238:37
2. Kudo A, Miseki Y (2009) Heterogeneous photocatalyst materials for water splitting. Chem Soc Rev 38:253
3. Connelly KA, Idriss H (2012) The photoreaction of TiO$_2$ and Au/TiO$_2$ single crystal and powder surfaces with organic adsorbates.

Emphasis on hydrogen production from renewable. Green Chem 14:260–280
4. Nadeem MA, Connelly KA, Idriss H (2012) The photoreaction of TiO$_2$ and Au/TiO$_2$ single crystal and powder with organic adsorbates. Int J Nanotechnol (special edition on nanotechnology in Scotland) 9:121–162
5. Connelly KA, Wahab AK, Idriss H (2012) Photoreaction of Au/TiO$_2$ for hydrogen production from renewables: a review on the synergistic effect between anatase and rutile phases of TiO$_2$. Mater Renew Sustain Energy 1:1–12
6. Walter MG, Warren EL, McKone JR, Boettcher SW, Mi Q, Santori EA, Lewis NS (2010) Solar water splitting cells. Chem Rev 110:6446–6473
7. Yang YZ, Chang CH, Idriss H (2006) Photo-catalytic production of hydrogen from ethanol over M/TiO$_2$ catalysts (M = Pd, Pt or Rh). Appl Catal B: Environ 67:217–222
8. Wahab AK, Al-Oufi M, Bashir S, Al-Salik Y, Katsiev H, Idriss H (2013) Photocatalyst, method of preparation, photolysis system. World Patent 13T&I0037-WO-PCT (2013); Gulf Cooperation Council (13T&I0037-GC-NP), Serial Number GCC/P/2013/25576 International Procedure (13T&I0037-WO-PCT), Serial Number PCT/IB13/59406
9. Jovic V, Al-Azria ZHN, Sun-Waterhouse D, Idriss H, Waterhouse GIN (2013) Photocatalytic H$_2$ production from bioethanol over Au/TiO$_2$ and Pt/TiO$_2$ photocatalysts under UV irradiation—a comparative study. Top Catal 56:1139–1151
10. Waterhouse GIN, Wahab AK, Al-Oufi M, Jovic V, Anjum DH, Sun-Waterhouse D, Llorca J, Idriss H (2013) Hydrogen production by tuning the photonic band gap with the electronic band gap of TiO$_2$. Scientific Rep 3:1–5
11. Jovic V, Chen WT, Blackford MG, Idriss H, Waterhouse GIN (2013) Effect of gold loading and TiO$_2$ support composition on the activity of Au/TiO$_2$ photocatalysts for H$_2$ production from ethanol–water mixtures. J Catal 305:307–317
12. Bowker M, Millard L, Greaves J, James D, Soares J (2004) Photocatalysis by Au nanoparticles: reforming of methanol. Gold Bull 37:3–4
13. Subramanian V, Wolf EE, Kamat PV (2004) Catalysis with TiO$_2$/gold nano composites: effect of metal particle size on the fermi level equilibration. J Am Chem Soc 126:4943–4950
14. Du L, Furube A, Yamamoto K, Hara K, Katoh R, Tachiya M (2009) Plasmon-induced charge separation and recombination dynamics in gold TiO$_2$ nanoparticles systems: dependence on TiO$_2$ particle size. J Phys Chem C 113:6454–6462
15. Seh ZW, Liu S, Low M, Zhang SY, Liu Z, Mlayah A, Han MY (2012) Janus Au–TiO$_2$ photocatalysts with strong localization of plasmonic near-fields for efficient visible-light hydrogen generation. Adv Mater 24:2310–2314
16. Zhang Z, Zhang L, Hedhili MN, Zhang H, Wang P (2013) Plasmonic gold nanocrystals coupled with photonic crystal seamlessly on TiO$_2$ nanotube photoelectrodes for efficient visible light photoelectrochemical water splitting. Nano Lett 13:14–20
17. Bashir S, Wahab AK, Idriss H (2014) Synergism and photocatalytic water splitting to hydrogen over Pt/TiO$_2$ catalysts: effect particle size. Catal Today (submitted)
18. http://webbook.nist.gov/Massfragmentationethanol
19. Kruse N, Chenakin S (2011) XPS characterization of Au/TiO$_2$ catalysts: binding energy assessment and irradiation effects. Appl Catal A 391:367–376
20. Yi CW, Luo K, Wei T, Goodman DW (2005) The composition and structure of Pd–Au surfaces. J Phys Chem B 109:18535
21. Li Z, Gao F, Wang Y, Calaza F, Burkholder L, Tysoe WT (2007) Formation and characterization of Au/Pd surface alloys on Pd (111). Surf Sci 601:1898–1908

22. Moulder JF, Stickle WF, Sobol PE, Bomben KD (1992) Handbook of X-ray photoelectron spectroscopy, Perkin-Elmer Corporation

23. Idriss H, Barteau MA (1994) Reactions of p-benzoquinone on TiO$_2$(001) single crystal: oligomerization and polymerization by reductive coupling. Langmuir 10:3639–3700

24. Idriss H, Barteau MA (1994) Characterization of TiO$_2$ surfaces active for novel organic synthesis. Catal Lett 26:123–139

25. Doniach S, Sunjic M (1970) Many-electron singularity in X-ray photoemission and X-ray line spectra from metals. J Phys C Solid 3:285

26. Montini T, Marelli M, Minguzzi A, Gombac V, Psaro R, Fornasiero P, DalSanto V (2012) H$_2$ production by renewables photoreforming on Pt–Au/TiO$_2$ catalysts activated by reduction. ChemSusChem 5:1800–1811

27. Nadeem MA, Murdoch M, Waterhouse GIN, Metson JB, Keane MA, Llorca J, Idriss H (2010) Photoreaction of ethanol on Au/TiO$_2$ anatase. Comparing the micro to nano particle size activities of the support for hydrogen production. J PhotoChem PhotoBiol A: Chem 216:250–255

28. Muir JN, Choi YM, Idriss H (2012) DFT study of ethanol on TiO$_2$ (110) rutile surface. Phys Chem Chem Phys 14:11910–11919

29. Idriss H, Kim KS, Barteau MA (1993) Carbon–carbon bond formation via aldolization of acetaldehyde on single crystal and polycrystalline TiO$_2$ surfaces. J Catal 139:119–133

30. Reztova T, Chang C-H, Koersh J, Idriss H (1999) Dark and photoreactions of ethanol and acetaldehyde over TiO$_2$/carbon molecular sieve fiber. J Catal 185:223–235

31. Kundu S, Vidal AB, Nadeem A, Senanayake SD, Stacchiola D, Idriss H, Rodriguez J (2013) Ethanol photo reaction on ruthenium metal/metal oxide modified rutile TiO$_2$ (110) single crystal surface. J Phys Chem C 117:11149–11158

11

Sustainable world through sustainable materials and integrated biorefineries

Said Salah Eldin Elnashaie · Firoozeh Danafar ·
Fakhru'l-Razi Ahmadun

Abstract The present world, with all its advancement, is not a sustainable world, simply because it is based on non-renewable raw materials (non-RRMs). Sustainable world is best expressed in terms of sustainable development and non-RRMs must be replaced by sustainable materials. The sustainable materials needed by modern society are very wide and the main pillars are biofuels and bioproducts. Both pillars are best related through integrated biorefineries (IBRs) formed of related concepts which are very important for the economic development and sustainability of all countries on our planet. Integrated biorefineries include the production of biofuels and bioproducts and utilizing novel technologies. An IBR contains at least two routes: a biochemical route based on a sugar platform and a thermo-chemical-catalytic route based on a syngas platform. It is a multiple inputs–multi outputs (MIMO) system with design flexibility to accept a wide range of biofeedstock, especially wastes. If the biorefinery consists of only one route/platform, or is limited with regard to MIMO or biofuels/bioproducts produced, then it should be considered an elementary biorefinery (EBR). Sustainable development engineering which is a subsystem of sustainable development (and sustainable world) is more general than Environmental Engineering; and Clean and Green Technology, because it also includes the utilization of RRMs to achieve not only sustainability but also clean environment. An integrated system approach based on system theory is used to analyse sustainable development, sustainable world, sustainable materials, IBRs, EBRs and their interactions.

Keywords Sustainable · Integrated biorefineries (IBR) · Biofuels · Renewable raw material (RRM)

S. S. E. Elnashaie (✉) · F.-R. Ahmadun
Department of Chemical and Environmental Engineering,
Faculty of Engineering, Universiti Putra Malaysia,
43400 Serdang, Selangor, Malaysia
e-mail: selnashaie@gmail.com

F. Danafar
Department of Chemical Engineering, College of Engineering,
Shahid Bahonar University of Kerman, 7618891167 Kerman,
Iran

S. S. E. Elnashaie
Chemical and Biological Engineering Department, University
of British Columbia (UBC), Vancouver, Canada

Introduction

System theory [1–17] is the basis of the integrated system approach, which is the most efficient methodology for knowledge classification, organization, transfer, and exchange [18]. The integrated system approach is very valuable in both research and education. In research, it is one of the most important tools for the development of new knowledge and novel processes, especially in areas where multi-disciplinary research and development is a must for innovative solutions [19, 20]. Sustainable development is one of those areas that are multi-disciplinary by their very nature [21]. It is formed by a number of sub-systems, each of which is formed by its own elements (or subsystems of subsystems depending upon the level of analysis). Sub-systems of sustainable development include both technical and non-technical categories, for example, technology, socioeconomic, political, ethical/moral, and so on. Focusing on any one of the sustainable development subsystems can only be successful within a framework that has the other subsystems as a background. Within the technological subsystem of sustainable development, a structural hierarchy of subsystems, followed by subsystems of sub-systems, down to elements gives the structure and boundaries of this important subsystem, especially from an engineering point of view. It is useful in this regard to use terminologies and classifications of system theory coupled with terminology of non-linear dynamics and stability theorem. As an example, we can consider efficient engineering as a subsystem of environmental engineering, representing a necessary but not sufficient condition for Clean and Green Technology (C&GT). This is due to the simple fact that applying efficient engineering without taking environmental constraints into consideration can achieve maximum productivity that would not necessarily be environmentally clean. C&GT will need efficient engineering as a prerequisite. Also, we can consider Environmental Engineering as a subsystem of sustainable development engineering, representing a necessary but not sufficient condition for sustainability. This is due to the simple fact that Environmental Engineering without using RRMs can achieve C&GT and maximum productivity but is not necessarily sustainable.

Sustainable development engineering will have both efficient and environmental engineering as prerequisites. The utilization of RRMs is at the heart of sustainability; this leads to the crucial importance of biofuels at one level and IBRs at higher levels, as discussed in this article, using the integrated system approach as an efficient tool. Although, biofuels from RRMs are very useful for a cleaner environment, it is not sufficient for sustainable development, which needs both biofuels and bioproducts produced from RRMs, therefore, requiring IBRs. This need of sustainable development and its coupling to IBRs and RRMs suggests other needs for new technologies and innovative solutions to old and new challenges as well as novel technologies for RRMs. It is also important to take into consideration the fact that RRMs are geographically usually spread over wide areas while classical non-RRMs are usually concentrated in certain areas; this will have important effect on the technology of getting the raw materials to the processing plant and the optimal layout of the processes and the plant. Novel technologies will need to be utilized to the most to make packages of RRMs with their novel technologies compete with classical non-RRMs and their well established technologies. For the first package to win the competition it has to use research extensively and utilize optimal coupling between experimental techniques and mathematical/computer modelling in the development and scaling up of novel technologies. The theme of this article is to stress the importance of the system theory when dealing with the issue of sustainability, biofuels (BFs), and IBRs. In this respect, an introduction will be presented about system theory, and

then the focus is on the how sustainable development (SD) relates to the system theory (ST) and integrated system approach (ISA).

System theory

System theory

System theory is a basic tool for dealing with sustainability and sustainable development, and hence it should be more widely used in engineering education and research. In this respect, definitions of some important terms are essential. First of all, what is a system? The word system derives from the Greek word systema and means an assemblage of objects united by some form of regular interaction or interdependence [17, 19]. A simpler, more pragmatic description regarding systems includes the following:

- A system is a whole composed of parts (elements or subsystems).
- The concepts of a system, subsystem, and element are relative and depend upon the degree of analysis.
- The parts of the system can be parts in the physical sense or they can be processes. A system can be formed of both (i.e. different parts of the system; a reactor and a regenerator combined to form a fluid catalytic cracking unit) [16], each part having a number of processes taking place within its boundaries.
- The properties of the system are not necessarily the sum of the properties of its components (elements or subsystems), although they are, of course, affected by those components. Instead, the properties of the system result from non-linear interaction (synergy) between elements or subsystems [17, 19].

The state of the system and state variables

The term state of the system, rigorously defined through the state variables of the system, is used extensively in discussing and modelling/simulation of systems. These state variables are chosen according to the nature of the system.

Input variables (parameters)

Input variables are not state variables. Instead, they are external to the system but affect the system (i.e. work on the system). For example, the feed temperature and composition of the feed stream to a distillation tower or a chemical reactor or the feed temperature to a heat exchanger are input variables.

Design variables (parameters)

They are associated with the design of the system and are usually fixed. Examples are the diameter and height of a continuous stirred tank reactor (CSTR) or of a tubular reactor.

Boundaries of system

A system has boundaries distinguishing it from the surroundings or environment. The relation between the system and its environment leads to one of the most important classifications of systems:

Isolated systems They do not exchange matter or energy with the environment (surroundings). They tend to the state of thermodynamic equilibrium (maximum entropy). An example is a batch adiabatic reactor.

Closed systems They do not exchange matter with the environment (surroundings), but they do exchange energy. Such systems, again, tend to thermodynamic equilibrium (maximum entropy). A batch non-adiabatic reactor is an example.

Open systems They exchange matter and energy with the environment (surroundings). They do not tend to thermodynamic equilibrium but to steady state or what should better be called a stationary non-equilibrium state, characterized by minimum entropy generation. A CSTR is an example.

The above shows that the term steady state commonly used in chemical/biological engineering and other disciplines is not precise enough. A more accurate term should be stationary non-equilibrium state, which is a characteristic of open systems, distinguishing it from stationary equilibrium state, associated with isolated and closed systems.

Steady and unsteady states and thermodynamic equilibrium of systems

Steady state occurs when the state of the system does not change with time, but the system is not at thermodynamic equilibrium. This steady state of lumped systems is a point in a space having the same dimensions as the problem (number of components + temperature + pressure, etc.), whereas that for distributed systems is a profile in the space coordinate(s) as additional dimension(s). Unsteady state of an open system starts at an initial condition and tends with time towards a steady state when the system is stable (a point for lumped system and profile for distributed systems).

Integrated system approach

Sustainable development in a changing global environment will require resilience at many levels, including human communities and economic enterprises. In the face of ever-increasing global complexity and volatility, it is essential to move beyond a simplistic steady state model of sustainability. Instead, we need to develop adaptive policies and strategies that enable societal and industrial institutions to cope with unexpected challenges, balancing their need to be able to achieve an efficient sustainable development. The current lack of success in improving industrial sustainability, coupled with the challenges of biocomplexity and resilience, indicates that sustainability is a system's problem requiring collaborative solutions with a cross-disciplinary nature [12, 22, 30, 44, 46, 48, 51–54, 58, 59]. A number of technical advances will likely improve the usefulness of models, including rigorous methodologies for dealing with missing and uncertain information; improved methods for interpretation of multivariate data sets and for multi-objective decision making involving trade-offs among conflicting goals; and novel modelling methods as alternatives to traditional mathematical models. More generally, there is a great need for operational definitions and metrics for sustainability and resilience in economic, ecological, and societal systems.

Basic principles of sustainable development

The simple analysis in the introduction and some other components discussed in this article highlight the following basic principles:

1. Sustainable development is a system formed of technological and non-technological subsystems with the following components as the principle ones:

    (a) Political (e.g. legislations and strategic decisions…).
    (b) Economical (e.g. investment in novel new technologies).
    (c) Social, ethical/moral (e.g. consumption trends, acceptance of novel clean technologies and products, moral/ethical factors).
    (d) Technological (e.g. novel efficient clean technologies, clean fuels, efficient utilization of renewable feedstock, new environmentally friendly products, in-process modification for minimum pollution maximum production (MPMP), efficient waste treatment) with special emphasis on the technological subsystem with the other subsystems in the background.

2. Sustainable development engineering is a subsystem of the technological subsystem of sustainable development.

3. Sustainable development can also be divided into the following [24, 27, 29, 34, 39, 41].

(a) Sustainable development with respect to production, which is the main emphasis of this article.

(b) Sustainable development with respect to consumption, which is in the background of this article.

Sustainable production and consumption (SP&C) emerged as a key issue on the sustainable development agenda at the UN Conference on Environment and Development (UNCED) in Rio de Janeiro in 1992. The agenda called on governments, businesses, and others to implement measures to promote efficiencies in production and encourage sustainable patterns of consumption. It went on to say that the developed countries should take the lead in introducing those measures. An international agenda had been introduced by the UN Commission on Sustainable Development (UNCSD), in cooperation with national governments; the Organization for Economic Cooperation and Development (OECD) and others responded to this mandate with an international work programme and recommendations for action. Proposed action items included the following:

- Pricing reforms to internalize environmental costs and remove subsidies that generate unsustainable consumption
- "Green" public procurement policies

Main actions taken for sustainable development include:

- Extending producer responsibility for the lifecycle environmental impacts of goods and services
- Eco-labelling programmes

The UNCSD has called specifically upon businesses to do the following:

- Integrate environmental criteria into their purchasing policies
- Design more efficient products and processes
- Increase the life-spans for durable goods
- Improve their after-sales services, reuse, and recycle
- Promote sustainable consumption through advertising, marketing, and product information

4. Efficient engineering is a subsystem of environmental engineering, and environmental engineering is a subsystem of sustainable development engineering. In other words, efficient engineering is necessary but not sufficient for environmental engineering, and environmental engineering is necessary but not sufficient for sustainable development engineering.

5. Metrics are necessary for measuring sustainability [31, 55].

According to the extensive study by the IChemE, metrics for sustainability can be divided into three groups:

environmental indicators, economic indicators, and social indicators. This emphasizes, however, the three categories do not show explicitly the importance of RRMs for sustainability, shown very simply and briefly above. RRMs represent a crucial component of sustainable development engineering and, thus, sustainable development and C&GT based on non-RRMs may be satisfactory from an environmental engineering point of view but is not sustainable. Biofuels [15] are an important subsystem of renewable clean energy, others are wind energy, solar energy, etc. There are a large number of biofuels; each one of them can be produced through different routes. Figure 1 shows some of these routes for biodiesel and biohydrogen. All biofuels and all technologies will have their positions in the clean fuels matrix of the future.

Figure 1 shows two types of biodiesels, the strategic Fischer–Tropsch biodiesel from syngas [4, 36, 45] and the non-strategic biodiesel from the transesterification of vegetable oils. A third type of biodiesel, which is proving to be more strategic than both, is the algae biodiesel [2, 6, 13, 14, 23, 35, 47, 50]. Among the most photosynthetically efficient plants are those various types of algae. Some species of algae are ideally suited to biodiesel production due to their high oil content (some well over 50 % oil) [7, 13, 35] and extremely fast growth rates, 200 times faster than soya bean [20]. Extensive research is carried out now on the growth of specially chosen strains of microalgae [3, 28, 40, 49] in optimally designed and operated photobioreactors [3].

Integrated bioreactors

An IBR is a complex facility that integrates biomass conversion processes and equipment to produce fuel, power, and chemicals/biochemicals from the biomass. It is analogous to today's petroleum refineries and its integrated petrochemical complexes, which produce multiple fuels and products from petroleum. IBRs have been identified as the most promising route to the creation of a new domestic and distributed bio-based industry. Sustainability does not depend only on sustainable fuels but also on sustainability of other commodities for modern societies. This important simple fact leads to the important concept of IBRs [39] producing not only sustainable fuels but also other sustainable commodities and energy. The National Renewable Energy Laboratory, which is a part of the USA Department of Energy (DOE), defines biorefinery as "a facility that integrates biomass conversion processes and equipment to produce fuels, power, and chemicals from biomass." The present view about IBRs is based on two platforms, the sugar platform and syngas platform, as discussed later. Production facilities for sustainable biofuels are a subsystem of IBRs. The implications of this are as follows:

Fig. 1 Some routes to biodiesel and biohydrogen

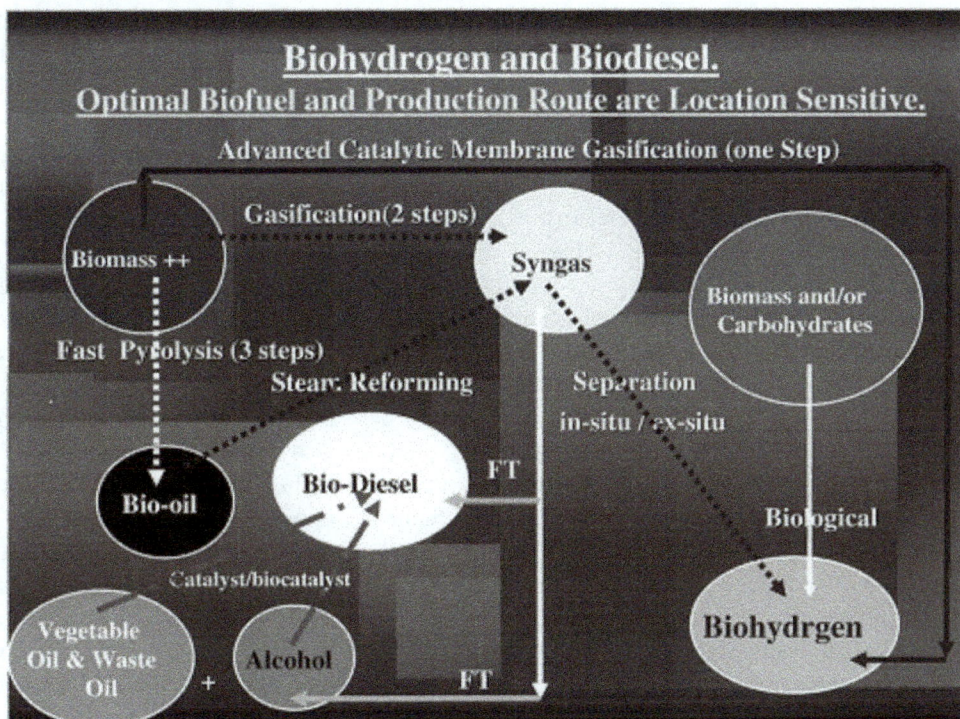

- A sustainable biofuels facility built today should be planned with its growth into an IBR in mind.
- Sustainable biofuel production will almost always be a part of IBRs.
- An advanced definition of IBRs and their subsystems should be developed together with a clear definition of biofuels and bioproducts.
- A clear definition of sustainability and quantification of these definitions into suitable metrics should be developed.

Both the National Science Foundation (NSF) and DOE are putting large research funds into development of IBRs [1, 5, 8–10, 25, 26, 32, 33, 42, 43, 61].

IBRs represent an integral critical subsystem of sustainable development, which is a multi-disciplinary system by its very nature, as discussed earlier. It is best to use the integrated system approach to study this complex multi-disciplinary system and its subsystems. Engineers are most interested in the technology part (which is a subsystem of the sustainable development system) but with a background understanding of other subsystems and collaboration with other disciplines, as discussed earlier. Also discussed earlier, sustainable development engineering is the most important subsystem of the technology part, which is itself a subsystem of the sustainable development system. As discussed earlier, the efficient utilization of RRMs is at the heart of sustainability. RRMs should be defined very clearly, for this is strongly related to the cycle of renewability, that is, fossil fuels are renewable, but over a cycle of hundreds of thousands (or even millions) of years. A RRM should be renewable over a period of 6–18 months. RRMs can be any kind of renewable waste, for example, agricultural waste, municipal waste, and so on, or special (energy) crops produced specifically to be used for this purpose, for example, switchgrass cultivated and produced especially for biofuels. On the other hand, useful agricultural products used today for biofuels, for example, corn for ethanol, vegetable oil for transesterification to biodiesel, etc., are not acceptable as RRMs because of the consumption of important edible products at a time of food shortage everywhere, especially in developing nations.

Renewable biomass, the main group of RRMs, is a storage tank for solar energy through biosynthesis, CO_2, and other nutrients. The ultimate aim for biofuels is to produce as much CO_2 as consumed in the biosynthesis of the biomass that produced it. This, with maximum efficiency, may lead to approaching zero net CO_2 emission. We cannot really dispose of any CO_2 resulting from fossil fuel; it only keeps circulating from one form/place to the other, except with sequestration by injection under the bottom of the ocean, which is expensive. Its side effects are not known and can only be practised by very large companies. All other techniques just move the CO_2 from one location/to the other without reducing the earth CO_2 added inventory from the carbon source that came from under the ground.

Hydrogen is a good clean fuel and will occupy its part in the clean fuels matrix and is, therefore, one of the potential products of IBRs. However, although the claim that it is 100 % clean is locally true, it is actually not globally true if its source is fossil fuel, whether directly through catalytic steam reforming (CSR) 95 % of hydrogen produced in United States is through CSR or indirectly through production of electricity (90 % of electricity in the United States is from coal) followed by electrolysis. It can be globally clean only if the source is bio, wind, hydro, nuclear, etc.

The range of fermentable sugars is expanding due to the development in microbiology (specially genetic engineering) and the discovery of efficient mutated microorganisms capable of fermenting the wide range of sugars produced from cellulose/hemicellulose hydrolysis. A strong challenge with lots of intensive research nowadays is in the field of efficient and clean hydrolysis of cellulose/hemicellulose to sugars suitable for fermentation with classical microorganisms and/or mutated ones. Important improvements are achieved in the enzymatic hydrolysis of cellulose/hemicellulose. The process of fermentation is improving continuously using novel membrane immobilized fermenters and novel modes of operation. Lignin from lignocelluloses is used as a cheap fuel to improve the energy efficiency of the IBRs; however, the future potential is to use it as a platform for a wide range of products to replace certain petroleum refining and petrochemical products. Important improvements are introduced to syngas production from biomass through both a one-step process (biomass gasification to syngas) and a two-step process (biomass fast pyrolysis to bio-oil followed by CSR of bio-oil to syngas). The two-step process seems to be better with regard to the percentage of H2 in the syngas. Certain difficulties associated with CSR of bio-oil are solvable through novel reformers configurations such as circulating fluidized bed (CFB) CSR. An important challenge in the biomass to FT biofuels process is the integration of the endothermic CSR process with the exothermic catalytic FT process into one integrated membrane catalytic autothermal process. This auto-thermic process can also be integrated to a fluidized bed catalytic chemical vapour deposition (FBCCVD) that will utilize CO2 as the feedstock and produce the very valuable carbon nanotubes and oxygen to be used in the catalyst regeneration for the CFB CSR. This part of a process can be considered an elementary biorefineries (EBR) and is described very shortly below.

Novel membrane catalytic reactor for catalytic steam reforming coupled to catalytic Fischer–Tropsch process and the novel over all reactor coupled to a novel process to crack CO2 to carbon nanotubes and oxygen coupled membrane reactors could also make a positive contribution to meet the challenge of sustainable development. This example is dedicated to treatment of one or more

intermediates, produced thermochemically from biomass, to produce biofuels that could replace blends of gasoline, diesel, or jet fuels, as well as bioproducts and other liquid hydrocarbons suitable to be fed to existing petroleum refineries (biocrude oil). These intermediates and products include fast pyrolysis bio-oil, syngas, ethanol, mixed alcohols, mixed oxygenates (C2 and higher), olefins, ethers, biofuels, biofuels, biocrude oil, etc. This combination of integrated thermo-chemical-catalytic (TCC) processes can be considered EBRs in contradistinction to IBRs which include both TCC as well as biochemical processes refineries. IBRs are large-scale complex systems addressing not only efficient production of a range of biofuels, but also numerous bioproducts. Moving beyond ethanol to higher hydrocarbons, biofuels are becoming increasingly important as blend barriers are reached. Renewable refinery feedstocks biocrude oil could reduce dependence on imported oil by replacing fossil fuels. This could allow progress beyond renewable gasoline-only blending components, and it could also redirect oxygen present in thermo-chemically-derived intermediates to value-added co-products with economic benefits. The oxygen issue is critical and can be addressed through different pathways. The pathway in this example consists of:

1. Extracting oxygen in fast pyrolysis bio-oil and syngas as CO2.
2. Using catalytic fluidized bed chemical vapour deposition reactors to re-carbonize CO2 into carbon nanotubes and oxygen.
3. Using oxygen to oxidize intermediates to biofuels and other bioproducts, as well as in the regeneration of catalyst(s).

Advancing these technologies includes choosing process operating parameters to achieve maximum products yields, integrity of process materials of construction, and long-term long life catalyst performance to fit the biomass feedstock. This is the best achieved by optimal coupling of experimental techniques, reactor modelling and computer simulation in novel configurations and catalytic processes. Coupled membrane catalytic reactors represent the heart of many of these processes, as shown below, in Fig. 2. This example focuses on a relatively simple, but novel, integrated biorefinery IBR to produce a wide range of products and advanced materials. Reactor modelling and computer simulation need to be combined with experimental results to achieve system viability. A simplified schematic diagram of the proposed process is shown in Fig. 2, where the starting feed is the bio-oil obtained from biomass through fast pyrolysis. Biomass such as corn stover, woody materials or rice straw is fast pyrolysed, and the resulting bio-oil (fast pyrolysis bio-oil, FPBO) is transformed into syngas in a novel catalytic Circulating fluidized bed membrane steam

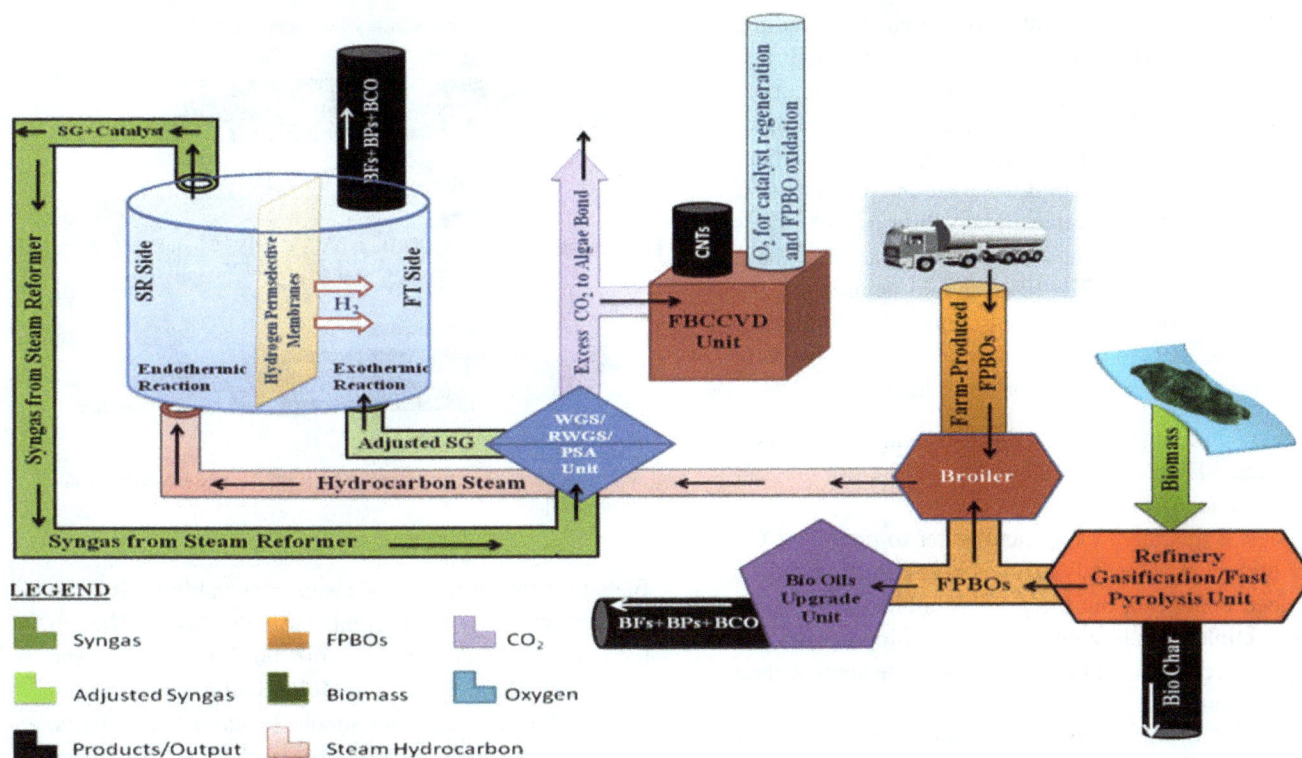

Fig. 2 Simplified diagram for the proposed units at the refinery

reformer (CFBMSR). The syngas then goes to reaction using novel efficient catalytic Fischer–Tropsch processes and catalyst(s) (e.g. nanoparticle catalysts). Other sources of syngas are also considered as shown in Fig. 2. Both CO and H2 from the steam reforming process transfer to the Fischer–Tropsch reactor through the selective membrane coupling them. The remaining CO2 is treated together with that resulting from catalyst regeneration as described below. Pre-treatment other external syngas is pre-treated processes, to optimize its composition. This will include to be done using a catalytic reverse water–gas-shift reactor to transform CO2 to CO, and CO2-absorption–desorption in a pressure swing adsorption unit to remove any excess CO2 in the feed to the Fischer–Tropsch unit. This excess CO2 together with all the CO2 from the CFBMSR will be treated in a novel catalytic fluidized bed chemical vapour deposition reactors unit to produce CNTs and oxygen.

It is worth mentioning, here, that the extraordinary properties of CNTs and its wide range of applications have encouraged considerable efforts to synthesize this valuable material. CO2 is a cheap, abundantly available and renewable feedstock that can be converted into highly added-value carbon nanostructures with high carbon yields. Besides that, CO2 shows great potential as oxidizing agent for improving the quality of CNT synthesis [11, 37, 38, 56, 60, 62] and enables to catalytically decompose amorphous carbon by the reverse

Boudouard reaction. Taking into account that for most applications, highly pure CNTs are required dictates that impurities should be removed during a post-synthesis step. As purification procedures may affect the structure of CNTs [57], much effort is still put into the development of a direct low temperature synthesis route of high-quality CNTs. Using the excess CO_2 to produce CNTs and/or bioproducts through photobiosynthesis will depend upon many economical and technological factors.

The O_2 will in turn be partly used to oxidize fast pyrolysis bio-oil FPBO to additional biofuels and bio-products, and partly to regenerate the steam reforming catalyst in the CFBMSR. As shortly mentioned above the CFBMSR is coupled to the Fischer–Tropsch reactor through membranes which allow the transfer of CO and H2 from the CFBMSR to the Fischer–Tropsch reactor as well as heat in the opposite direction from the exothermic reactions in the Fischer–Tropsch reactor to the endothermic reactions in the CFBMSR and the unreacted syngas SG from the Fischer–Tropsch unit reactor will be recycled to the reverse water–gas-shift unit. Use of fluidized beds for the catalytic processes will increase efficiency and facilitate on-line sampling and characterization of the different catalysts during the different stages of process development to reach the commercial scale.

The EBR needs to be started up by a special procedure. Each catalytic reactor must then operate at steady state for at least 2,000 h (3 months) with a single catalyst loading. This will require a good control system for each unit and for the overall biorefinery. In addition to on-line measurement of state variables, small samples will need to be withdrawn and analysed during operation to follow changes in state variables, including catalyst characteristics. If this EBR is integrated to other lignocelluloses to biofuels EBR based on the sugar platform, this integration of these two EBRs results in an IBR. IBRs should be able to contribute to sustainability and not only renewable biofuels. It is important for IBRs to include all that is bio, whether with regard to feedstock, processes, or both, for instance,

- Biomass utilizing sequential thermal catalytic processes to produce FT biodiesel (the processes are not bio, but the feedstock is bio)
- Biological treatment of CO and water to produce CO2 and hydrogen (biocatalysed water–gas shift) reaction; the feed is not bio, but the process is bio)
- Lignocelluloses utilizing sequential bioprocesses to produce cellulosic bioethanol (both feedstock and processes are bio)
- Utilization of bioprocesses to transfer one form of energy to another, for example, integrated bioenergy to biofuel cells (IBE-BFCs) to change intermittent solar energy into continuous electric energy, for many applications including auto-thermal housing IBRs can either grow with time as parts of existing plants (e.g. pulp and paper industry) or start as an IBR based on renewable fuels production and grow into a complete IBR. Other forms of renewable classical energy (RCE) (when suitable, e.g. wind energy) can be integrated with the biofuels to generate the total renewable energy profile of an IBR. The central intermediates (platforms) for the biofuels part of the IBRs seem to be fermentable sugars suitable for fermentation to bioethanol and/or biobutanol and syngas suitable for both hydrogen extraction and use in the Fischer–Tropsch process for producing fuels from methanol up to diesel. It is important, also, to notice that it is possible to combine the gasification of biomass (or fast pyrolysis followed by catalytic steam reforming (CSR) of the product bio-oil) to produce biosyngas with the fermentation of this biosyngas to produce bioethanol and or biobutanol. Direct utilization of solar energy through different direct techniques (e.g. photocells) without passing through the biosynthesis stage is a possibility, but it has its own bottlenecks and should be integrated with other technologies (e.g. IBE-BFCs) to fully utilize it as a part of auto-thermal housing and IBRs. From the above, it is clear that neither single biofuel nor one technology will dominate for any single one of these biofuels. Humanity will move from today's matrix of dirty, non-renewable

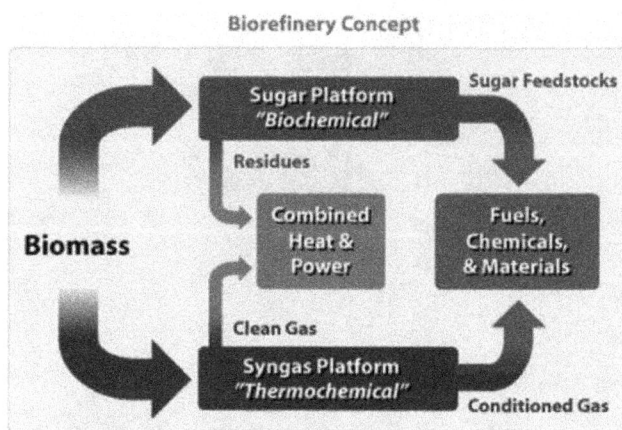

Biorefinery Concept

Fig. 3 Preliminary biorefinery structure with the main two sugar and syngas platforms

fuels to a future matrix of clean, renewable fuels, with the sub-matrix of different biofuels from different technologies occupying a large portion of the clean fuels matrix, and that sustainability will not be achieved through biofuels only but should be extended to IBRs, combining biofuels and other bioproducts with efficient bioenergy to achieve sustainable development. Figure 3 shows a simple schematic diagram of an IBR formed of only two platforms; the sugar and syngas platforms as the bioenergy is an integral part of any IBR. It can be considered an integration of two EBRs, and should be optimized with regard to RRMs utilizations and useful biofuels and bioproducts production with minimum to zero emissions.

Conclusions

Today's matrix of dirty, non-renewable fuels is moving to a future matrix of clean, renewable fuels, with the sub-matrix of different biofuels from different technologies occupying a large portion of the clean fuels matrix. To achieve sustainable development, biofuels processes should be extended to IBRs, combining biofuels and other bioproducts with efficient bioenergy. IBRs represent an integral critical subsystem of sustainable development, which is a multi-disciplinary system by its very nature. It is best to use the ISA to study this complex multi-disciplinary system and its subsystems. Engineers are most interested in the technology part (which is a subsystem of the sustainable development system) but with a background understanding of other subsystems and collaboration with other disciplines.

The utilization of RRMs is at the heart of sustainability for biofuel production. It leads to the crucial importance of biofuels at one level and IBRs at higher levels using the integrated system approach as an efficient tool. Although, biofuels from RRMs are very useful for a cleaner environment, it is not sufficient for sustainable development which needs both biofuels and bioproducts produced from RRMs, therefore, requiring IBRs. This need of sustainable development and its coupling to IBRs and RRMs suggests other needs for new technologies and innovative solutions to old and new challenges as well as novel technologies for RRMs. Novel technologies will need to be utilized to the most to make packages of RRMs with their novel technologies compete with classical non-RRMs and their well established technologies. For the first package to win the competition it has to use research extensively and utilize optimal coupling between experimental techniques and mathematical modelling in the development and scaling up of novel technologies.

The current lack of success in improving industrial sustainability, coupled with the challenges of biocomplexity and resilience, indicates that sustainability is a system's problem requiring collaborative solutions with a multi-disciplinary nature. There is a great need for operational definitions and metrics for sustainability and resilience in economic, ecological, and societal systems. In this respect, the importance of the system theory is evident that emphasis it should be more widely used in engineering education and research.

References

1. Aguilar A, Bochereau L, Matthiessen-Guyader L (2008) Biotechnology and sustainability: the role of transatlantic cooperation in research and innovation. Trends Biotechnol 26(4): 163–165
2. Alvarado P (2007) Biodiesel from algae and the biofuels discussion in Argentina. http://www.treehugger.com/files/2007/03/biodiesel_from_1.php. Retrieved Aug 2007
3. Arudchelvam Y, Nirmalakhandan N (2012) Optimizing net energy gain in algal cultivation for biodiesel production. Bioresour Technol 114:294–302
4. Bermu'dez V, Lujan JM, Pla B, Linares WG (2011) Comparative study of regulated and unregulated gaseous emissions during NEDC in a light-duty diesel engine fuelled with Fischer Tropsch and biodiesel fuels. Biomass Bioenergy 35:789–798
5. Bohlmann GM (2005) Biorefinery process economics. In: Paper presented at the World Congress on Industrial Biotechnology and Bioprocessing
6. Briggs M (2004) Widescale biodiesel production from algae. http://www.unh.edu/p2/biodiesel/article_alge.html. Retrieved Aug 2007
7. Chen L, Liu T, Zhang W, Chen X, Wang J (2012) Biodiesel production from algae oil high in free fatty acids by two-step catalytic conversion. Bioresour Technol 111:208–214
8. Chen S, Wen Z, Liao W, Liu C, Kincaid RL, Harrison JH (2005) Studies into using manure in a biorefinery concept. Appl Biochem Biotechnol 124(1–3):999–1016
9. Christensen BH, Nielsen C (2005) Biorefineries with optimal carbon utilization for liquid bio-fuels. In: Paper presented at the First International Biorefinery Workshop
10. Clark JH, Budarin V, Deswarte FEI, Hardy JJE, Kerton FM, Hunt AJ (2006) Green chemistry and the biorefinery: a partnership for a sustainable future. Green Chem 8:853–860
11. Corthals S, Noyen JV, Geboers J, Vosch T, Liang D, Ke X, Hofkens J, Tendeloo GV, Jacobs P, Sels B (2012) The beneficial effect of CO2 in the low temperature synthesis of high quality carbon nanofibers and thin multiwalled carbon nanotubes from CH4 over Ni catalysts. Carbon 50:372–384
12. Delft University of Technology (2003) Annual report of Delft University of Technology (TUDelft). Technology Assessment Section, Delft, Netherlands
13. Demirbas A, Demirbas MF (2011) Importance of algae oil as a source of biodiesel. Energy Convers Manag 52:163–170
14. Dewulf J, Langenhove HV (eds) (2006) Renewables-based technology: sustainability assessment. Wiley, Chichester. Directory: biodiesel from algae oil (n.d.). Retrieved Aug 2007
15. Ellison K (2007) Biodiesel boom heading toward Wall street. Business 2.0 Magazine. http://money.cnn.com/2007/09/25/technology/biodieselboom.biz2/?postversion07092609. Retrieved Sept 2007
16. Elnashaie SSEH, Elshishini SS (1996) Dynamic modelling, bifurcation and chaotic behaviour of gas-solid catalytic reactors. Gordon and Breach, Amsterdam
17. Elnashaie SSEH, Garhyan P (2003) Conservation equations and modelling of chemical and biochemical processes. Marcel Dekker, New York
18. Elnashaie SSEH, Grace JR (2007) Complexity, bifurcation and chaos in natural and man-made lumped and distributed systems. Chem Eng Sci 62(13):3295–3325
19. Elnashaie SSEH, Uhlig F (2007) Numerical techniques for chemical and biological engineers using MATLAB. Springer, New York
20. Ferrell J, Sarisky-Reed V (2008) National Algal Biofuels Technology Roadmap. In: A technology roadmap resulting from the National Algal Biofuels Workshop. December 9–10, 2008, College Park, Maryland (Publication date May 2010)
21. Fiksel J (2006) Sustainability and resilience: toward a systems approach. Sci Pract Policy 2(2):14–21
22. Food and Agriculture Organization of the United Nations (1997) A system approach to biogas technology. http://www.fao.org/sd/EGdirect/EGre0022.htm. Retrieved Aug 2007
23. Gallagher BJ (2011) The economics of producing biodiesel from algae. Renew Energy 36:158–162
24. Geist HJ (2011) Linkages of sustainability. Bioscience 61(4):328–330
25. Governmental and industry partnerships for developing biorefineries (2004) Biobased Fuels, Power and Products Newsletter, pp 1–2
26. Halog A, Mao H (2011) Assessment of hemicellulose extraction technology for bioethanol production in the emerging bioeconomy. Int J Renew Energy Technol 2(3):223–239
27. Hens L (ed) (2005) The World Summit on Sustainable Development: the Johannesburg Conference. Springer, Dordrecht
28. Ho Sh, Chen Ch, Lee DJ, Chang JS (2011) Perspectives on microalgal CO2-emission mitigation systems—a review. Biotechnol Adv 29:189–198
29. Horbach J (ed) (2005) Indicator systems for sustainable innovation (sustainability and innovation). Physica-Verlag, Heidelberg
30. Hramov AE, Koronovskii AA (2005) Generalized synchronization: a modified system approach. Physics Review E (Stat Nonlinear Soft Matter Phys) 71(6):1–4

31. Institution of Chemical Engineers (2002) The sustainability metrics: sustainable development progress metrics recommended for use in the process industries. http://nbis.org/nbisresources/metrics/triple_bottom_line_indicators_process_industries.pdf

32. Integrated biorefineries (n.d.) (2007) http://www1.eere.energy.gov/biomass/integrated_biorefineries.html. Retrieved Aug 2007

33. Kochergin V, Kearney M (2006) Existing biorefinery operations that benefit from fractal-based process intensification. Appl Biochem Biotechnol 130(1–3):349–360

34. Koltun P (2010) Materials and sustainable development. Prog Nat Sci: Mater Int 20:16–29

35. Krohn BJ, McNeff CV, Yan B, Nowlan D (2011) Production of algae-based biodiesel using the continuous catalytic Mcgyan process. Bioresour Technol 102:94–100

36. Lapuerta M, Armas O, Hernández JJ, Tsolakis A (2010) Potential for reducing emissions in a diesel engine by fuelling with conventional biodiesel and Fischer–Tropsch diesel. Fuel 89:3106–3113

37. Lou Z, Chen C, Huang H, Zhao D (2006) Fabrication of Y-junction carbon nanotubes by reduction of carbon dioxide with sodium borohydride. Diamond Relat Mater 15(10):1540–1543

38. Magrez A, Seo JW, Kuznetsov VL, Forro L (2007) Evidence of an equimolar C2H2–CO2 reaction in the synthesis of carbon nanotubes. Angew Chem Int Ed 46(3):441–444

39. Makower J, Fleischer D (2003) Sustainable consumption and production: Strategies for accelerating positive change. Environmental Grantmakers Association Nanotechnology: Opportunities and challenges (n.d.), New York. http://nano.nd.edu/ESTS40403/index.html. Retrieved Aug 2007

40. Odlare M, Nehrenheim E, Ribé V, Thorin E, Gavare M, Grube M (2011) Cultivation of algae with indigenous species—potentials for regional biofuel production. Appl Energy 88:3280–3285

41. Omer AM (2008) Energy, environment and sustainable development. Renew Sustain Energy Rev 12:2265–2300

42. Realff MJ, Abbas C (2004) Industrial symbiosis, refining the biorefinery. J Indust Ecol 7(3–4):5–9 (Bull Sci Technol Soc)

43. Regalbuto JR (2011) The sea change in US biofuels' funding: from cellulosic ethanol to green gasoline. Biofuels Bioprod Biorefin 5(5):495–504

44. Rogers PL, Eon YJ, Svenson CJ (2005) Application of biotechnology to industrial sustainability. Process Saf Environ Protect 83(B6):499–503

45. Romar H, Lahti R, Tynjälä P, Lassi U (2011) Co and Fe catalysed Fischer–Tropsch synthesis in biofuel production. Top Catal 54(16–18):1302–1308

46. Rotmans J, van Asselt MBA, de Bruin AJ, den Elzen MGJ, de Greef J, Hilderink HBM (1994) Global change and sustainable development: A modelling perspective for the next decade. National Institute for Public Health and the Environment, Bilthoven

47. Sazdanoff N (2006) Modeling and simulation of the algae to biodiesel fuel cycle. Unpublished honors undergraduate thesis, Ohio State University

48. Seok H, Nof ShY, Filip FG (2012) Sustainability decision support system based on collaborative control theory. Annu Rev Control 36:85–100

49. Shahid EM, Jamal Y (2011) Production of biodiesel: a technical review. Renew Sustain Energy Rev 15:4732–4745

50. Sheehan J, Dunahay TG, Benemann JR, Roessler PG, Weissman JC (1998) A look back at the US Department of Energy's Aquatic Species Program: biodiesel from algae. National Renewable Energy Laboratory, Golden

51. Sheppard AW, Gillespie L, Hirsch M, Begley C (2011) Biosecurity and sustainability within the growing global bioeconomy. Curr Opin Environ Sustain 3:4–10

52. Subhadra BG (2011) Macro-level integrated renewable energy production schemes for sustainable development. Energy Policy 39(4):2193–2198

53. Subramanian K (2000) The system approach. Hanser Gardner, Cincinnati

54. Taniquchi I, Tabata D, Kusuda H, Koga T, Sotomura T (2003) Construction of bio-fuel cells using catalytic electrodes for oxidation of sugars and reduction of oxygen. In: Paper presented at the 204th meeting of the Electro Chemical Society

55. Trazil D, Ma G, Beloff BR (2003) Sustainability metrics. In: Paper presented at the 11th International Conference of Greening of Industry Network, Innovation for Sustainability

56. Tsang SC, Harris PJF, Green MLH (1993) Thinning and opening of carbon nanotubes by oxidation using carbon dioxide. Nature 362:520–522

57. Verdejo R, Lamoriniere S, Cottam B, Bismarck A, Shaffer M (2007) Removal of oxidation debris from multi-walled carbon nanotubes. Chem Commun 5:513–515

58. Von Bertalanffy L (1976) General system theory: foundations, development, applications. George Braziller, New York

59. Way LA, Wilson G (eds) (2005) Managing for tomorrow: resource-based communities & the environment. University of Northern British Columbia, Prince George

60. Wen Q, Qian W, Wei F, Liu Y, Ning G, Zhang Q (2007) CO2-assisted SWNT growth on porous catalysts. Chem Mater 19(6):1226–1230

61. Worden RM, Grethlein AJ, Jain MK, Datta R (1991) Production of butanol and ethanol from synthesis gas via fermentation. Fuel 70:615–619

62. Xu X, Huang S (2007) Carbon dioxide as a carbon source for synthesis of carbon nanotubes by chemical vapor deposition. Mater Lett 61:4235–4237

12

Preface for the special issue of the 1st Saudi–Chinese Oil Refinery Forum (1st SCORF 2013)

Zi-Feng Yan · Hamid A. Al-Megren

The continuous increasing of heavier crude oil supply and regulatory pressure for environmentally more acceptable oil products puts forwards ever more critical requirements on oil refining industry. The fast economic development in China leads to ever-increasing consumption of refined oil products and refining capacity and makes China world's second largest oil refiner, consuming 488 million tons of crude oil in 2013. As one of the most important oil processing, fluid catalytic cracking (FCC) is of great significance in converting heavy oil into high-octane gasoline, diesel and liquefied petroleum gas (LPG). In China, about 75–80 % of gasoline and more than 40 % of diesel come from FCC products. Currently, the FCC modified processes provide approximately one-third of the world's propylene.

In Saudi Arabia, Saudi Aramco is a fully integrated, global petroleum enterprise and a world leader in exploration and producing, refining, distribution, shipping and marketing. The company manages the largest proven reserves of conventional crude oil, 260.1 billion barrels, and the fourth-largest gas reserves in the world, 275.2 trillion cubic feet. Saudi Arabia, the world's biggest crude exporter, also expects to become the top single refiner producer of refined products such as fuel and petrochemicals. Saudi Arabia oil refinery capacity is at a current level of 2.122 Mbpd, up from 2.107 Mbpd 2 years ago. Saudi Arabia could consume nearly a tenth of the kingdom's current production capacity of about 12.0 million bpd when they are all fully operational in 2017.

Environmental protection nowadays has been a general consensus worldwide and environmentally driven regulations are also requiring significant improvement in the quality of gasoline and diesel in many parts of the world. Reducing organic sulfur content in gasoline has been recognized as one way of reducing emission of sulfur oxides. In China, the sulfur content in gasoline is required as low as 150 ppm since 2006 and further reduced to be less than 50 ppm since 2011. In the case of gasoline pool, nearly 90 % of sulfur content comes from FCC gasoline in China and about 33 % in USA. Therefore, refineries' efforts focus on effectively reducing the sulfur species coming from FCC unit by already existing technologies or developing more efficient and economical methods.

Saudi refineries are also busy improving products' quality by reducing sulfur in gasoline and diesel and reducing benzene in gasoline to improve local environment and to maintain competitiveness in international markets. Saudi Arabia has been leading the regional shift by working to implement projects that will improve gasoline and diesel qualities. To comply with mandatory sulfur specifications for gasoline and diesel between 2013 and 2016, the Kingdom of Saudi Arabia plans to construct multiple clean-fuel projects that will improve gasoline and diesel qualities. Saudi Arabia is seeking to reduce sulfur content in diesel and gasoline to 10 parts per million (ppm) and to lower benzene content in gasoline to 1 %. With these new fuel specifications, Saudi Arabian refining operations will comply with international standards. This will represent a dramatic reduction in sulfur levels from 2012, when Saudi Arabia's maximum sulfur level for diesel was greater than 500 ppm. The planned upgrades

Z.-F. Yan
State Key Laboratory of Heavy Oil Processing, CNPC Key Laboratory of Catalysis, China University of Petroleum, Qingdao, China

H. A. Al-Megren (✉)
Petrochemical Research Institute, King Abdulaziz City for Science and Technology, Riyadh, Saudi Arabia
e-mail: almegren@kacst.edu.sa

and revamps are necessary to meet future market demand for higher-value, lower-sulfur transportation fuels.

As the crude oils are getting heavier and the demand for high value petroleum products is increasing, FCC technologies are developing under the following directions: (a) reducing the yields of dry gas and coke, and increasing the yield of light oil; (b) increasing the diesel to gasoline ratio of the product; (c) increasing the processing capacity of inferior feedstock; (d) improving the product quality; and (e) increasing the yield of light olefins. Catalysts technology is the core for oil refining industry. In the year 2013, the estimated value of worldwide refining catalysts is approximately US\$ 3.13 trillion while the value of FCC catalysts is close to US\$ 0.91 trillion. In the past decades, refineries and catalyst companies in China have developed scores of new FCC catalysts (categories such as Orbit, LB, LVR-60, CC-20D) for heavy oil conversion to more light oil yields and resisting coking and metal contamination, FCC catalysts (categories such as GOR, LGO, LBO, LHO-1) special for decreasing olefin production, and FCC catalyst additives (categories such as CHO, CA, LBO-A, MS-011) special for more propylene, improving octane numbers of FCC gasoline or decreasing sulfur content of FCC gasoline.

Besides, the FCC unit technologies have also been developed to adapt the request of FCC catalysts and feedstock. One of the most successful examples is the two-stage riser (TSR) FCC technology developed by China University of Petroleum. In 2002, the TSR FCC technology was firstly commercially applied in a 100 kt/a industrial unit. After the technological renovation and transformation, the dry gas and coke yields decreased 2.7 wt % while the liquid products yield increased 2.7 wt %. Moreover, the cetane number of diesel increased 7 units. At present, there are 12 industrial units, including that in the stage of transformation or new construction, applied the TSR technology. The accumulative processing capacity has reached 9 Mt/a, the processing capacity of the largest unit is 1.6 Mt/a.

The 1st Saudi–Chinese Oil Refinery Forum (1st SCORF 2013), organized by King Abdul Aziz City for Science and Technology (KACST) in Riyadh with cooperation with China University of Petroleum (CUP), provided an opportunity for oil refinery key players to meet and discuss the latest developments in refinery technologies, innovative catalysts and sulfur management in refinery process, FCC process, development, and upgrading of crude oil and heavy residues, integrations of refining and petrochemical processes and hydro-processing of crude oil fractions.

This special issue presents selected papers from the conference (1st SCORF 2013) pertinent to catalyst preparation, modification and characterization, catalytic reactions, new advances in environmental technologies, and

refining technologies and new processing plants. The papers have been subject to an initial screening process by the conference organizers and then a normal peer-review process. The following highlights may be as a general summary of the topics in this issue.

In China, the needs for effective and efficient conversion of heavy oils have necessitated persistent high interest in research and development of refining catalysts, both focused on FCC catalyst active components and matrix materials. The Y zeolite is a single largest catalyst employed worldwide as FCC active component and 30–50 % of all motor fuels in the last decades have been produced using Y zeolite catalysts. Presently, one focus is the rare-earth modification of Y to modulate its acidity and cracking activity. The papers on "Research on the high activity of REY zeolite in fluid catalytic cracking reaction" and "The development of FCC catalysts for producing FCC gasoline with high octane numbers" present the importance of rare earth in modification of the acidity and pore properties of USY and improvement of VGO conversion. Besides, another paper titled "The application of mesoporous alumina with rich Brönsted acidic sites in FCC catalysts" offers an innovative idea to reduce coke yield and increase FCC feed conversion by modification of alumina matrix, on which the cracking active sites of Brönsted acidic sites are introduced while coking inclined Lewis acidic sites are reduced by a simple sol–gel method.

Besides, zeolites are also used in other refining processes such as hydroisomerization and hydrocracking. Coincidentally, facing the same thorny problems of diffusion and effective contact of bigger hydrocarbon molecules on active sites in microporous zeolites, an innovation is to improve the accessibility by preparing micro-/mesoporous composite zeolites. From the laboratory of China University of Petroleum, efforts directed to the reconstruction and/or post-treatment of commercial zeolites for hierarchical zeolites SAPO-11 and USY are described, respectively, contributed by H. Song et al. and X. Li, et al.

About 75–80 % of gasoline and 30–40 % of diesel comes from FCC products in China. Therefore, it has enormous economic and social benefits to convert heavier crude oils into high value petroleum products. One of the greatest progresses in the past decade in China is the development and intensively industrial applications of two-stage riser (TSR) FCC process, which can be operated in different modes for market demands. The paper "Multifunctional two-stage riser fluid catalytic cracking process" describes the TSR FCC process for maximizing light oil and its development for increasing diesel yield, enhancing the conversion of heavy oil, and reducing the olefin content of gasoline. In this special issue, another innovation for FCC unit is described by Dr. R. Partha, who concentrated in his scientific paper on "HS-FCC™ high-severity fluidized catalytic cracking: A newcomer to the FCC family" that is considered a breakthrough technology in the refining and petrochemical industry. HS-FCC produces four times more light olefins with minimum gasoline loss and produces high-octane gasoline as well. Full-scale commercial unit study for 30,000 BPSD HS-FCC plant was completed in 2012.

In China, the critical requirements for community health and pollution control and prevention have gained national attention in the light of increasingly serious air pollution caused by the use of fossil fuel. Sulfur abatement and sulfur managements have been a critical issue of laboratory and refinery for several years. In this special issue, the paper on "Additives for in situ reduction of sulfur in fluid catalytic cracking (FCC) gasoline" elaborates the FCC additives prepared by impregnation and gives examples of such additives and their expected functions role in the in reducing sulfur in gasoline catalytic cracking processes. Another paper titled "Development of hydro-treating catalysts with improved performance" underlines the major improvements of the catalysts for deep HDS of gas oils derived from heavy crudes, indicative of overwhelming influence of refractive sulfur species on the overall HDS. Besides, another innovative work done by J. Zhao et al. is the bifunctional catalysts for upgrading FCC gasoline by coupling reactive adsorption desulfurization and aromatization of olefins (RADS-Ar).

It is known that industrial FCC plants always run with high operating costs and at the fast-fluidized bed regime, concerning intricate two-phase or three-phase flow behaviors and complicated catalytic cracking reactions. Thereby successful modeling simulation relying on the advanced computer makes it possible to estimate sophisticated flow reactions with higher accuracy, by which preliminary test costs might be greatly reduced in the foreseeable future. Paper titled "An Integrated Methodology for the Modeling of FCC Riser Reactor." by Y. Du described a novel methodology for modeling FCC riser reactor named equivalent reactor network model (ERN).

It is hoped that readers will find the papers in this special issue of interest, usefulness and stimulation in their relevant research. We appreciate all authors and referees for their scientific contributions to the publication of this special issue. Last but not least, we would like to express our gratitude to the editors of Applied Petrochemicals Research and the journal manager for their strong support and wonderful assistance.

13

Study of feed temperature effects on performance of a domestic industrial PSA plant

Ehsan Javadi Shokroo · Mohammad Shahcheraghi ·
Mehdi Farniaei

Abstract The Parsian N_2-PSA industrial plant, situated in the southern pars zone of Iran, was studied numerically by mathematical modeling and numerical simulation. The model coupled PDEs are solved using fourth order Runge–Kutta scheme. In this work, we are dealing with the feed temperature and investigating its effect on the N_2 purity and recovery, which is known as an operating variable. Finally, the results of simulations showed that the feed temperature near to 25 °C is well suited to N_2 production with respect to its purity and recovery. In addition, as the feed temperature increases N_2 productivity decreases.

Keywords Pressure swing adsorption · Industrial plant · N_2 production · Numerical simulation · Carbon molecular sieve

Nomenclature

A_w	Wall cross-sectional area (cm^2)
AD	Adsorption step
B	L-F parameter (atm^{-1})
BD	Blow down step
cp_g	Gas heat capacities (cal/g.K)
cp_s	Pellet heat capacities (cal/g K)
cp_w	Wall heat capacities (cal/g K)
D_L	Axial dispersion coefficient (cm^2/s)
ED	Equalization to depressurization step
EP	Equalization to pressurization step
h_i	Internal heat-transfer coefficient (cal/cm^2 K s)
h_o	External heat-transfer coefficient (cal/cm^2 K s)
ΔH	Average heat of adsorption (cal/mol)
ID	Idle step
K	Parameter for the LDF model
K_L	Axial thermal conductivity (cal/cm s K)
L	Bed length (cm)
P	Total pressure (atm)
PP	Providing purge step
Pr	Reduced pressure (—)
PG	Purge step
PR	Pressurization step
q	Amount adsorbed (mol/g)
q^*	Equilibrium amount adsorbed (mol/g)
q_m	Saturated amount adsorbed (mol/g)
R	Gas constant (cal/mol K)
R_p	Radius of the pellet (cm)
R_{Bi}	Inside outside radius of the bed (cm)
R_{Bo}	Outside outside radius of the bed (cm)
t	Time (s)
T	Gas phase temperature (k)
T_{atm}	Temperature of the atmosphere (K)
T_w	Wall temperature (K)
u	Interstitial velocity (cm/s)
y_i	Mole fraction of species i in gas phase
Z	Axial distance (cm)

Greek letters

α	Particle porosity
ε	Voidage of the adsorbent bed
ε_t	Total void fraction
ρ_g	Gas density (g/cm^3)

E. J. Shokroo (✉) · M. Shahcheraghi · M. Farniaei
PART-SHIMI Company, Martyr Fahmideh Talent Foundation,
Campus of Knowledge Based Companies, 87, Mirzaye Shirazi,
71 888 41111 Shiraz, Fars, Iran
e-mail: ehsan.javadi@hotmail.com

ρ_p Pellet density (g/cm^3)
ρ_B Bulk density (g/cm^3)
ρ_w Wall density (g/cm^3)

Subscripts
B   Bed
i Component i
p   Pellet
g   Gas phase
s   Solid
w   Wall

Introduction

In general, three commercial methods are available for nitrogen purification, namely cryogenic, membrane technologies and pressure swing adsorption (PSA). PSA process is a wide operating unit for separation and purification of gases that operates based on capability of solid adsorption and selective separation of gases. The important operational parameter in this system is the pressure, and most industrial units operate at\or vicinity of the surrounding temperature. Today, the PSA process completely is known in a wide region of the processes, and this process was preferred in contrast to other conventional separation methods, especially for lower capacity and higher purity [23, 25, 33]. The nitrogen purification by pressure swing adsorption (N_2-PSA) system is well suited to rapid cycling, in contrast to other cyclic adsorption separation processes, and this has the advantage of reducing the absorbent inventory and therefore the capital costs of the system [23]. Use of PSA process to gas separation took place for the first time in 1958 by Skarstrom. He provided his recommended PSA cycles to enrich oxygen and nitrogen in air under the subject of heatless drier [30]. Thus, Skarstrom invented a two-bed PSA cycle with equalization step for oxygen production from air using zeolite 13X adsorbent in 1966 [31]. The main reasons for the success of this technology are many reforms that reached in this field and also is the new design and configuration for the cycles and devices [10, 16, 17, 29].

PSA process performance was strongly influenced by design parameters (such as: bed size, adsorbent physical properties, configuration and number of beds) and operating variables (such as: pressurization time, production time, purge time, feed flow rate, purge flow rate, production flow rate, temperature and/or pressure variations). So, this may be the maximum performance to obtain in terms of the best process variables. Thus, it is important that the behavior of the PSA operating variables was undertaking a review to know the optimum operating conditions. In recent years, use of this method was followed by researchers as a more important separation technique in air separation.

In modeling a kinetically controlled PSA process, the key requirement is an adequate representation of the mass transfer kinetics. For the systems showing kinetic selectivity, the mass transfer resistance is usually in the micropores and both linear driving force (LDF) [11], 1987; [1, 14] and micropore diffusion [7, 28] models. Shin and Knaebel [28] assumed that the diffusivities remain constant, while Farooq and Ruthven [7] allowed for the concentration dependence of the micropore diffusivity in accordance with the chemical potential gradient as the driving force. Detailed studies of diffusion in microporous adsorbents reveal that, for both zeolites [5, 26] and carbon molecular sieves [3, 15], the micropore diffusivity varies strongly with sorbate concentration.

Fernandez and Kenney [9] provided a theoretical analysis and adaptation to the experimental results for the separation of oxygen and nitrogen from air by a single bed pressure swing adsorption. They have compared the analytical approximations with the results of the experiments, and concluded that the approximately analytical solution can predict the adsorption bed dynamics for short cycle time. These authors also have reported that the exact numerical solution is an efficient method in the modeling and numerical simulation of the adsorption bed dynamics, especially in the case of long cycle time. Hassan et al. [11] proposed a simple dynamic model for the O_2-PSA process based on linear mass transfer rate relations and Langmuir equilibrium equation. These authors have assumed that the pressure remains constant during the adsorption and desorption steps. Farooq et al. [8] proposed a kinetically dynamic model for the O_2-PSA process, in which the adsorption is controlled by equilibrium instead of kinetic. Their testing system was a Skarstrom cycle [30]. The advantage of the kinetically proposed model by these researchers for the PSA systems is that the effects of mass transfer resistance and axial dispersion can easily be evaluated. The considered system by the authors is a simple two-bed PSA Skarstrom system, but utilizing the model is not limited for the multi-bed systems which are commonly used in large-scale units. Farooq & Ruthven [6] suggested a linear driving force model (LDF) assuming frozen solid concentration during pressurization and blowdown steps to numerical simulation of air separation process by adsorption method using carbon molecular sieve (CMS). A fix bed under a four-step cycle was used in the PSA experiments. The cycle steps were the same as the steps of Skarstrom fundamental cycle: (1) high-pressure production, (2) blowdown, (3) purge with product and, (4) pressurization with feed. The authors have concluded that a good agreement is between the simple LDF model and the experimental results and, also the LDF model has a high compatibility with the more complicated pore-diffusion model. Farooq & Ruthven [24] modeled the PSA process dynamic behavior for the nitrogen recovery using zeolite 5A, zeolite 13X, and CMS using the theoretical and experimental investigation. The LDF and pore-diffusion models were also used to compare the rate of mass transfer in the adsorption beds. The experiments were performed in a dual-bed system with two configurations: (1) the Skarstrom cycle and (2) the Self-purging modified cycle. The comparison between simulation results and experiments showed that the simple LDF model can predict the effect of process variables, but the complicated pore-diffusion model indicated a better adaptation to the experimental results. Farooq & Ruthven [7] suggested a pore-diffusion model for modeling the bulk two-component gas separation process using a PSA system based on Langmuir equilibrium and also considering concentration dependency for diffusion coefficient. The PSA system used by the authors was investigated in two configurations: (1) Skarstrom cycle and modified Skarstrom cycle with pressure equalization step and self-purging. The concentration dependency for diffusion coefficient of micropore was also examined. The concentration dependency has a large effect on cyclic steady state (CSS) performance. Budner et al. [2] improved a thermal non-equilibrium model for the multicomponent adsorption process. They also have designed a software for the computation and simulation air separation process based on vacuum swing adsorption (VSA) technique using molecular sieve zeolite adsorbent. They have claimed that the developed and studied mathematical model using its software is able to design and optimize the VSA units for oxygen production. Mendes et al. [18] investigated a PSA system using zeolite 5A and Skarstrom cycle through simulations and experiments. They showed that the pressure increasing during pressurization step causes increasing dispersion in the PSA beds and finally results in decrease of product purity and recovery. Time reduction of depressurization in blowdown step down to 4 s does not affect on product purity and recovery. Taking the pressure equalization step in the Skarstrom cycle causes improvement of product purity and recovery. Mendes et al. [19] examined the PSA unit for oxygen separation from air using zeolite 5A through simulations and experiments. The mentioned PSA system was studied in two configurations as Skarstrom cycle and Skarstrom cycle with pressure equalization step. They showed that, as the production flow rate increases the product purity decreases; as the production step time increases the product purity reduces while its recovery increases. Shin et al. [27] found the optimal conditions for oxygen recovery and productivity using a two-bed PSA system with incomplete equalization step. According to their findings, use of the incomplete equalization step for PSA cycle is the leading cause of

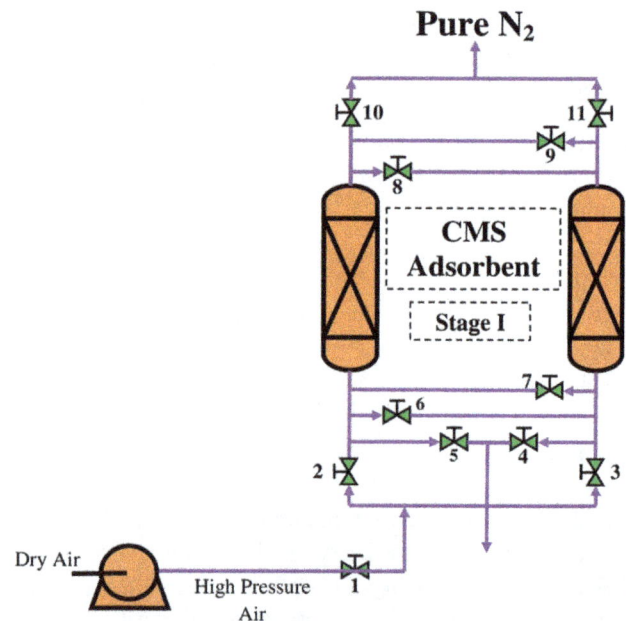

Fig. 1 Schematic diagram of Parsian N₂-PSA industrial plant [22]

productivity improvement in the case of feed pressurization. The maximum productivity occurs using the incomplete equalization step, but its level is less than that for the case of feed pressurization. On the other hand, the maximum recovery always is obtained using complete equalization step. Cruz et al. [4] proposed a heuristic method to optimizing cyclic adsorption separation processes (PSA/VSA), in which the system acts based on Skarstrom cycle with pressure equalization step. They studied the various configurations for equalization step, and they found that bottom-to-bottom configuration is just efficient when adsorbent productivity is low and the Skarstrom cycle is used. The top-to-top configuration always is the most efficient in terms of product recovery and power consumption. Moghadaszadeh et al. [21] investigated different process variables such as production recovery, cycle time, production and purge flow rates on oxygen purity. They examined a four-bed PSA pilot consisting of seven steps to oxygen separation from air using zeolite 13X. These authors showed that the increasing purge flow rate maximized the product purity. Also, increasing cycle time is the leading cause of power consumption decrease, while power requirement will be increased. Mofarahi et al. [20] investigated a four-bed PSA consisting seven steps to oxygen separation from air using zeolite 5A. They reported that as long as the adsorption pressure increases from 4 to 6 bar, the oxygen purity and recovery increase. The PSA unit performance is improved at higher cycle times. When production flow rate increases oxygen recovery increases, while oxygen purity will be reduced.

The Parsian N₂-PSA industrial plant, a two-bed PSA process, operates as modified Skarstrom cycle [23, 33].

Table 1 Model equations [23, 25, 33]

$$\frac{\partial C_i}{\partial t} - D_L \frac{\partial^2 C_i}{\partial z^2} + \frac{\partial (C_i.u)}{\partial z} + \rho_p.\left(\frac{1-\varepsilon}{\varepsilon}\right).\frac{\partial \bar{q}_i}{\partial t} = 0 \tag{1}$$

$$\frac{\partial C}{\partial t} - D_L \frac{\partial^2 C}{\partial z^2} + \frac{\partial (C.u)}{\partial z} + \rho_p.\left(\frac{1-\varepsilon}{\varepsilon}\right).\sum_{i=1}^{N} \frac{\partial \bar{q}_i}{\partial t} = 0 \tag{2}$$

$$-K_l \frac{\partial^2 T}{\partial z^2} + \varepsilon.\rho_g.c_{p,g}.\left(u\frac{\partial T}{\partial z} + T\frac{\partial u}{\partial z}\right) + \left(\varepsilon_t.\rho_g.c_{p,g} + \rho_B.c_{p,s}\right).\frac{\partial T}{\partial t} - \rho_B.\sum_{i=1}^{N}\left(\frac{\partial \bar{q}_i}{\partial t}.(-\Delta\bar{H}_i)\right) + \frac{2h_i}{R_{B,i}}(T - T_w) = 0 \tag{3}$$

$$\rho_w.c_{p,w}.A_w\frac{\partial T_w}{\partial t} = 2\pi R_{B,i}h_i(T - T_w) - 2\pi R_{B,o}h_o(T_w - T_{atm}); A_w = \pi\left(R_{B,o}^2 - R_{B,i}^2\right) \tag{4}$$

$$P(t) = a.t^2 + b.t + c \tag{5}$$

$$-\frac{dP}{dz} = a.\mu.u + b.\rho.u.|u|; a = \frac{150}{4R_p^2}.\frac{(1-\varepsilon)^2}{\varepsilon^2}; b = 1.75\frac{(1-\varepsilon)}{2R_p\varepsilon} \tag{6}$$

$$q_i = \frac{q_{m,i}.B_i.P_i}{1+\sum\limits_{j=1}^{s} B_j.P_j}; q_{m,i} = k_1 + k_2.T; B_i = k_3.\exp(k_4/T); n_i = k_5 + k_6/T \tag{7}$$

$$\frac{\partial \bar{q}_i}{\partial t} = \omega_i.(\dot{q}_i - \bar{q}_i); \omega_i = \frac{15D_{ei}}{r_c^2}; \frac{15D_{ei}}{r_c^2} = C_i.P_r^{0.5}.(1 + B_i.P_i)^2 \tag{8}$$

Figure 1 shows a schematic diagram of the process. The six-step process used is as follows: (1) co-current feed pressurization (PR) of a partially pressurized bed by a previous pressurizing pressure equalization step (EP); (2) high-pressure adsorption (AD) step; (3) counter-current depressurizing pressure equalization (ED) step; (4) counter-current blow down (BD) step; (5) counter-current purge with a light product (PG) step; (6) co-current EP step.

In this work, the Parsian N_2-PSA industrial plant is studied by mathematical modeling and numerical simulation. Furthermore, the effect of feed temperature on the plant performance (N_2 recovery and productivity) is investigated.

Mathematical model

To develop a mathematical model for a PSA system, the below main assumptions were given:

a. Gas behaves an ideal gas.
b. The flow pattern is described by the axially dispersed plug-flow model.
c. Adsorbing properties throughout the tower would remain constant and unchanged.
d. Radial gradient is to be negligible.
e. Equilibrium equations for the components of the feed (N_2, O_2, Ar) can be expressed by three-component Langmuir–Freundlich isotherm.
f. Mass transfer rate is expressed by a linear driving force equation.
g. Thermal equilibrium between gas and solid phases is assumed.
h. Pressure drop along the bed is calculated by the Ergun's equation.

The model equations for the bulk phase in the adsorption bed are written in Table 1 [12, 13, 32].

For the coupled PDEs problem, the well-known Danckwerts boundary conditions are applied [23, 25, 33]. The adsorption isotherm parameters and diffusion rate constant of N_2, O_2 and Ar over CMS are shown in Table 2. In Table 3, the characteristics of adsorbent and adsorption bed are shown.

Results and discussion

The fourth order Runge–Kutta scheme (for the time derivatives) and the implicit finite difference scheme (for the space derivatives) were used to solve the mathematical model that consisted of coupled partial differential equations.

The plant operating conditions data supplied by Parsian Co. can be found in Table 4 [22]. In this simulation study, the feed (air) components were assumed to be N_2, O_2 and Ar. To validate the simulation results, the results of this work first were compared with the plant data. Figure 2 shows the simulated N_2 purity as a function of feed flow rate, with N_2-PSA industrial plant data. This figure also shows as the feed flow rate increases the nitrogen purity decreases, which are in correspondence with other PSA simulations results [23, 25, 33]. In addition, as obvious in this figure, the simulation and presented model in this work predict the results of the plant data with a fairly high accuracy. The effect of feed temperature on N_2 purity and recovery is indicated in Fig. 3. It can be seen from this figure that the increasing of the feed temperature is the leading cause of a reduction in the N_2 purity and an increment in the N_2 recovery. The adsorption is an inherently exothermic phenomenal, so the decreasing of

Table 2 Equilibrium\rate parameters and heat of adsorption of N_2, O_2 and Ar on CMS [12]

Parameter	Component		
	N_2	O_2	Ar
$k_1 \times 10^3$ (mol/g)	23.63	15.27	20.42
$k_2 \times 10^3$ (mol/g k)	−0.0638	−0.00323	−0.00530
$k_3 \times 10^4$ (1/atm)	361	22.9	239.7
k_4 (k)	1,444	9,66.1	324.6
k_5 (−)	1.692	1.187	1.646
k_6 (k)	−270	−106	−238.2
Heat of adsorption, $-\Delta H_i$ (cal/mol)	3,197.532	3,297.828	3,398.124

Table 3 The characteristics of adsorbent and adsorption bed

Adsorbent [12]		Adsorption bed [22]	
Adsorbent	CMS	Length L (cm)	245
Type	Sphere	Inside radius, $R_{B,i}$ (cm)	80
Micropore diameter, R_p (cm)	3 A	Outside radius, $R_{B,o}$ (cm)	80.50
Particle density, ρ_p (g/cm³)	0.9	Heat capacity of column, Cp_w (cal/g k)	0.12
Bulk density, ρ_B (g/cm³)	0.633	Density of column, ρ_w (g/cm³)	7.83
Heat capacity, Cp_s (cal/g k)	0.22	Internal heat-transfer coefficient, h_i (cal/cm² k s)	9.2×10^{-4}
Particle porosity, α	0.30	External heat-transfer coefficient, h_o (cal/cm² k s)	3.4×10^{-4}

Table 4 Industrial N_2-PSA operating condition [22]

Composition, mol (%)	Feed	Product	Tail
N_2	79.00	99.99	N/A
O_2	20.00	0.01	N/A
Ar	1.00	N/A	N/A
Pressure (bar)	7.00	6.50	1.00
Temperature (°C)	35.00	40.00	25.00
Flow rate (N m³/h)	319.800	N/A	N/A

the N_2 purity with the feed temperature has occurred in the normal way. In other hand, the incremental behavior of the N_2 recovery through the increasing of the feed temperature is caused by gas molar compression in the adsorption (AD) step, which means that the N_2 volume increases in the product stream. This figure shows that the feed temperature near to 25 °C is well suited to N_2 production in terms of its purity and recovery. Figure 4 depicts as the feed temperature increases N_2 productivity decreases. Productivity is defined as the ratio of moles of N_2 in the product stream to the kg of the adsorbent per cycle time. The cyclic partial concentration of N_2 at the top of the bed during a whole cycle is shown in Fig. 5. It is evident also from this figure that the process

performance (N_2 purity) is modified to correspond to lower feed temperature (25 °C) because of the cyclic concentration of N_2 placed at an upper level. In other words, it can reach the high pure N_2 (as the current purity in the plant) using higher feed flow rates when the feed temperature is near to 25 °C, which causes the increasing of N_2 recovery.

Conclusions

The Parsian N_2-PSA, located in the southern pars zone of Iran, has been simulated. The effect of feed temperature on the process performance is studied by a mathematical

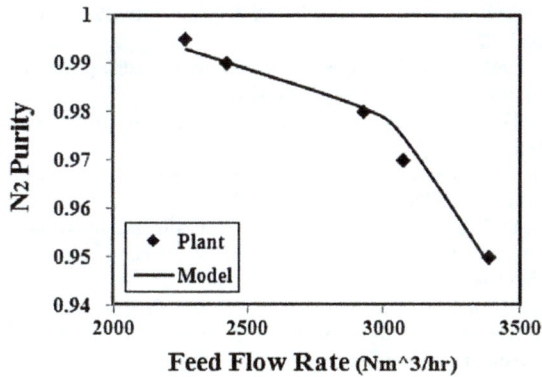

Fig. 2 Simulated purity and plant data of N_2-PSA industrial process as a function of feed flow rate

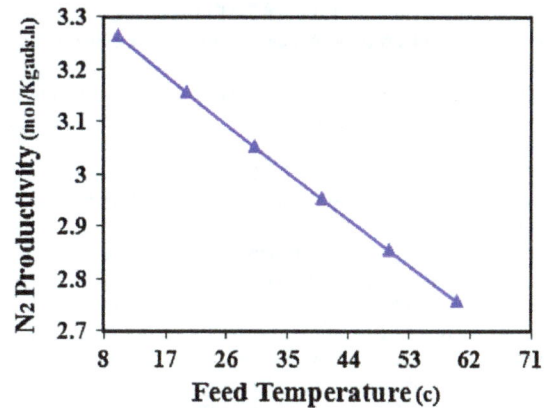

Fig. 4 Effect of feed temperature on the N_2 productivity

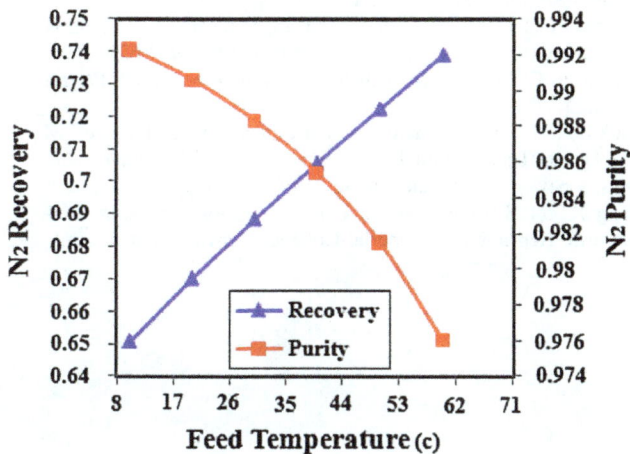

Fig. 3 N_2 purity and recovery as a function of feed temperature

that the unit performance (in terms of N_2 purity and recovery) is in may well conditions when the feed temperature is near to 25 °C. Furthermore, as the feed temperature increases N_2 productivity decreases.

Acknowledgments The authors gratefully acknowledge the Parsian Gas Refining Co. for agreement with this research.

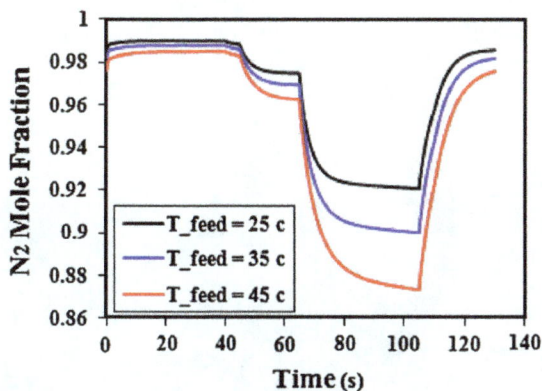

Fig. 5 Cyclic partial concentration of N_2 during a whole cycle at the top of the bed

modeling numerical simulation. The mathematical model in the gas phase takes into account the balances in energy, mass and momentum and the coupled PDEs are solved using fourth order Runge–Kutta scheme. Results showed

References

1. Ackley MW, Yang RT (1990) Kinetic separation by pressure swing adsorption: method of characteristics. A1ChE J 36:1229–1238
2. Budner Z, Dula J, Podstawa W, Gawdzik A (1999) Study and modeling of the Vacuum Swing Adsorption (VSA) process employed in the production of oxygen. Ins Chem Eng, 77, Part A, 405–412
3. Chihara K, Suzuki M, Kawazoe K (1978) Concentration dependence of micropore diffusivities-diffusion of propylene in molecular sieving carbon 5A. J Chem Eng Japan 11:153–155
4. Cruz P, Magalhaes FD, Mendes A (2005) On the optimization of cyclic adsorption separation processes. AIChE J 51:1377–1395
5. Doetsch IH, Ruthven DM, Loughlin KF (1974) Sorption and diffusion of n-Heptane in 5A Zeolite. Can J Chem 52:2717–2724
6. Farooq S, Ruthven DM (1990) A comparison of linear driving force and pore diffusion models for a pressure swing adsorption bulk separation process. Chem Eng Sci 45:107–115
7. Farooq S, Ruthven DM (1991) Numerical simulation of a kinetically controlled pressure swing adsorption bulk separation process based on a diffusion model. Chem Eng Sci 46:2213–2224
8. Farooq S, Ruthven DM, Boniface HA (1989) Numerical simulation of a pressure swing adsorption oxygen unit. Chem Eng Sci 44(12):2809–2816
9. Fernandez G, Kenney CN (1983) Modeling of the pressure swing air separation process. Chem Eng Sci 38(6):827–834
10. Fuderer A, Rudelstorfer E (1976) Selective adsorption process, US Patent 3,986,849

11. Hassan MM, Raghavan NS, Ruthven DM (1986) Air Separation by pressure swing adsorption on a carbon molecular sieve. Chem Eng Sci 41:1333–1343

12. Jee JG, Kim MB, Lee CH (2005) Pressure swing adsorption to purify oxygen using carbon molecular sieve. Chem Eng Sci 60:869–882

13. Jee JG, Park HJ, Haam SJ, Lee CH (2002) Effects of nonisobaric and isobaric steps on O_2 pressure swing adsorption for an aerator. Ind Eng Chem Res 41:4383

14. Kapoor A, Yang RT (1989) Kinetic separation of methane-carbon dioxide mixture by adsorption on molecular sieve carbon. Chem Eng Sci 44:1723–1733

15. Kawazoc K, Suzuki M, Chihara K (1974) Chromatographic study of diffusion in molecular-sieving carbon. J Them Eng Jpn 7:151–157

16. Malek A, Farooq S (1985) Hydrogen purification from refinery fuel gas by pressure swing adsorption. AIChE J 44:1985

17. Malek A, Farooq S (1997) Study of a six-bed pressure swing adsorption process. AIChE J 43:2509

18. Mendes AMM, Costa CAV, Rodrigues AE (2000) Analysis of nonisobaric steps in nonlinear bicomponent pressure swing adsorption systems: application to air separation. Ind Eng Chem Res 39:138–145

19. Mendes AMM, Costa CAV, Rodrigues AE (2001) Oxygen separation from air by PSA: modeling and experimental results part I: isothermal operation. Sep Purif Tech 24:173–188

20. Mofarahi M, Towfighi J, Fathi L (2009) Oxygen separation from air by four-bed pressure swing adsorption. Ind Eng Chem Res 48:5439–5444

21. Moghadaszadeh Z, Towfighi J, Mofarahi M (2008) Study of a four-bed pressure swing adsorption for oxygen separation from air. Int J Chem Bio Eng 1(3):140–144

22. Parsian GAS Refinement Co., N2-PSA plant, http://www.nigc-parsian.ir/

23. Ruthven DM (1984) Principle of adsorption and adsorption processes. John Wiley & Sons Inc, New York

24. Ruthven DM, Farooq S (1990) Air separation by pressure swing adsorption. Gas Sep Purif 4:141–148

25. Ruthven DM, Farooq S, Knaebel KS (1994) Pressure swing adsorption. VCH Publishers Inc, New York

26. Ruthven DM, Loughlin KF (1971) The sorption and diffusion of n-butane in linde 5a molecular sieve. Chem Eng Sci 26:1145–1154

27. Shin HS, Kim DH, Koo KK, Lee TS (2003) Performance of a two-bed pressure swing adsorption process with incomplete pressure equalization. Adsorption 6:233–240

28. Shin HS, Knaebel KS (1988) An experimental study of diffusion-induced separation of gas mixtures by pressure swing adsorption. AIChE J 34:1409–1416

29. Sircar S, Golden TC (2000) Purification of hydrogen by pressure swing adsorption. Sep Sci Technol 35:667

30. Skarstrom CW (1960) Method and apparatus for fractionating gaseous mixture by adsorption, US Patent, No. 2,944,627

31. Skarstrom CW (1966) Oxygen concentration process, US Patent, No. 3,237,377

32. Shokroo EJ, Shahcheraghi M, Farrokhizadeh A, Farniaei M (2014) The Iranian Jam Petrochemical's H_2-PSA enhancement using a new steps sequence table. Pet Coal 56(1):13–18

33. Yang RT (1987) Gas separation by adsorption processes. Butterworth, Reprinted by Imperial College Press, London

14

Esterification of cooking oil for biodiesel production using composites $Cs_{2.5}H_{0.5}PW_{12}O_{40}$/ionic liquids catalysts

Jianxiang Wu · Yilong Gao · Wei Zhang · Yueyue Tan · Aomin Tang · Yong Men · Bohejin Tang

Abstract Here, two ionic liquids 1-methyl-3-propane sulfonic-imidazolium (PSMIM) and 1-methyl-3-propane sulfonic-imidazolium hydrosulfate (PSMIMHSO$_4$) are synthesized, and these ionic liquids mixed with different of heteropolyacid $Cs_{2.5}H_{0.5}PW_{12}O_{40}$ have been used as catalysts for esterification of cooking oil for preparation of biodiesel. Then those catalysts are characterized by Infrared spectrometer, X-ray diffractometer, nuclear magnetic resonance, elemental analyses and high-performance liquid chromatography. PSMIM and PSMIMHSO$_4$ mixed with $Cs_{2.5}H_{0.5}PW_{12}O_{40}$ at the mass ratio of 1:1 are able to effectively catalyze esterification, using cooking oil as starting material at ratio of 1:20 (catalyst/cooking oil) and cooking oil to methanol at mass ratio of 1:6 for preparation of biodiesel with 3.5 h at 343 K. The result showed that 97.1 % yield of biodiesel could be obtained at optimized operation using PSMIMHSO$_4$ mixed with $Cs_{2.5}H_{0.5}PW_{12}O_{40}$ at the mass ratio of 1:1 as catalyst.

Keywords $Cs_{2.5}H_{0.5}PW_{12}O_{40}$ · Ionic liquid · Biodiesel · Acidic catalyst

Introduction

In the past decades, the global energy shortage and environment deterioration continue to grow rapidly. The quality of life today is dependent upon access to a bountiful supply of affordable and low-cost energy. For a sustainable development, the energy should be derived from non-fossil sources, which are ideally reliable and safe, affordable, and limitless. Therefore, development of clean alternative fuels and renewable energy has become an important subject for world-wide researchers. Because of its environmentally friendliness, renewability and other advantages, biodiesel is expected to replace traditional diesel fuel to meet the needs of sustainable development of society [1, 2]. At present, biodiesel production mainly uses a variety of oils and lower alcohol as raw materials, and acid and alkalinity can be used as catalyst [3] such as homogeneous acid or base, lipase, or heterogeneous acid or base catalysts. Conventionally, this reaction is carried out using homogeneous acid or base catalysts, such as sulfonic acid, potassium hydroxide, sodium hydroxide, or their alkoxides [4–8]. However, these catalysts are corrosive and are not easily recovered, leading to the release of environmentally unfriendly effluents, which inevitably leads to a series of environmental problems. Because of the detrimental effects of these catalysts, great efforts directed toward the development of environmentally friendly catalysts have been made. The use of lipases as biocatalysts for biodiesel production has been of great interest due to its environmental friendliness [9–12]. However, some alcohols such as methanol deactivate the lipase to some extent and the enzyme stability was poor. Moreover, glycerol, which is a byproduct, easily adsorbs on the surface of the lipase and inhibits the enzyme activity. Although heterogeneous acid or base catalysts are environmentally friendly, they also have drawbacks, such as low catalytic activity and deactivation [3, 13–15]. Due to the disadvantages of equipment corrosion and separation difficulty, concentrated sulfuric acid is much maligned in the esterification stage [16–20].

Keggin heteropoly acids (HPAs) have been reported to have the potential for biodiesel synthesis [14, 21–27]. Solid

J. Wu · Y. Gao · W. Zhang · Y. Tan · A. Tang · Y. Men · B. Tang (✉)
College of Chemistry and Chemical Engineering, Shanghai University of Engineering Science, Shanghai 201620, China
e-mail: tangbohejin@sues.edu.cn

super acid $Cs_{2.5}H_{0.5}PW_{12}O_{40}$ is not only BrÖnsted acid, but also Lewis acid, which has good catalytic activity, selectivity, and corrosion resistance [28, 29]. The activity of $Cs_{2.5}H_{0.5}PW_{12}O_{40}$ relative to the other catalysts became higher than conventional acids such as H_2SO_4, Al_2O_3–SiO_2, zeolites and acidic resins [15, 30–32]. $Cs_{2.5}H_{0.5}PW_{12}O_{40}$ is a widely applicable water-tolerant solid acid. As an important green solvent, acidic ionic liquids (ILs) can dissolve a variety of solute and favor the dispersion of catalyst in catalytic phase transfer reaction. Ionic liquids have attracted much interest as relatively clean and promising catalysts and alternative solvents that possess important attributes, such as wide liquid range, negligible vapor pressure, high catalytic activity, excellent chemical and thermal stability, potential recoverability, design possibilities, and ease of separation of the products from reactants [33–37]. Meanwhile acidic ionic liquid could also improve the acid catalytic performance [38]. Due to the booming demand of biodiesel and the limitations of traditional method, a novel method for the synthesis of biodiesel has been highly desirable. The method of mixing phosphotungstic salt with ionic liquid presents the remarkable advantage over the high yield of esterification.

In this study, 1-methyl-3-propane sulfonic-imidazolium (PSMIM), 1-methyl-3-propane sulfonic-imidazolium hydrosulfate (PSMIMHSO$_4$) and $Cs_{2.5}H_{0.5}PW_{12}O_{40}$ were prepared. Performances of two ionic liquids mixed with $Cs_{2.5}H_{0.5}PW_{12}O_{40}$ were, respectively, evaluated as catalysts in the esterification. Catalysts were characterized by nuclear magnetic resonance (NMR), infrared (IR) spectroscopy, elemental analyses, X-ray diffractometer (XRD) and high-performance liquid chromatography (HPLC). The effects of varying reaction conditions on the production of biodiesel were extensively studied and compared to different catalysts.

Experimental

Preparation of PSMIM

1,3-propyl sulfonic acid (40.0 g) was completely dissolved in 300 mL toluene. N-methyl imidazole (27.0 g) was slowly added while being stirred under an ice bath. Then the temperature was slowly raised to 298 K and maintained for 2 h. The product was then filtered from the mixture liquid and washed with diethyl ether and ether acetate for three times, respectively. The solid was obtained after heating at 343 K for 5 h [38].

Preparation of PSMIMHSO$_4$

PSMIM (23.3 g) was dissolved in 100 mL deionized water. Then sulfuric acid (10.9 g) was slowly added while being stirred at room temperature. After that, the temperature was raised to

363 K and then maintained for 2 h. The deionized water was removed from the product using a rotary evaporator [38].

Preparation of $Cs_{2.5}H_{0.5}PW_{12}O_{40}$

An aqueous solution of Cs_2CO_3 (0.3528 g) was added drop-wise to an $H_3PW_{12}O_{40}\cdot19H_2O$ (2.8000 g) solution under vigorous stirring. The fine suspension was held at room temperature overnight and subsequently to dryness at 323 K. The ratio polyacid–cesium carbonate was regulated in such a way that the final stoichiometry corresponded to $Cs_{2.5}H_{0.5}PW_{12}O_{40}$ [39].

Preparation of mixture catalysts

PSMIM was mixed with $Cs_{2.5}H_{0.5}PW_{12}O_{40}$ with deionized water as catalyst at room temperature and dried at 353 K, while mixing at the mass ratio of 1:1 named PSC. Meanwhile, PSMIMHSO$_4$ was mixed with $Cs_{2.5}H_{0.5}PW_{12}O_{40}$ with the same method, while mixing at the mass ratio of 1:1 named PSSC.

Preparation of biodiesel

Cooking oil (10 g), methanol and the catalyst [40, 41] were taken into a flask with stirring and refluxing. Then the catalyst was separated from flask by centrifuge. Subsequently, methanol (5.0 g) and potassium hydroxide (0.25 g) were also added into the flask with stirring and reflux for 0.5 h. The methanol was removed from the product using a rotary evaporator at 313 K, followed by cooling to room temperature. After that, the supernatant was separated from product by centrifuge and quickly subjected for HPLC analysis,

Characterization and evaluation of the catalyst

Infrared spectrometer (AVATAR 370 from Thermo Nicolet) was used to analyze the ionic liquid and heteropoly acid and its salt at 500–4,000 cm^{-1}.

X-ray Diffractometer (Deutschland RUKER D2 PHASER) was scanned with CuKα in the range of 10°–80° at a rate of 0.02°s^{-1}.

Nuclear magnetic resonance (NMR) was used to characterize the ionic liquid.

PSMIM^{1}HNMR (400 MHz, D$_2$O): δ2.138 (m, 2H, $J = 7.2$ Hz), 2.822 (t, 2H, $J = 7.2$ Hz), 3.806 (s, 3H), 4.270 (t, 2H, $J = 7.2$ Hz), 7.365 (s, 1H), 7.437(s, 1H), 8.655 (s, 1H). ^{13}CNMR (400 MHz, D$_2$O): δ25.11, 35.73, 47.24, 47.72, 122.16, 123.76, 136.17.

PSMIMHSO$_4^1$HNMR (400 MHz, D$_2$O): δ2.135(t, 2H, $J = 7.2$ Hz), 2.745 (s, 2H), 3.717 (s, 3H), 4.181 (t, 2H, $J = 7.2$ Hz), 7.271 (s, 1H), 7.339 (s, 1H), 8.560 (s, 1H). ^{13}CNMR(400 MHz, D$_2$O): δ24.99, 35.65, 47.15, 47.65, 122.10, 123.71, 136.06.

Elemental analyses (EA) were performed on a Perkin-Elmer 2,400 elemental analyzer for C, H, N and S. Anal. calc. for PSMIM: C, 41.18 %; H, 5.88 %; N, 13.73 %; S, 15.69 %; found: 41.24 %; H, 5.83 %; N, 13.74 %, S, 15.65 %. Anal. calc. for PSMIMHSO$_4$: C, 27.81 %; H, 4.64 %; N, 9.27 %; S, 21.19 %. found: C,27.83 %; H, 4.60 %; N, 9.33 %; S, 21.14 %.

The purity of the ionic liquid was analyzed by HPLC (Agilent1100). The conditions of HPLC: separation column is the ODS C18. Water and methanol were used as a mobile phase, the ratio of water and methanol was 1:9 and the speed was 1.0 mL/min. The detectable wavelength of UV was 215 nm at room temperature.

Analysis

The yield was analyzed by HPLC (Agilent1100). The conditions of HPLC: separation column is the ODS C18. Acetonitrile was used as a mobile phase at a flow rate of 1.0 mL/min. The volume of each sample was 15 µL. The detectable wavelength of UV was 205 nm at room temperature.

The calculation formula is:

Yield of biodiesel = (Conversion of cooking oil) × (Selectivity of cooking oil to biodiesel formation)

Result and discussion

FT-IR of ionic liquids

Figure 1 shows FT-IR of sulfonic acid ionic liquids. The peaks at 3156, 3116 and 1575 cm^{-1} represent –N–H, –C–H and –C=N– stretching vibration of the imidazole ring,

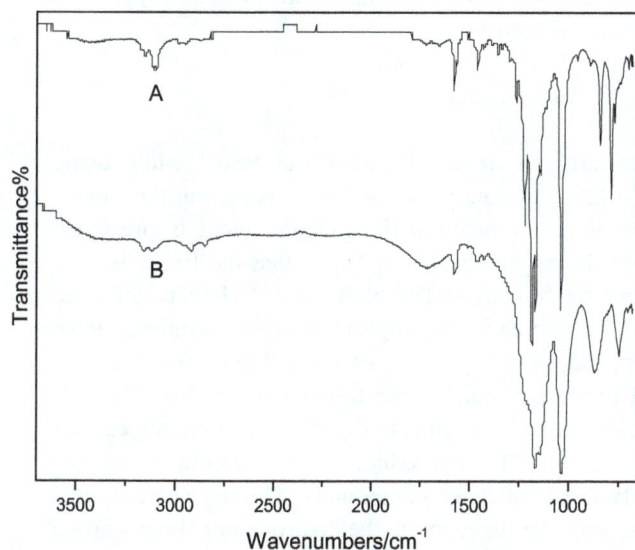

Fig. 1 FT-IR of ionic liquids *A* PSMIM, *B* PSMIMHSO$_4$

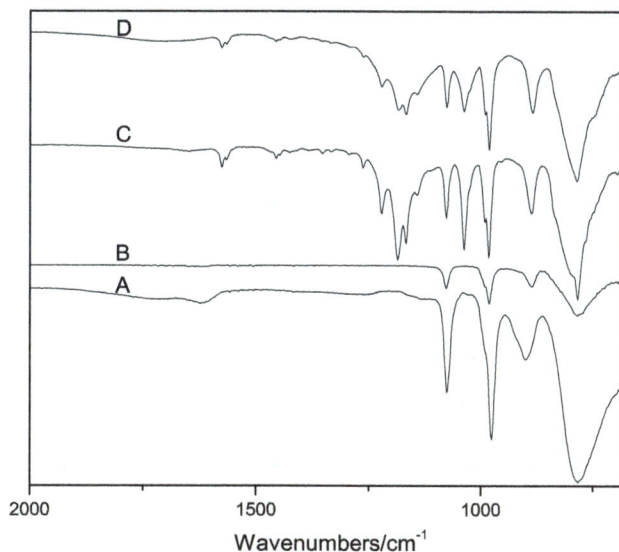

Fig. 2 FT-IR of phosphotungstic acid, its salt and compound *A* H$_3$PW$_{12}$O$_{40}$·19H$_2$O, *B* Cs$_{2.5}$H$_{0.5}$PW$_{12}$O$_{40}$, *C* PSC *D* PSSC

respectively. The peaks from 2,965 to 2,992 cm^{-1} are ascribed to the C–H stretching vibration of –CH$_2$ and N–CH$_2$, and the peaks at 1,487 and 1,398 cm^{-1} are formation vibration of CH$_2$ and CH$_3$. The peak at 749 cm^{-1} is the bending vibration of the imidazole ring. Meanwhile, a strong peak appears at 1,171 cm^{-1}, which is the S=O stretching vibration of HSO$_4$- and -SO$_3$ [38].

FT-IR of phosphotungstic acid and its salt

Figure 2 shows FT-IR of phosphotungstic acid and its salt. The peaks at 1,075, 976, 904 and 787 cm^{-1} represent H$_3$PW$_{12}$O$_{40}$·19H$_2$O. The peaks at 1078, 983, 889 and 785 cm^{-1} represent Cs$_{2.5}$H$_{0.5}$PW$_{12}$O$_{40}$. It can be seen that H$_3$PW$_{12}$O$_{40}$·19H$_2$O and Cs$_{2.5}$H$_{0.5}$PW$_{12}$O$_{40}$ keep the Keggin structure of phosphotungstic acid from FT-IR. The peak at 1,078 cm^{-1} belongs to P-O$_a$ (Oa-oxygen atoms bound to three W atoms and to P) stretching vibration of the center of tetrahedron. The peak at 983 cm^{-1} is W-O$_b$ (O$_b$-terminal oxygen atom) vibration. The peak at 889 cm^{-1} is W-O$_c$-W (O$_c$-corner sharing bridging oxygen atom) vibration. The peak at 785 cm^{-1} is W-O$_e$-W (O$_e$-edge sharing bridging oxygen atom) vibration [39, 42]. When Cs$_{2.5}$H$_{0.5}$PW$_{12}$O$_{40}$ mixed with ionic liquids, compounds also keep the Keggin structure of phosphotungstic acid and the structure of ionic liquids.

XRD of phosphotungstic acid and its salt

XRD patterns of t phosphotungstic acid and its salt are shown in Fig. 3. XRD patterns of H$_3$PW$_{12}$O$_{40}$·19H$_2$O at 8.5°, 10.6°, 20.4° and 27.9° are the characteristic peaks. XRD patterns of Cs$_{2.5}$H$_{0.5}$PW$_{12}$O$_{40}$ at 10.4°, 18.3°,

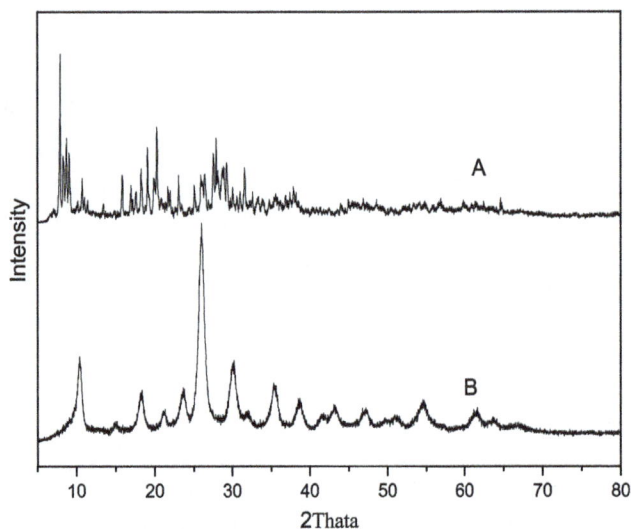

Fig. 3 XRD of phosphotungstic acid and its salt *A* H₃PW₁₂O₄₀·19H₂O, *B* Cs₂.₅H₀.₅PW₁₂O₄₀

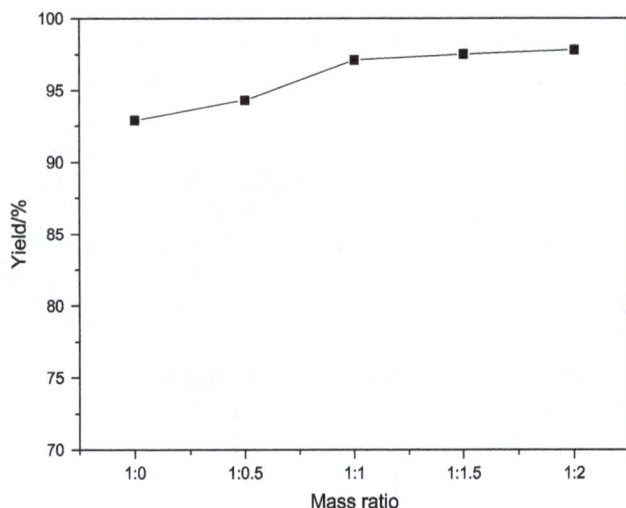

Fig. 4 Yield of biodiesel at different mass ratio at 343 K (The weigh of $Cs_{2.5}H_{0.5}PW_{12}O_{40}$ precedes over PSMIMHSO₄ in the mass ratio)

26.0°, 30.1° and 35.4° are the characteristic peaks [39, 42]. As $H_3PW_{12}O_{40} \cdot 19H_2O$ and $Cs_{2.5}H_{0.5}PW_{12}O_{40}$ remain the same, characteristic peaks of XRD patterns, it is concluded that $Cs_{2.5}H_{0.5}PW_{12}O_{40}$ keep the Keggin structure.

Effect of mass ratio of mixed catalytic system

$Cs_{2.5}H_{0.5}PW_{12}O_{40}$ was mixed with PSMIMHSO₄ at different mass ratio to catalyze the reaction at 343 K for 3.5 h.

Figure 4 shows that the yield of biodiesel rises, as the ratio of PSMIMHSO₄ is increasing. After the mass ratio of 1:1, the yield of biodiesel rises slowly. Thus, the optimized mass ratio is 1:1 in the reaction.

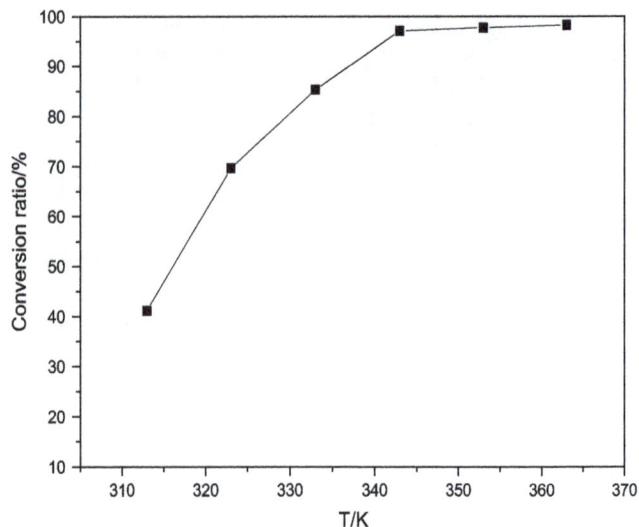

Fig. 5 Effect of reaction temperature on PSSC-catalyzed esterification

Effect of reaction temperature

The experiment of esterification was tested while using PSSC (0.5 g) as catalyst from 313 to 363 K for 3.5 h, cooking oil was 10 g and methanol was 60 g. By adjusting the temperature of water bath kettle to what it is needed.

Figure 5 shows that the yield of biodiesel significantly increases with the rising of the temperature. From Fig. 5, it is observed that the higher activity is favored at high temperature, which is likely related to that the more H⁺ released from the ionic liquid at higher temperature, thus leading to the stronger acidity, and the higher yield [38]. Meanwhile, the acidic activity of $Cs_{2.5}H_{0.5}PW_{12}O_{40}$ also is increasing while rising the temperature. After 343 K, the yield of biodiesel is nearly full and yield incremental very slow. Thus, 343 K is the temperature to obtain the satisfactory esterification yield.

Effect of reaction time

The experiment of esterification was tested while using PSSC (0.5 g) as catalyst at 343 K by adjusting the time.

The effect of reaction time on the yield is presented (Fig. 6). It can be seen from Fig. 6 that the PSSC is very efficient for biodiesel. The high yield 97.1 % is achieved after 3.5 h, which is very high for biodiesel synthesis from cooking oil [43, 44]. It is generally known that the biodiesel reaction is carried out through three steps. The triglycerides are transformed to diglycerides, monoglycerides and glycerol. The byproduct content would cause the hydrolysis of the ester compounds. The diglycerides and monoglycerides have both the hydrophobic long carbon chain and the hydrophilic hydroxyl, which can bring the

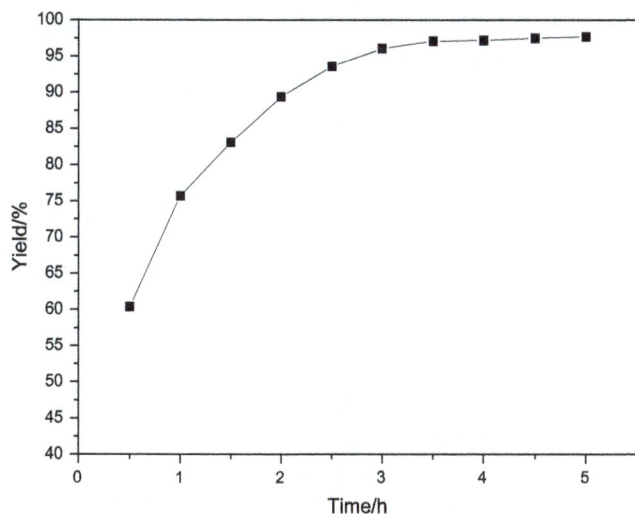

Fig. 6 Effect of reaction time

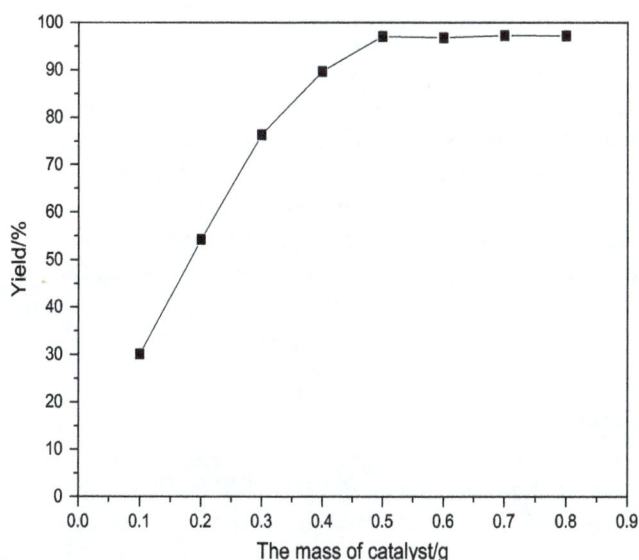

Fig. 7 Effect of the mass of catalyst

water to the acid sites. The yield varies little after 3.5 h, which remained nearly constant, indicating a nearly equilibrium yield. Therefore, the appropriate reaction time is 3.5 h.

Effect of the mass of catalyst

The experiment of esterification was tested while using PSSC as catalyst at different mass at 343 K for 3.5 h.

In Fig. 7, the influence of the mass of catalyst is presented. The optimum mass of catalyst was found to be 0.5 g in this reaction. The excessive of the catalyst did not lead to an obvious further increase in yield, indicating a nearly equilibrium yield was achieved.

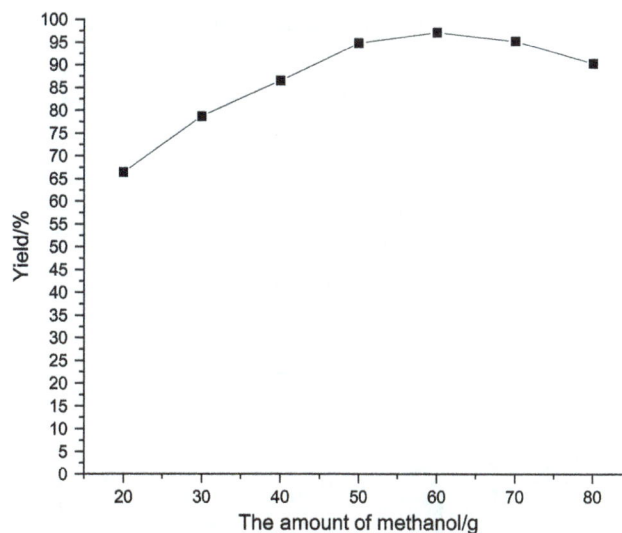

Fig. 8 Effect of the amount of methanol

Effect of the methanol amount

The experiment of esterification was tested while using PSSC as catalyst at different mass at 343 K for 3.5 h.

The mass ratio of the cooking oil and methanol is investigated (Fig. 8). The methanol amount is vital for the reaction. Since the reaction involved in biodiesel is reversible, one would expect that increasing the amount of methanol would shift the reaction equilibrium toward the products [45]. The biodiesel would produce water and the byproduct water would cause the hydrolysis of the ester, which reduced the yield. More methanol promotes the reaction equilibrium forward to obtain the high yield. On the other hand, too much methanol may cause the dilute effect and decrease yield [38]. When 60 g methanol is added, the maximum yield is reached. Therefore, the mass of methanol is optimized as 60 g.

Compared with traditional acid catalysts

By comparing the yield of PSMIM and its functionalized PSMIMHSO$_4$ from Table 1, it is found that the yield of PSMIMHSO$_4$ improves more than PSMIM. It is explained that the presence of the anionic group HSO$_4^-$ enhanced the acidity of BrÖnsted acid, which improve the activity of ionic liquid and have the positive of acid catalyst in the process of esterification. But the acidity of two ionic liquid is lower than H$_2$SO$_4$, so their yields are lower than H$_2$SO$_4$ [38]. However, the high byproduct content causes the serious hydrolysis of ester products to decrease the yield of reaction. The traditional homogeneous catalyst H$_2$SO$_4$ has relatively low activities for the production of biodiesel. Although H$_2$SO$_4$ has a strong acidity, materials and

Table 1 Yield of biodiesel in different catalytic (0.5 g) at 343 K for 3.5 h

The sample number	Component	Yield (%)
1	PSMIM	70.2
2	PSMIMHSO$_4$	73.1
3	H$_2$SO$_4$	88.9
4	H$_3$PW$_{12}$O$_{40}$·19H$_2$O	89.1
5	Cs$_{2.5}$H$_{0.5}$PW$_{12}$O$_{40}$	92.9
6	PSC	94.3
7	PSSC	97.1

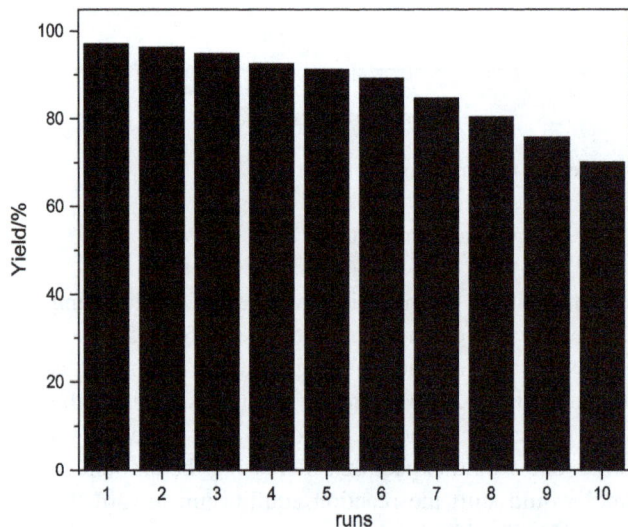

Fig. 9 Catalyst reusability at 343 K for 3.5 h

products are carbonized partly during the process, which reduced the biodiesel quality.

Per equal catalyst weight, H$_3$PW$_{12}$O$_{40}$·19H$_2$O is almost as effective as H$_2$SO$_4$ [46] and the yield of H$_3$PW$_{12}$O$_{40}$·19H$_2$O is almost the same as H$_2$SO$_4$. H$_3$PW$_{12}$O$_{40}$·19H$_2$O is more active than H$_2$SO$_4$ in the reaction and exhibits the great advantage of these H$_3$PW$_{12}$O$_{40}$·19H$_2$O as catalyst compared to the traditional catalyst in view of corrosivity. However, H$_3$PW$_{12}$O$_{40}$·19H$_2$O is soluble in methanol and difficult to separate from water or methanol. Therefore, salts of alkaline cation Cs$^+$ for H$_3$PW$_{12}$O$_{40}$·19H$_2$O are introduced in the study. The salts with large monovalent ions exhibited an impressive performance that they revealed similar activities compared to the parent acid in acid-catalyzed reactions [47]. On the other hand, Cs$_{2.5}$H$_{0.5}$PW$_{12}$O$_{40}$, which are both Brönsted acid and Lewis acid, exhibits larger surface area and higher activity than H$_3$PW$_{12}$O$_{40}$·19H$_2$O [46], and the acid activity of Cs$_{2.5}$H$_{0.5}$PW$_{12}$O$_{40}$ is about 15 times higher than H$_2$SO$_4$ [28]. That is, the yield of Cs$_{2.5}$H$_{0.5}$PW$_{12}$O$_{40}$ is higher than H$_3$PW$_{12}$O$_{40}$·19H$_2$O and H$_2$SO$_4$.

Cs$_{2.5}$H$_{0.5}$PW$_{12}$O$_{40}$ and ionic liquid are mixed to form new catalytic system to improve the yield. As a result, the

yield has been obviously improved. Because ionic liquid has a high viscosity, it could disperse Cs$_{2.5}$H$_{0.5}$PW$_{12}$O$_{40}$ uniformly so as to increase the contact area of Cs$_{2.5}$H$_{0.5}$PW$_{12}$O$_{40}$. In addition, the ionic liquid could dissolve the carbon deposition, which could probably cover the acid site of Cs$_{2.5}$H$_{0.5}$PW$_{12}$O$_{40}$. So it could be used as a phase transfer catalyst. Thus, the yield of mixed catalytic system is improved.

Catalyst reusability

Catalytic stability of Cs$_{2.5}$H$_{0.5}$PW$_{12}$O$_{40}$/PSMIMHSO$_4$, which is presented in Fig. 9, is evaluated by performing consecutive batch runs with the same catalyst sample, at same conditions. It is found that the yield of biodiesel kept unchanged, even slightly decreased with runs at the first five runs. This may be because the more acids of catalyst are formed under mechanical agitation, which increases the surface area of catalyst to enhance the catalytic property. However, the yield of biodiesel obviously decreased after the sixth run mainly due to the loss of acids in the process of taking sample or separating product. Nevertheless, about 70.1 % yield can still be obtained even when esterification reaction of the 10th run is finished. Therefore, the advantage of esterification using Cs$_{2.5}$H$_{0.5}$PW$_{12}$O$_{40}$/PSMIMHSO$_4$ as phase transfer catalyst is obvious.

Conclusions

In the present investigation, PSMIM, PSMIMHSO$_4$ and Cs$_{2.5}$H$_{0.5}$PW$_{12}$O$_{40}$ were successfully prepared. Then two ionic liquids mixed with Cs$_{2.5}$H$_{0.5}$PW$_{12}$O$_{40}$ were used as the phase transfer catalysts for biodiesel production. And those were characterization by NMR, IR, EA, XRD and HPLC. PSSC was used as a catalyst to test the optimum conditions, which was the mass ratio of 1:1(PSMIMHSO$_4$/ Cs$_{2.5}$H$_{0.5}$PW$_{12}$O$_{40}$). The best reaction temperature was determined at 343 K, the best reaction time was 3.5 h, the optimum mass of catalyst was 0.5 g and the optimum amount of methanol was 60 g. Meanwhile, through the comparison for the catalytic activity of PSMIM, PSMIMHSO$_4$, H$_2$SO$_4$, H$_3$PW$_{12}$O$_{40}$·19H$_2$O and Cs$_{2.5}$H$_{0.5}$PW$_{12}$O$_{40}$, the acid activity of Cs$_{2.5}$H$_{0.5}$PW$_{12}$O$_{40}$ was higher than others under the same conditions. After mixing, the yield was improved, so the acid activity of the mixed catalytic system was higher than any others. In addition, because of the presence of anionic group HSO$_4^-$, the acid activity of Cs$_{2.5}$H$_{0.5}$PW$_{12}$O$_{40}$ mixed with PSMIMHSO$_4$ was higher than the acid activity of Cs$_{2.5}$H$_{0.5}$PW$_{12}$O$_{40}$ mixed with PSMIM.

Acknowledgments Research was sponsored by The Program for Professor of Special Appointment (Eastern Scholar) at Shanghai Institutions of Higher Learning and the Scientific Research Foundation for the Returned Overseas Chinese Scholars, State Education Ministry.

References

1. Ma F, Hanna MA (1999) Biodiesel production: a review. Bioresour Technol 70:1–15
2. Gerpen JV (2005) Biodiesel processing and production. Fuel Process Technol 86:1097–1107
3. Lopez DE, Goodwin JG Jr, Bruce DA, Lotero E (2005) Transesterification of triacetin with methanol on solid acid and base catalysts. Appl Catal A 295:97–105
4. Crabbe E, Nolasco-Hipolito C, Kobayashi G, Sonomoto K, Ishizaki A (2001) Biodiesel production from crude palm oil and evaluation of butanol extraction and fuel properties. Process Biochem 37:65–71
5. Al-Widyan MI, Al-Shyoukh AO (2002) Experimental evaluation of the transesterification of waste palm oil into biodiesel. Bioresour Technol 85:253–256
6. Felizardo P, Neiva Correia MJ, Raposo I, Mendes JF, Berkemeier R, Bordado JM (2006) Production of biodiesel from waste frying oils. Waste Manag 26:487–494
7. Tapasvi D, Wiesenborn D, Gustafson C (2005) Process model for biodiesel production from various feedstocks. Transact ASAE 48:2215–2221
8. Vicente G, Martínez M, Aracil J (2005) Optimization of *Brassica carinata* oil methanolysis for biodiesel production. J Am Oil Chem Soc 82:899–904
9. Shimada Y, Watanabe Y, Sugihara A, Tominaga Y (2002) Enzymatic alcoholysis for biodiesel fuel production and application of the reaction to oil processing. J Mol Catal B Enzym 17:133–142
10. Wu H, Zong M, Lou W (2003) Transesterification of fste oil to biodiesel in solvent free system catalyzed by immobilized lipase. Chin J Catal 11:903–908
11. Xu Y, Du W, Liu D, Zeng J (2003) A novel enzymatic route for biodiesel production from renewable oils in a solvent-free medium. Biotechnol Lett 25:1239–1241
12. Du W, Xu Y, Liu D, Zeng J (2004) Comparative study on lipase-catalyzed transformation of soybean oil for biodiesel production with different acyl acceptors. J Mol Catal B Enzym 30:125–129
13. Gryglewicz S (1999) Rapeseed oil methyl esters preparation using héterogeneous catalysts. Bioresour Technol 70:249–253
14. Suppes GJ, Dasari MA, Doskocil EJ, Mankidy PJ, Goff MJ (2004) Transesterification of soybean oil with zeolite and metal catalysts. Appl Catal A 257:213–223
15. Jitputti J, Kitiyanan B, Rangsunvigit P, Bunyakiat K, Attanatho L, Jenvanitpanjakul P (2006) Transesterification of crude palm kernel oil and crude coconut oil by different solid catalysts. Chem Eng J 116:61–66
16. Harmer MA, Sun Q (2001) Solid acid catalysis using ion-exchange resins. Appl Catal A 221:45–62
17. Sharma M (1995) Some novel aspects of cationic ion-exchange resins as catalysts. React Funct Polym 26:3–23
18. Chakrabarti A, Sharma M (1993) Cationic ion exchange resins as catalyst. React Polym 20:1–45
19. Tejero J, Cunill F, Iborra M, Izquierdo J, Fité C (2002) Dehydration of 1-pentanol to di- < i > n </i > -pentyl ether over ion-exchange resin catalysts. J Mol Catal A Chem 182:541–554
20. Siril P, Cross HE, Brown D (2008) New polystyrene sulfonic acid resin catalysts with enhanced acidic and catalytic properties. J Mol Catal A Chem 279:63–68
21. Schuchardt U, Vargas RM, Gelbard G (1995) Alkylguanidines as catalysts for the transesterification of rapeseed oil. J Mol Catal A Chem 99:65–70
22. Furuta S, Matsuhashi H, Arata K (2004) Biodiesel fuel production with solid superacid catalysis in fixed bed reactor under atmospheric pressure. Catal Commun 5:721–723
23. Guerreiro L, Castanheiro J, Fonseca I, Martin-Aranda R, Ramos A, Vital J (2006) Transesterification of soybean oil over sulfonic acid functionalised polymeric membranes. Catal Today 118:166–171
24. Aranda DA, Santos RT, Tapanes NC, Ramos ALD, Antunes OAC (2008) Acid-catalyzed homogeneous esterification reaction for biodiesel production from palm fatty acids. Catal Lett 122:20–25
25. Shu Q, Yang B, Yuan H, Qing S, Zhu G (2007) Synthesis of biodiesel from soybean oil and methanol catalyzed by zeolite beta modified with La^{3+}. Catal Commun 8:2159–2165
26. Mbaraka IK, Radu DR, Lin VS-Y, Shanks BH (2003) Organosulfonic acid-functionalized mesoporous silicas for the esterification of fatty acid. J Catal 219:329–336
27. Takagaki A, Toda M, Okamura M, Kondo JN, Hayashi S, Domen K, Hara M (2006) Esterification of higher fatty acids by a novel strong solid acid. Catal Today 116:157–161
28. Okuhara T, Mizuno N, Misono M (2001) Catalysis by heteropoly compounds—recent developments. Appl Catal A 222:63–77
29. Kozhevnikov IV (1998) Catalysis by heteropoly acids and multicomponent polyoxometalates in liquid-phase reactions. Chem Rev 98:171–198
30. Lotero E, Liu Y, Lopez DE, Suwannakarn K, Bruce DA, Goodwin JG (2005) Synthesis of biodiesel via acid catalysis. Ind Eng Chem Res 44:5353–5363
31. Arzamendi G, Campo I, Arguiñarena E, Sánchez M, Montes M, Gandía LM (2008) Synthesis of biodiesel from sunflower oil with silica-supported NaOH catalysts. J Chem Technol Biotechnol 83:862–870
32. Narasimharao K, Brown D, Lee AF, Newman A, Siril P, Tavener S, Wilson K (2007) Structure–activity relations in Cs-doped heteropolyacid catalysts for biodiesel production. J Catal 248:226–234
33. Seddon KR (1997) Ionic liquids for clean technology. J Chem Technol Biotechnol 68:351–356
34. Welton T (1999) Room-temperature ionic liquids. Solvents for synthesis and catalysis. Chem Rev 99:2071–2084
35. Wasserscheid P, Keim W (2000) Ionic liquids-new solutions for transition metal catalysis. Angew Chem 39:3772–3789
36. Wilkes JS (2002) A short history of ionic liquids—from molten salts to neoteric solvents. Green Chem 4:73–80
37. Dupont J, de Souza RF, Suarez PA (2002) Ionic liquid (molten salt) phase organometallic catalysis. Chem Rev 102:3667–3692
38. Wu Q, Chen H, Han M, Wang D, Wang J (2007) Transesterification of cottonseed oil catalyzed by Brønsted acidic ionic liquids. Ind Eng Chem Res 46:7955–7960
39. Kimura M, Nakato T, Okuhara T (1997) Water-tolerant solid acid catalysis of Cs2.5H0.5PW12O40 for hydrolysis of esters in the presence of excess water. Appl Catal A 165:227–240
40. Sharma Y, Singh B (2008) Development of biodiesel from karanja, a tree found in rural India. Fuel 87:1740–1742

41. Fang L, Xing R, Wu H, Li X, Liu Y, Wu P (2010) Clean synthesis of biodiesel over solid acid catalysts of sulfonated mesopolymers. Sci China Chem 53:1481–1486

42. Zhu Z, Yang W (2009) Preparation, characterization and shape-selective catalysis of supported heteropolyacid salts K2. 5H0. 5PW12O40,(NH4) 2.5 H0. 5PW12O40, and Ce0. 83H0. 5PW12O40 on MCM-41 mesoporous silica. J Phys Chem C 113:17025–17031

43. Jacobson K, Gopinath R, Meher LC, Dalai AK (2008) Solid acid catalyzed biodiesel production from waste cooking oil. Appl Catal B 85:86–91

44. Guo F, Fang Z, Tian X-F, Long Y-D, Jiang L-Q (2011) One-step production of biodiesel from Jatropha oil with high-acid value in ionic liquids. Bioresour Technol 102:6469–6472

45. Liang X, Gong G, Wu H, Yang J (2009) Highly efficient procedure for the synthesis of biodiesel from soybean oil using chloroaluminate ionic liquid as catalyst. Fuel 88:613–616

46. Timofeeva M, Maksimovskaya R, Paukshtis E, Kozhevnikov I (1995) Esterification of 2, 6-pyridinedicarboxylic acid with n-butanol catalyzed by heteropoly acid $H_3PW_{12}O_{40}$ or its Ce(III) salt. J Mol Catal A Chem 102:73–77

47. Zhang S, Zu Y-G, Fu Y-J, Luo M, Zhang D-Y, Efferth T (2010) Rapid microwave-assisted transesterification of yellow horn oil to biodiesel using a heteropolyacid solid catalyst. Bioresour Technol 101:931–936

15

Development of improved catalysts for deep HDS of diesel fuels

Syed Ahmed Ali

Abstract Deep hydrodesulfurization (HDS) of gas oils continues to attract research interest due to environmental-driven regulations which limit its sulfur content to 10–15 ppm in several countries. This paper highlights some of the recent studies conducted at King Fahd University of Petroleum and Minerals to develop improved HDS catalysts. The first study was focused on the effect of Co/(Co + Mo) ratio in CoMo/Al$_2$O$_3$ catalysts on HDS pathways of benzothiophene (BT) and dibenzothiophene (DBT). Co/Co + Mo ratio exhibited significant influence on the direct desulfurization (DDS) pathway, but showed no influence on the hydrogenation pathway. A Co/Co + Mo ratio of 0.4 exhibited optimum promotion effect of Co for HDS by DDS route and hence overall HDS. The second study investigated the effect of phosphorus addition on simultaneous HDS reactions and their pathways. The results indicate that phosphorus modification of CoMo/γ-Al$_2$O$_3$ catalysts resulted in enhancement of HDS due to increased dispersion of MoO$_3$ and the maximum enhancement was achieved with 1.0 wt % P$_2$O$_5$. Enhancement of HDS rates was in the following order: 4,6-DMDBT (51 %) > 4-MDBT (38 %) > DBT (26 %). In the third approach, a series of NiMo catalysts supported on Al$_2$O$_3$-ZrO$_2$ composites containing 0–10 wt % ZrO$_2$ was synthesized, characterized and evaluated for deep desulfurization of gas oil. An increase of 1.3–2.5 times increase in HDS activity at 320–360 °C was observed due to reduced interaction between Al$_2$O$_3$ and the active metals. A correlation was found between the enhancement of hydrogenation activity of sulfided catalysts and the reducibility of their oxide precursors.

Keywords Arabian gas oils · Hydrodesulfurization · HDS pathways · CoMo/Al$_2$O$_3$ · NiMo/Al$_2$O$_3$-ZrO$_2$

Introduction

Global trend in diesel sulfur limit is towards 10–15 ppm levels, which is required to adopt advanced pollution control technologies. With increase in heavy/sour crude production, the requirement for deep desulfurization will increase all over the world, especially in fast growing regions of Asia–Pacific and Middle East. Advanced processes/catalysts are required to meet the specifications and increasing demands of ultra-low sulfur diesel [5, 15]. Ultra-low sulfur levels (10–15 ppm) can be achieved by deep hydrodesulfurization (HDS) of middle distillate streams. Many factors such as the catalysts, process parameters, and feedstock quality have a significant influence on the degree of desulfurization of diesel feeds.

Detailed analysis of gas oils obtained from Arabian Light (AL-GO), Arabian Medium (AM-GO) and Arabian Heavy (AH-GO) crude oils in terms of reactive and refractory sulfur, nitrogen, as well as aromatic species has been reported [7]. The sulfur, nitrogen and aromatic contents were considerably higher in AH-GO as presented in Table 1. Refractory sulfur and the alkyl-carbazole content in the AH-GO, which are significantly (3–4 times) higher than that in AL-GO, hinders the deep HDS of AH-GO. Major sulfur species were alkyl-benzothiophenes (alkyl-

S. A. Ali (✉)
Center of Research Excellence in Petroleum Refining and Petrochemicals, King Fahd University of Petroleum and Minerals, Dhahran 31261, Saudi Arabia
e-mail: ahmedali@kfupm.edu.sa

Table 1 Aromatics, sulfur and nitrogen species in arabian gas oils [7]

Component	Type of gas oil hydrotreated		
	AL-GO	AM-GO	AH-GO
Total aromatics (%)	21	25	32
Sulfur (ppm)			
Reactive	4,134	4,741	6,500
Refractive	3,567	5,859	10,500
Total	7,701	10,600	17,000
Nitrogen (ppm)			
Carbazole	1.4	3.1	4.3
Alkyl carbazole	84.0	95.0	308.0
Total	85.4	98.1	312.3

Reproduced with permission from Elsevier

BTs) comprising C2-C5 alkyl chain, dibenzothiophene (DBT), as well as the considerable amounts of alkyl-DBTs, such as 4-DBT, 4,6- DMDBT and 4,6,x-TMDBT. AH-GO contained the largest amount of alkyl-DBTs with two or more alkyl carbon atoms, which are the refractory sulfur species. The results also show presence of significantly higher nitrogen content in AH-GO, which are mainly alkyl-carbazoles.

The conventional HDS process is usually conducted over sulfided CoMo/γ-Al$_2$O$_3$ or NiMo/γ-Al$_2$O$_3$ catalysts. A variety of metal contents, promoters, support properties, etc., have been studied and used for the development of versatile deep HDS catalyst system. Despite their robust nature, conventional catalysts, however, do not have sufficient activity to desulfurize diesel feed streams to ultra-low sulfur levels under normal operating conditions. They require severe operating conditions such as high temperature, low space velocity and high hydrogen partial pressure. Such severe processing conditions generally lead to rapid catalyst deactivation, shorter cycle lengths and reduced throughput.

Hence, the development and application of highly active and stable catalysts are among the most desired options for reducing the sulfur content of diesel to ultra-low levels by deep desulfurization. Substantial improvement in catalyst activity is necessary when the sulfur content is to be reduced to ultra-low levels (<15 ppm). This was achieved by one or more of the following approaches: (a) nature of active species; (b) support choice and modification; and (c) preparation procedures. Intensive efforts have been devoted to develop highly active hydrodesulfurization (HDS) catalysts [15].

The paper highlights the results from some of the recent studies conducted at the Center for Refining and Petro-chemicals of the King Fahd University of Petroleum and Minerals (KFUPM) to develop improved HDS catalysts.

The studies covered include (a) effect of Co/(Co + Mo) ratio in CoMo/Al$_2$O$_3$ catalysts on HDS pathways of benzothiophene and dibenzothiophene; (b) role of phosphorus addition on simultaneous HDS of dibenzothiophene and alkyl dibenzothiophenes over CoMo/Al$_2$O$_3$ catalysts; and (c) deep HDS of gas oil over NiMo catalysts supported on Al$_2$O$_3$-Zr$_2$O$_3$ composites.

Effect of Co/(Co + Mo) ratio in Como/γ-Al$_2$O$_3$ catalysts on HDS pathways

The HDS of dibenzothiophenes generally takes place by two routes: (a) a hydrogenation (HYD) pathway involving aromatic ring hydrogenation, followed by C–S bond cleavage; and (b) a direct desulfurization (DDS) or hydrogenolysis pathway via direct C–S bond cleavage without aromatic ring hydrogenation [13]. The direct desulfurization route is favorable as it consumes less H$_2$, making this route more economical. Therefore, many researchers have focused their research to enhance the DDS selectivity [8, 14, 17].

The objective of this study was to investigate the influence of Co/(Co + Mo) ratio on the HDS of benzo-thiophene (BT) and dibenzothiophene (DBT). The study covered overall HDS as well the DDS and HYD pathways for HDS [1].

Catalyst preparation and evaluation

A series of CoMo/γ-Al$_2$O$_3$ catalysts was prepared with Co/(Co + Mo) ratio of 0.3, 0.4 and 0.5 while maintaining a total metal oxide content of 19 wt %. For the sake of simplicity, the catalysts were denoted as aCM, in which a represents ten times the Co/(Co + Mo) molar ratio. The catalysts were prepared by co-impregnation of (NH$_4$)$_6$Mo$_7$O$_{24}$•4H$_2$O and Co(NO$_3$)$_2$•6H$_2$O on calcined γ-Al$_2$O$_3$ at room temperature. The actual MoO$_3$ loadings were 15.4, 14.0 and 12.4 wt % for 3, 4 and 5, respectively. The surface areas of 3, 4 and 5 catalysts were 166, 155, 146 m^2/g whereas their pore volumes were 0.33, 0.28 and 0.24 ml/g, respectively. The oxide catalysts were sulfided prior to their performance tests by light kerosene spiked with dimethyl disulfide (2.5 wt % sulfur) in a tubular reactor under flowing hydrogen (7.5 NL/h) at 750 psig and 593 K for 16 h.

The catalysts were tested in a 100 ml batch autoclave reactor which was loaded with 50 g decalin and 0.105 g of BT and 0.144 g of DBT. Hence, the model compounds contributed 500 ppm each of sulfur content, which resulted in a total sulfur content of 1,000 ppm in the feed. Half a gram of fresh catalyst was used for each experiment, which amounted to a catalyst-to-feedstock ratio of 1 wt %.

Table 2 Pseudo-first order reaction rate constants and activation energies for the HDS of BT and DBT [1]

Catalyst	Co/Co + Mo molar ratio	Temperature (°C)	$1,000 \times k$ (min^{-1})		k_{BT}/k_{DBT}
			k_{BT}	k_{DBT}	
3CM	0.310	300	101	1	101
		325	139	5	28
		350	150	24	6.3
4CM	0.407	300	108	10	10.8
		325	142	17	8.4
		350	166	48	3.5
5CM	0.505	300	90	5	18
		325	106	8	13.3
		350	152	24	6.3

Reproduced with permission from Springer

Experiments were carried out for 2 h under a hydrogen pressure (6.1 MPa) and at 573, 598 and 623 K. The product samples taken during the course of the experimental run were analyzed by gas chromatograph equipped with sulfur chemiluminscence detector (GC-SCD).

Results and discussion

HDS kinetics BT and DBT were determined by assuming a pseudo first-order reaction and the results are summarized in Table 2 [1]. Comparison of first-order rate constants obtained at different temperatures and catalysts provides information about the effectiveness of cobalt addition on BT and DBT desulfurization. The HDS rate of BT was much higher than DBT, especially at lower temperature. However, the HDS of DBT was more sensitive to Co/(Co + Mo) ratio than the HDS of BT. Overall HDS rates of BT and DBT were higher over 4CM catalyst than either 3CM or 5CM. Activation energies of the HDS of DBT were 3–4 times higher than the HDS of BT. Lowest activation energy for DBT was for 4CM.

HDS of DBT takes place via direct desulfurization (DDS) and/or Hydrogenation (HYD) routes. Cyclohexyl benzene content was almost same for products obtained over all catalysts indicating no influence of Co/(Co + Mo) ratio on the HYD route. Biphenyl content was highest over 4CM which signifies that DDS is enhanced when the Co/(Co + Mo) ratio is 0.4.

The results of this study have clearly demonstrated that Co/Co + Mo ratio has significant influence on the overall HDS of BT and DBT as well as on the DDS pathway, but showed no influence on the HYD pathway. A Co/Co + Mo ratio of 0.4 was found to be optimum for both overall HDS as well as the HDS by DDS pathway.

Role of P addition on simultaneous HDS of DBT and alkyl DBTs over CoMo/Al$_2$O$_3$ catalysts

Previous studies have shown that modification of HDS catalysts by acidic species, such as phosphorus, improves the activities of Mo/γ-Al$_2$O$_3$ catalysts [9, 10]. However, most of the earlier studies focused on influence of phosphorus on HDS catalysts using single model compound. In the present study, simultaneous HDS of DBT and 4-methyl dibenzothiophene (4-MDBT) as well as DBT and 4,6-DMDBT were studied over a series of phosphorus promoted CoMo/γ-Al$_2$O$_3$ catalysts with the aim of investigating the effect of phosphorus addition on simultaneous HDS reactions. The study covered overall HDS as well as the DDS and HYD pathways for HDS [3].

Catalyst preparation and evaluation

A series of catalysts was prepared in which the total metal oxide (MoO$_3$ + CoO) content was kept constant at 19.0 wt % with a Co/(Co + Mo) ratio of 0.4. γ-alumina support was modified by addition of 0.5, 1.0 or 1.5 wt % of P$_2$O$_5$ before impregnation of the active metals. The catalysts were prepared by co-impregnation of ammonium heptamolybedate tetrahydrate and cobalt nitrate heaxahydrate on calcined γ-alumina at room temperature. The catalysts were then dried at 373 K for 12 h followed by calcination at 773 K for 1 h. The catalysts were denoted as CMP(a), in which a represents P$_2$O$_5$ content in wt %.

Figure 1 shows the pore size distribution of phosphorus modified and unmodified catalysts. Phosphate is adsorbed on the walls of the pores and blocks small pores initially resulting in stronger reduction in pore volume (up to 35 %) than surface area (up to 26 %). The apparent average size of unblocked pores increases due to phosphate adsorption. However, in reality, there was slight decrease in pore size that was originally present in CMP(0) catalyst due to adsorption of P$_2$O$_5$. This phenomenon results in decrease in surface area, pore volume and average pore size with increase in P$_2$O$_5$ content.

Catalyst evaluation procedures used were similar to those presented in Sect. 2.1. Two sets of experiments: (a) DBT and 4-MDBT; and (b) DBT and 4,6-DMDBT were carried out. The quantities of model compounds added in the decalin feedstock were controlled so that each model compound contributed 500 ppm of sulfur in the feedstock.

Results and discussion

X-ray diffraction patterns of γ-Al$_2$O$_3$ and CMP catalysts show that the crystallinity of CoMoO$_4$ phase increased with

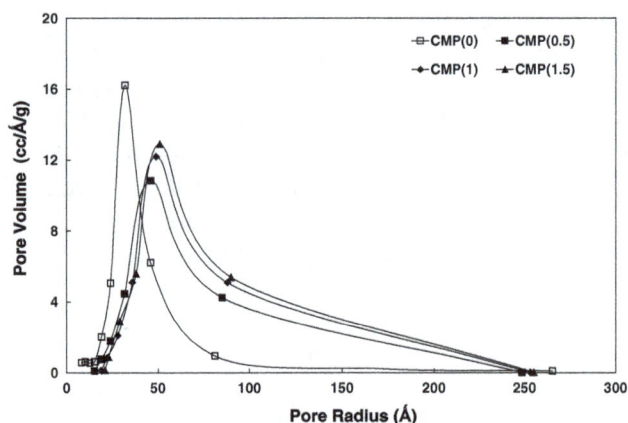

Fig. 1 Pore-size distribution of CMP catalysts [3]. (Reproduced with permission from Elsevier)

P addition. It also resulted in increased dispersion of MoO_3 phase, which is catalytically active phase. Modification of $CoMo/\gamma\text{-}Al_2O_3$ catalysts by addition of phosphorus strongly increased the HDS activity and the maximum activity enhancement was achieved with 1.0 wt % P_2O_5. However, further increase in phosphorus content reduces the activity which may be due to enhanced $CoMoO_4$ formation, as noticed from XRD results, causing loss of Mo dispersion and formation of relatively stable Co-Mo-P compounds.

Comparison of reaction rates for DDS and HYD pathways provide an insight into the reactivities of model compounds and effect of phosphorus addition on catalyst. In the simultaneous HDS of DBT and 4-MDBT both molecules mainly react by the DDS pathway and thus compete for the same DDS sites (Fig. 2). However, in the case of simultaneous HDS of DBT and 4,6-DMDBT, the latter preferably reacts over the HYD sites, while DBT reacts mainly over DDS sites (Fig. 3). Thus, DBT faces less competition from 4,6-DMDBT than from 4-MDBT for the DDS sites resulting in higher conversion of DBT in the presence of 4,6-DMDBT than 4-MDBT.

Figure 4 summarizes the enhancement in catalyst activity due to addition of 1 wt % P_2O_5 by DDS and HYD pathways as well as the overall HDS at 623 K. The results show that about 90 % enhancement in HDS of DBT and 4-MDBT was via DDS route. This is in contrast to about 47 % enhancement in 4,6-DMDBT enhancement was via DDS route. On the absolute basis, however, the overall enhancement in HDS rates by phosphorus addition was in the following order: 4,6-DMDBT (51 %) >4-MDBT (38 %) >DBT (26 %). Incidentally, steric hindrance by methyl substituents also increases in the same order. Since 4,6-DMDBT is one of the most refractive sulfur compound in the diesel fuel, enhancement in its HDS rate by about 50 % by phosphorus addition will substantially contribute in achieving the 10–15 ppm sulfur level.

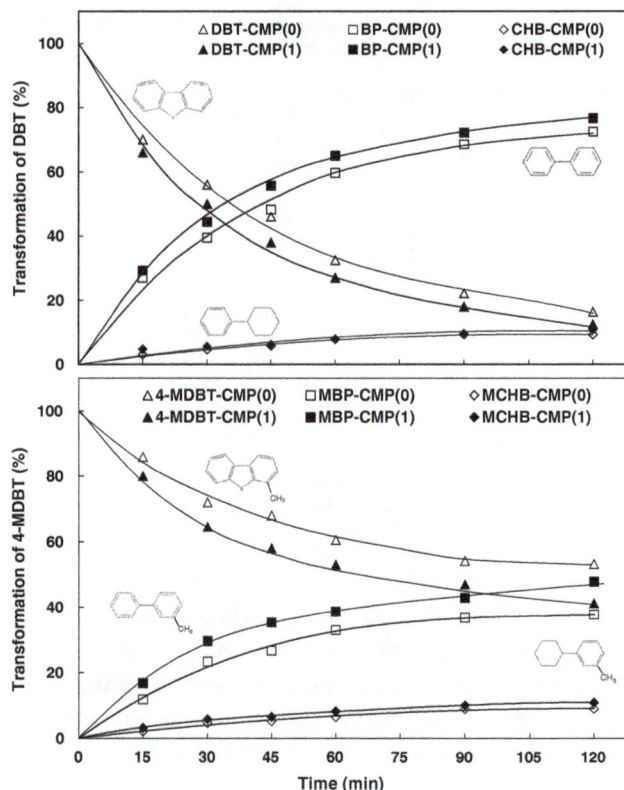

Fig. 2 Product distribution during simultaneous HDS of DBT and 4-MDBT over CMP(0) and CMP(1) catalysts at 623 K [3]. (Reproduced with permission from Elsevier)

Deep HDS of gas oil over nimo catalysts supported on Alumina–Zirconia composites

Alumina is the most widely used support material in HDS catalysts because it is highly stable, contains acidic and basic sites, has reasonably high surface area and porosity, can be easily formed into desired shapes, and is relatively inexpensive. However, alumina undergoes a direct interaction with the active metal species, and has an indirect role in determining the promotional effects of the secondary metal species. These effects have generated an immense interest in new supports for deep HDS catalysts, such as TiO_2, ZrO_2, MgO, C, SiO_2, zeolites, etc. [6], [11], [12] [16]). Our study focused on the synthesis and characterization of a series of $ZrO_2\text{-}Al_2O_3$ composite oxides containing 0–10 wt % ZrO_2 [2].

Catalyst preparation, characterization and evaluation

A series of $ZrO_2\text{-}Al_2O_3$ composite supports having a ZrO_2 content of 0.0, 2.5, 5.0 and 10.0 wt % were prepared. Zirconium(IV) oxynitrate hydrate $[ZrO(NO_3)_2 \bullet xH_2O]$ (99 %), ammonium heptamolybdate pentahydrate $[(NH_4)_6 Mo_7O_{24} \bullet 5H_2O]$ (99.98 %), and nickel(II) nitrate nonahydrate $[Ni(NO_3)_2 \bullet 9H_2O]$ (98.5 %) were used as sources of

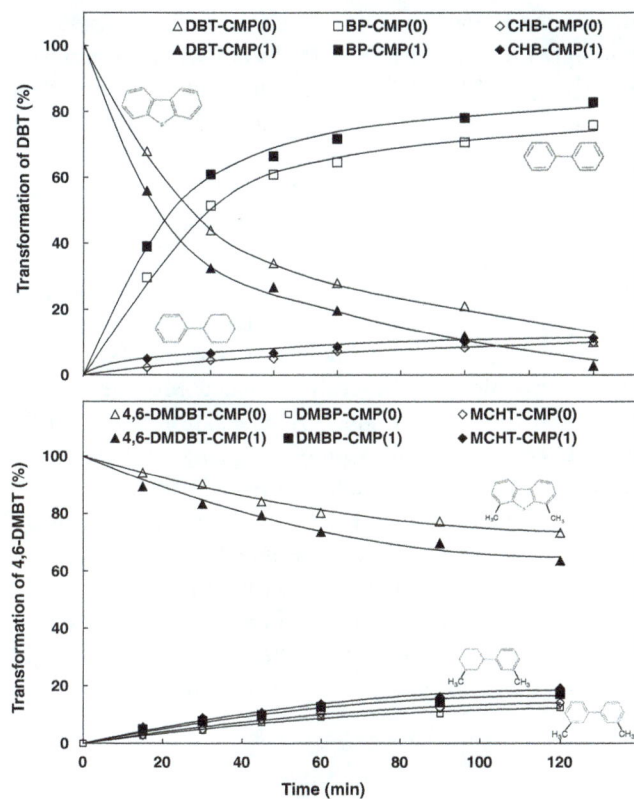

Fig. 3 Product distribution during simultaneous HDS of DBT and 4,6-DMDBT over CMP(0) and CMP(1) catalysts at 623 K [3] (Reproduced with permission from Elsevier)

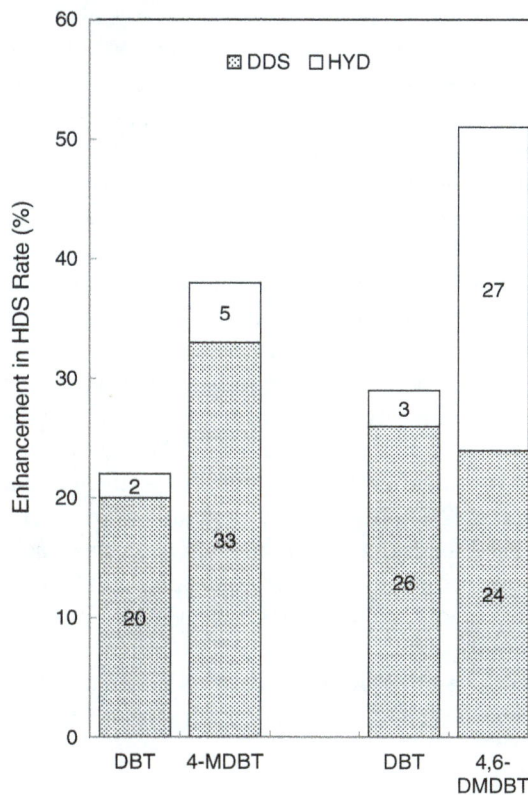

Fig. 4 Enhancement in HDS rates at 623 K due to addition of 1 wt % P_2O_5 [3] (Reproduced with permission from Elsevier)

zirconium, molybdenum and nickel, respectively. The composite supports were synthesized by peptizing Al_2O_3 together with $ZrO(NO_3)_2 \cdot xH_2O$. Typical synthesis procedure involved dispersion of 40.0 g AlO(OH)$\cdot H_2O$ in 200 ml deionized water at 70 °C while stirring for 3 h followed by addition of 105 ml aqueous solution of $ZrO(NO_3)_2 \cdot xH_2O$. The concentration of $ZrO(NO_3)_2 \cdot xH_2O$ was controlled in order to achieve ZrO_2 contents of 0.0, 2.5, 5.0 and 10.0 wt %. An aqueous solution containing Mo, Ni and citric acid in molar ratio of 1:0.4:1, respectively, was used to impregnate the supports with a loading of 18 wt % MoO_3 by pore-volume impregnation. The catalysts are referred to as NMAZ-0, NMAZ-2.5, NMAZ-5, and NMAZ-10, where the numbers denote the percentage of ZrO_2 content.

X-ray diffraction analysis of the supports was carried out on a Siemens-D5005® diffractometer using Cu-K$_\alpha$ radiation under 40 kV, 40 mA, and scan range from 10° to 80°. UV–Visible diffuse reflectance experiments were performed on a Thermo Evolution-600® high-performance spectrophotometer. A Quantachrome Autosorb-1C® system was used for temperature-programmed desorption of ammonia (NH3-TPD) of supports. H_2 temperature-programmed reduction (H_2-TPR) was conducted on the catalysts after calcination in air at 550 °C for 3 h. The

composition of metals in the ZrO_2-Al_2O_3 supports was quantitatively verified by inductively coupled plasma mass spectrometry (ICP-MS).

Catalytic activity for the HDS of gas oil was determined in a fixed-bed flow reaction system using gas oil feed (S content = 10,000 ppm) which was obtained from a local refinery. The tubular reactor (ID = 1.5 cm; L = 74 cm) was loaded with 10 ml of uncalcined catalyst as pellets of size between 0.50 and 0.85 mm. Prior to performance testing, the catalyst was presulfided in situ using white kerosene mixed with dimethyl disulfide (S content = 2.5 wt %). The performance of catalysts was evaluated at 320, 340 and 360 °C while the LHSV (1.0 h^{-1}), H_2 pressure (6 MPa) and H_2 flow rate (250 Nm^3/m^3) were kept constant. Total sulfur content in the feed and product was determined by pyrofluorescence detection method using Antek 7000S analyzer. The boiling-range distribution was determined by simulated distillation using Agilent GC 3800.

Results and discussion

X-ray diffraction patterns indicate homogenous dispersion of 2.5–10 wt % of ZrO_2 in bulk Al_2O_3. Mono-modal pore-size distribution—decrease in pore-size with increasing

ZrO_2 content. NH_3-TPD results show that incorporation of 5 wt % or more ZrO_2 neutralized the weak acid sites of Al_2O_3 and generated a different type of stronger acid sites. UV–Visible diffuse reflectance spectroscopy shows the presence of tetrahedral and octahedral Mo^{6+} ions species. An increase in octahedral species with addition of ZrO_2 was observed, possibly due to weaker interaction of active metals on composite support. This result is also supported by H_2-TPR measurements.

Acidity of the composite catalysts was assessed by NH_3-TPD and the results are presented in Table 3. It seems that the presence of 2.5 % ZrO_2 has neutralized the strong acid sites on the Al_2O_3 supported catalyst and at the same time generated greater density of the weak ones. However, the density and strength of the strong acid sites increased in the 10 % ZrO_2 supported catalyst. It could be postulated that the presence of ZrO_2 has initially neutralized the strong acid sites, but upon further addition of ZrO_2, stronger acid sites have evolved for the catalyst containing 10 % ZrO_2. The exact nature of these acid sites, however, could not be identified in this study.

Deep desulfurization of gas oil, carried out in a bench-scale flow reactor at 320, 340 and 360 °C, indicate that the addition of ZrO_2 increased the catalytic activity—especially at higher temperature. The results are summarized in Table 4. Compared to Al_2O_3-based catalyst, the 1.5 order HDS rate constant was about 1.3, 1.8 and 2.5 times higher for catalysts containing 2.5, 5 and 10 wt % ZrO_2, respectively. Apparent activation energies were found to be in the range of 32–36 kcal/mole, which are comparable to the reported values [4].

These trends confirm the results of H_2-TPD characterization that the addition of ZrO_2 reduces the interaction between Al_2O_3 support and the active metals, and led to the formation of the easily reduced Mo species, which formed more active sites. A correlation was also found between the enhancement of hydrogenation activity of sulfided catalysts and the reducibility of their oxide precursors, as determined by the amount of hydrogen consumed in TPR experiments followed by mass spectroscopy. The hydrogen consumption increased with increase in ZrO_2 content indicating enhanced hydrogenation activity. This is perhaps the only major drawback of the incorporation of ZrO_2 in the composite support.

Table 3 NH_3-TPD results of the of NiMo/Al_2O_3- ZrO_2 Catalysts [2]

Support	80–350 °C		350–550 °C		Total desorbed NH_3 (μmol/g)
	Peak Temperature (°C)	Desorbed NH_3 (μmol/g)[a]	Peak Temperature (°C)	Desorbed NH_3 (μmol/g)[a]	
AZ-0	165	15,250	435	3,327	18,577
AZ-2.5	166	64,139	–	–	64,139
AZ-5	170	23,675	380	1,112	24,787
AZ-10	187	8,752	482	1,531	10,283

Reproduced with permission from Elsevier

[a] Obtained from the calibrated integrated intensity of the signal

Table 4 Performance evaluation results of NiMo/Al_2O_3- ZrO_2 catalysts [2]

| | Feed | Catalyst | | | | | | | | | | |
		NMAZ-0			NMAZ-2.5			NMAZ-5			NMAZ-10		
Reaction temperature (°C)		320	340	360	320	340	360	320	340	360	320	340	360
Sulfur content (wt ppm)	10,000	1,390	579	114	1,180	399	66	767	213	30	389	117	20
Reduction in S content compared to NMAZ-0 (%)					15	31	42	45	63	75	72	80	82
H_2 consumption (NL/L)		19.8	28.9	36.7	23.2	33.9	46.7	26.5	42.6	53.2	29.5	47.8	62.8
Rate constant (1.5 order)		3.4	6.3	17.6	3.8	9.6	22.6	5.2	11.7	34.5	8.1	16.5	42.7
Apparent E_a (kcal/mol)			32.1			33.2			35.4			35.5	
Simulated distillation (°C)													
5 %	179.9	181.6	180.6	178.4	183.2	179.7	176.2	183.8	178.2	173.1	184.4	179.7	170.0
50 %	277.1	275.6	274.0	272.3	274.0	272.4	271.0	273.8	272.8	267.7	273.6	272.4	264.4
95 %	357.0	348.2	343.1	350.5	339.4	338.0	357.8	351.0	354.6	336.5	362.6	338.0	315.2

Reproduced with permission from Elsevier

Concluding remarks

It has been demonstrated that development of deep HDS catalysts with improved performance can be achieved by different approaches:

(1) Incorporation of proper ratio of active metals effects DDS route of HDS. Highest DDS activity was observed at Co/(Co + Mo) ratio of 0.4. Since the total metal oxide content is fixed at 19 wt %, higher Co content reduces Mo content and the catalytic activity. Lower Co content, on the other hand, results in not enough promotion of Mo catalytic activity.

(2) Addition of P_2O_5 as a second promoter weakens the interaction between Mo and γ-Al_2O_3 resulting in increased dispersion of MoS_2 particles. Maximum activity enhancement was achieved with 1.0 wt % P_2O_5. However, further increase in phosphorus content reduces the activity which may be due to enhanced $CoMoO_4$ formation causing loss of Mo dispersion.

(3) Application of ZrO_3-Al_2O_3 composite support reduces the interaction between Al_2O_3 support and the active metals, and led to the formation of the easily reduced Mo species, which formed more active sites. A correlation was found between the enhancement of hydrogenation activity of sulfided catalysts and the reducibility of their oxide precursors.

These studies have contributed in increasing the understanding of scientific reasons for achieving improved performance of deep HDS catalysts. Among the different approaches studied, it is difficult to identify a single most-effective method of catalyst improvement. It seems that several approaches need to be applied simultaneously to achieve the highest performing deep HDS catalysts.

Acknowledgments The author acknowledges the support of King Fahd University of Petroleum and Minerals (KFUPM). Acknowledgement is due to the Ministry of Higher Education, Saudi Arabia for establishing the Center of Research Excellence in Petroleum Refining and Petrochemicals (CoRE-PRP) at KFUPM. The author thanks Mr. Khurshid Alam for meticulously conducting the product analysis.

References

1. Ahmed K, Ali SA, Ahmed S, Al-Saleh MA (2011) Simultaneous hydrodesulfurization of benzothiophene and dibenzothiophene over CoMo/Al$_2$O$_3$ catalysts with different [Co/(Co + Mo)] ratios. React Kinet Mech Catal 103:123–133
2. Al-Daous MA, Ali SA (2012) Deep desulfurization of gas oil over NiMo catalysts supported on alumina-zirconia composites. Fuel 97:662–669
3. Ali SA, Ahmed S, Ahmed KW, Al-Saleh MA (2012) Simultaneous hydrodesulfuriz-ation of dibenzotiophene and substituted dibenzothiophenes over phosphorus modified CoMo/Al$_2$O$_3$ catalysts. Fuel Process Technol 98:39–44
4. Ancheyta J, Angeles MJ, Macias M, Marroquin G, Morales R (2002) Changes in apparent reaction order and activation energy in the hydrodesulfurization of real feedstocks. Energy Fuel 16:189–193
5. Barrow K (2009) More product-sulfur reduction on horizon. Oil Gas J 107:38–46
6. Breysse M, Afanasiev P, Geantet C, Vrinat M (2003) Overview of support effects in hydrotreating catalysts. Catal Today 86:5–16
7. Kim T, Ali SA, Alhooshani K, Park J-I, Al-Yami M, Yoon S-H, Mochida I (2013) Analysis and deep hydrodesulfurization reactivity of Saudi Arabian gas oils. J Ind Eng Chem 19:1577–1582
8. Kwak C, Kim MY, Choi K, Moon SH (1999) Effect of phosphorus addition on the behavior of CoMoS/Al$_2$O$_3$ catalyst in hydrodesulfurization of dibenzothiophene and 4,6-dimethyldi-benzothiophene. Appl Catal A Gen 185:19–27
9. Oyama ST, Gott T, Zhao H, Lee Y (2009) Transition metal phosphide hydroprocessing catalysts: a review. Catal Today 143:94–107
10. Prins R, Pirngruber G, Weber T (2001) Metal phosphides and zeolite-like mesophorus materials as catalysts. CHIMIA Int J Chem 55:791–795
11. Ramirez J, Sánchez-Minero F (2008) Support effects in the hydrotreatment of model molecules. Catal Today 130:267–271
12. Rayo P, Ramirez J, Rana MS, Ancheyta J, Aguilar-Elguézabal A (2009) Effect of the incorporation of Al, Ti, and Zr on the cracking and hydrodesulfurization activity of NiMo/SBA-15 catalysts. Ind Eng Chem Res 48:1242–1248
13. Shafi R, Hutchings GJ (2000) Hydrodesulfurization of hindered dibenzothiophenes: an overview. Catal Today 59:423–442
14. Shimada H, Sato T, Yoshimura Y, Hiraishi J, Nishijima A (1988) Support effect on the catalytic activity and properties of sulfided molybdenum catalysts. J Catal 110:275–284
15. Stanislaus A, Marafi A, Rana MS (2010) Recent advances in the science and technology of ultra-low sulfur diesel (ULSD). Catal Today 153:1–68
16. Trejo F, Rana MS, Ancheyta J (2008) CoMo/MgO-Al$_2$O$_3$ supported catalysts: an alternative approach to prepare HDS catalysts. Catal Today 130:327–336
17. Yao W, Zhongchao S, Anjie W (2004) Kinetics of hydrodesulfurization of dibenzothiophene catalyzed by sulfided Co-Mo/MCM-41. Ind Eng Chem Res 43:2324–2329

16

The energy challenge

Chris Llewellyn Smith

Abstract This paper provides an overview of the enormous challenge of meeting future energy demand in an environmentally responsible manner. A portfolio approach is required, which must include more solar, wind, hydro, bio and marine energy wherever sensible, and demand reduction (through better planning, especially in the world's expanding cities), increased efficiency, more nuclear power, and (if feasible, safe and economically competitive) carbon capture and storage. In the longer term, the world will need much more solar power, advanced nuclear fission, and fusion power—if it can be made to work reliably and competitively. The policy priority is to put a (high) price on carbon in the context of a global agreement (through a tax, or cap and trade with a floor price to provide certainty for investors), which is easily said, but very hard to do.

Keywords Energy demand · Energy supply · Energy efficiency · Fossil fuels · Renewable energy · Nuclear energy

Introduction

The biggest challenge of the twenty-first century is to provide sufficient food, water, and energy to allow everyone on the planet to live decent lives in decent environments, in the face of rising population, the threat of climate change, and (sooner or later) declining fossil fuels. Provision of sufficient energy is a necessary (but not sufficient) means to meet the overall challenge.

C. Llewellyn Smith (✉)
Oxford University, Oxford, UK
e-mail: c.llewellyn-smith@physics.ox.ac.uk

The world is using energy at a rate of 2.4 kW per person: that is the equivalent of 24 old-fashioned 100 W incandescent light bulbs burning continuously for every man woman and child on the planet. Average use per person is 10.4 kW in the USA while it is only 0.21 in Bangladesh. The average is 4.6 kW in the UK and in the Middle East as a whole, although it is much higher in the Gulf Cooperation Council (GCC) countries. For everyone to use energy at the same rate as the average person in the UK or the Middle East, total energy consumption would have to go up 1.9-fold today, or 2.4-fold when the population reaches 9 billion. This is probably impossible so changes in behaviour and expectations will be necessary.

Some 78 % of the world's primary energy is currently generated by burning fossil fuels (oil, coal, gas) which is causing potentially catastrophic climate change, and horrendous pollution, and is unsustainable as they won't last forever. The rest (in thermal equivalent terms) is provided by: burning combustible renewables and waste—10 %; hydropower—5.8 %; nuclear power—4.7 %; geothermal + solar + wind + marine power—1.2 %. The popular hope that the last category might replace fossil fuels must be tempered by the realisation that to do so its contribution would have to increase by a factor of 65, and the increase would have to be even greater in the future assuming world consumption continues to grow, driven by (very welcome) economic development in China, Africa and India.

The scale of the challenge is also shown by the International Energy Agency's 2011 'new policies scenario'. According to this scenario, which assumes the successful implementation of all agreed national policies and announced commitments designed to save energy and reduce use of fossil fuels, energy use will increase 33 % in the period 2010–2035, while the use of fossil fuels will increase 23 % (the increases come almost entirely from non-OECD countries). Although this falls far short of expectations, and of what is needed to temper climate change, the scenario requires very demanding measures, such as a 70 % increase in nuclear power and 60 % more hydropower (think of the civil engineering). Most commentators anticipate larger increases in energy use, e.g. the BP Energy Outlook 2030 (published in early 2012), which is based on BP judgement of what will actually happen and is not a business as usual extrapolation, projects a 39 % increase by the earlier date 2030, with fossil fuels up 31 %.

Future of fossil fuels

There is said to be a Saudi saying: "My father rode a camel. I drive a car. My son flies a plane. His son will ride a camel". Could this be true?

It is observed, and understood, that conventional oil production in a given region peaks when roughly half the primordial endowment of extractable oil has been produced. Predictions of when world oil production will peak depend on estimates of the primordial endowment and assumptions about how much it will be economic to extract, which in turn depend on assumptions about developments in technology and the oil price. A 2009 literature review by the UK Energy Research Centre concluded that the peak 'is likely to occur before 2030' and that there is a 'significant risk' that it will occur before 2020.

But there is plenty of oil in unconventional places (e.g. the deep ocean off Brazil and the arctic) and lots of unconventional oil (heavy oil, tight/shale oil, oil from oil shale or tar sands). Without significant technological developments, exploitation of most of these reserves will however be expensive and would be difficult to expand fast enough to compensate for an early peak in conventional oil production. The debate on the effect this will have on oil prices continues. Confidence in continued high prices would presumably lead to increased investments in turning coal and gas into oil.

There is plenty of coal and gas. According to the IEA, global coal reserves that are economically recoverable with current technology are enough for 150 years with current use, while there is enough recoverable gas for some 125 years, which potential unconventional resources could increase to 250 years. The fracking/shale gas revolution in the USA has had a dramatic effect on the world gas market, e.g. by freeing up liquefied natural gas (previously destined for the US) from Qatar, which in 2011 provided 27 % of the UK's gas, up from less than 0.15 % in 2008, and is a major contributor in the Far East. It is unclear, however, whether the expanding rate of production of shale gas is sustainable, and how much is really going to be accessible.

Climate change and carbon capture and storage

Climate scientists tell us that it would take thousands of years for the level of carbon dioxide in the atmosphere to drop to preindustrial levels were carbon emissions to stop abruptly. It follows that on time scales shorter than a thousand years, it is the cumulative emission of carbon-dioxide that drives climate change—not the rate of emission. Consequently, as long as power stations and certain large industrial plants continue to burn fossil fuels, the only measure that it is available to prevent the produced carbon dioxide driving additional climate change is to capture and bury as much as possible underground, where it must stay for thousands of years if the whole exercise is to have any point.

It is therefore very important to develop carbon capture and storage (CCS), which has not yet been demonstrated in a complete large-scale system, and to understand better what it will cost, and how long the carbon will stay underground in different geological conditions. CCS is likely to be expensive, and whether—assuming it is feasible—it should then be deployed on a large scale will depend on the cost in comparison to alternative low carbon energy sources (unless the public is willing to pay for these alternatives, and/or pay over the odds for CCS, it would seem likely that most of the carbon in the world's remaining 'cheap' fossil fuels will be emitted into the atmosphere on a time scale of a (few) hundred years, with severe consequences for the climate).

Necessary actions

In addition to developing CCS (and rolling it out on a large scale, if it this is feasible, safe and affordable), meeting the energy challenge requires:

- Reducing energy use/improving efficiency, which can have a large impact and save a lot of money, as discussed further below, but is unlikely to do more than curb the rising growth in global energy, which is driven by rising living standards in the developing world.
- Developing and expanding low carbon energy sources; we need everything we can sensibly get, but I will argue that without major contributions from solar and/ or nuclear (fission and/or fusion) it will not be possible to replace the 14 TW currently provided by fossil fuels.
- Devising economic tools and ensuring the political will to make the above happen.

Use of energy, demand reduction and energy efficiency

Globally, energy use is shared: 31 % industry, 31 % transport, 27 % residential buildings, 9 % commerce and public service, 2 % agriculture, forestry and fisheries, according to the Internal Energy Agency. Private use (in transport + buildings) is large, so although what we each do individually does not matter, what we do collectively is enormously important.

There is huge scope for reducing demand by, for example, designing buildings to make good use of natural light, and planning cities to encourage walking, bicycling or use of public transport. There are major opportunities in rapidly developing countries, where low-carbon development paths should be adopted as early as possible in planning expanding transport systems and cities (once a city ends up like Phoenix, going back is next to impossible). Changes in planning and procurement by cities and communities, designed to optimise energy use and reduce CO_2 emissions, are increasingly important drivers of demand and efficiency.

Managing demand and matching multiple sources of supply and demand through a smart grid (which collects information on grid conditions, use and costs across a large network, with multiple connections, and bidirectional flows) is becoming increasingly important. This should lead to more efficient energy use and reduce costs by lowering peak demand on the grid, and will be necessary when the penetration of intermittent sources (wind, solar) increases to much above the 20 % level.

Big efficiency gains are possible, from (e.g.) improved building insulation, more efficient lighting (which consumes 19 % of the world's electricity), and more efficient internal combustion engines or moving to hybrids or electric cars. Energy consumption per unit of GDP has been falling (in the UK it is now 40 % of the 1971 level) but GDP has been rising faster. Much more could be done while saving a lot of money (according to McKinsey, US CO_2 emissions could be reduced by nearly 20 % by actions which would all save money). It's not happening as fast might be expected because of:

- The 'rebound effect'—greater efficiency can lead to greater use. Similarly, trying to reduce demand by telling people how to save money and energy, e.g. by adjusting the thermostat, may lead them to conclude that they can afford to use more.
- Affluence in the developed world. Most of us don't care about relatively small savings, even if we know that collectively they could have a large effect—we need to be compelled to make them by regulations, e.g. on building construction and the performance of cars.
- Lack of capital in parts of the developing world, which may inhibit investment in efficient devices and solutions even if the pay-back time is short.

Reducing demand and improving efficiency are imperative. They can curb the 40 % increase in world energy consumption expected by 2030, but it would be fanciful to think that they could produce an overall decrease. It is therefore vital, in parallel, to radically expand the use of low carbon energy sources.

Low carbon energy sources

It is interesting to ask: what can replace the 14 TW (and rising) of primary power that the world derives from fossil fuels? The maximum additional potential of wind + hydro + bio + (enhanced) geothermal + marine energy appears to be no more than 6 or 7 TW [in detail: I think that the maximum additional practical, *thermal equivalent,* potentials are—wind 3 TW (35 × 2009; about half today's global electricity generation), hydro 2 TW (2 × 2009), bio 1 TW (2/3 of 2009), (enhanced) geothermal 1 TW (50 × 2009), marine 0.1 TW (600 × 2009)]. We should use as much of these sources of energy as we reasonably can, noting that (1) their potential contributions are very location dependent (none except wind seems to have much potential in the GCC countries), (2) this is easier said than done as most are currently more expensive than fossil fuels (ignoring externalities), and wind and marine (and solar) provide energy intermittently, and if used on a large scale need to be backed up by other sources, or supplemented by large scale energy storage systems which urgently need to be developed.

The conclusion is that if/when fossil fuels become unaffordable, or we renounce their use, solar and/or nuclear energy, which are not included in the list above, will have to play a major role.

Solar energy

Solar energy could *in principle* easily provide all the world's energy needs: with 15 % efficiency (which is readily available from Photo Voltaic and Concentrated Solar Power), 0.5 % of world's land surface could provide 20 TW of electricity. However, although solar capacity has grown at an average of some 40 % p.a. in the last decade, solar today only provides less than 0.01 TW (compared to world electricity consumption of 2.4 TW) so there is a very long way to go. High priority should be given to driving down costs (which is happening) and developing storage (for use at night and when the sun is not shining: for CSP thermal storage is appropriate, and being used; for PV hydrogen, or hydrocarbons synthesised from H and CO_2, could be used for storage and as an energy vector). Projects such as Desertec, which aims to supply Europe with solar and wind energy from N Africa, will require long distance high voltage DC transmission. The potential is of course large in the GCC countries, where solar energy can play a big role as a source of energy for desalination as well as electricity generation.

Nuclear

Nuclear and hydro are the only large-scale low-carbon sources of energy currently in use, apart from burning biomass that is renewed. However, while hydro-power could at most be expanded by a factor of about three globally, nuclear has a much bigger potential and in my opinion should be expanded now.

The first examples of a new improved generation of nuclear reactors are currently being built, but this is not happening as rapidly as might be hoped because of concerns about:

- the cost of nuclear power, which has been relatively high in the past (although less than that of coal or gas power if the environmental costs of burning fossil fuels are included). The cost depends strongly on the cost of borrowing the large sums needed to build reactors, which is generally higher than the cost of borrowing to invest in coal and gas plants because of the greater risks (of changes in government policy; being undercut by gas and coal in the future; delays in the planning process and/or construction; and cost over-runs). On paper,[1] the new 'generation 3' reactors look competitive with coal, but construction of the first units is running late and going way over budget, and more construction experience is needed to establish the real cost.
- safety, which is mainly a problem of perception: objectively the safety record of nuclear power is good compared to that of almost all other major sources of power (although there is no room for complacency);
- disposal of waste, which is technically not an issue although reduction in the volume of waste needing long-term storage (which is possible—see below) would be helpful;
- proliferation, which is mainly a political issue.

Technically the rate at which nuclear power can be expanded is limited by lack of capacity. There are relatively few suppliers, and there is a need to expand the skills-base, which has been largely lost in countries such as the UK. At one time there were concerns about the availability of uranium, but it now seems that there is quite probably enough to provide for ten times current use for

one hundred years at less than twice the current cost (which today only accounts for 2–4 % of the cost of nuclear electricity). This is long enough to further develop options that would prolong the nuclear age and could have some immediate benefits, including:

- re-cycling nuclear fuel (20 % more energy for a given amount of Uranium; less waste).
- fast breeder reactors (60 times as much energy/kg U, less waste, but more expensive, and slow to deploy—so development should not be delayed too long): fast breeders look promising as 'waste burners' to be used in conjunction with conventional reactors.
- thorium reactors (lots of it; less waste, but fuel fabrication much more demanding).
- fusion, which is intrinsically very attractive but extremely demanding.

Conclusions on the global energy challenge

Dealing with the enormous challenge will need a portfolio approach, which must include measures such as more solar, wind, hydro, bio and marine energy wherever sensible, and particularly: demand reduction, increased efficiency, more nuclear power, and (if feasible, safe and economically competitive) carbon capture and storage. In the longer term: the world will need much more solar power, advanced nuclear fission, and fusion power—if it can be made to work reliably and competitively.

There is a huge R&D agenda that requires more resources, which should be judged in comparison to the $400 billion p.a. which is currently spent subsidising fossil fuels (subsidies for renewables are currently about $65 billion p.a., while annual public funding of energy R&D is about $25 billion).

Above all there is a need for the political will to make a transition to a more energy efficient low carbon economy, which—as well as requiring the development of new and cheaper technologies—will have to be driven by financial incentives (especially a high carbon price) and tougher regulations of buildings, urban development and energy use.

---

[1] The Future of the Nuclear Fuel Cycle, MIT (ISBN 978-0-9828008-4-3) 2011.

Environmental impacts of ethylene production from diverse feedstocks and energy sources

Madhav Ghanta · Darryl Fahey · Bala Subramaniam

Abstract Quantitative cradle-to-gate environmental impacts for ethylene production from naphtha (petroleum crude), ethane (natural gas) and ethanol (corn-based) are predicted using GaBi® software. A comparison reveals that the majority of the predicted environmental impacts for these feedstocks fall within the same order of magnitude. Soil and water pollution associated with corn-based ethylene are however much higher. The main causative factor for greenhouse gas emissions, acidification and air pollution is the burning of fossil-based fuel for agricultural operations, production of fertilizers and pesticides needed for cultivation (in the case of ethanol), ocean-based transportation (for naphtha) and the chemical processing steps (for all feedstocks). An assessment of the environmental impacts of different energy sources (coal, natural gas and fuel oil) reveals almost similar carbon footprints for all the fossil fuels used to produce a given quantity of energy. For most of the environmental impact categories, the GaBi® software reliably predicts the qualitative trends. The predicted emissions agree well with the actual emissions data reported by a coal-based power plant (Lawrence Energy Center, Lawrence, KS) and a natural gas-based power plant (Astoria Generating Station, Queens, NY) to the United States Environmental Protection Agency. The analysis shows that for ethylene production, fuel burning at the power plant to produce energy is by far the dominant source (78–93 % depending on the fuel source) of adverse environmental impacts.

Keywords Ethylene · Environmental impact analysis · Fossil fuels · Corn ethanol · Life-cycle assessment

Introduction

Ethylene, with a worldwide consumption of 133 million tonnes/year, is the chemical industry's primary building block [1]. Major industrial uses of ethylene include (a) polymerization to polyethylene and other copolymers; (b) oligomerization to normal alpha-olefins; (c) oxidation to ethylene oxide and acetaldehyde; (d) halogenation and dehydrohalogenation to vinyl chloride; (e) alkylation of benzene to ethylbenzene; and (f) hydroformylation to propionaldehyde [1–3]. In the USA, 70 % of the total ethylene production capacity comes from steam cracking of naphtha and the remaining 30 % from the thermal cracking of ethane [4]. The increased availability of natural gas (and thus ethane) in the USA, as a result of hydraulic fracturing of shale rock, has stimulated feasibility studies of building new ethylene crackers by Chevron Phillips Chemical Company (1.5 million tonnes/year), LyondellBasell Industries (400,000 tonnes/year), Dow Chemical Company (900,000 tonnes/year), Shell Chemical Company (1 million tonnes/year) and Sasol (1 million tonnes/year) [5, 6].

An alternative source for ethylene is the dehydration of ethanol obtained from a renewable source, such as corn, sugarcane, and from cellulose or agricultural waste. Ethylene sourced from sugarcane is claimed to be greener than that produced from fossil fuel-based sources [7]. There is

M. Ghanta · D. Fahey · B. Subramaniam
Center for Environmentally Beneficial Catalysis, University of Kansas, Lawrence, KS 66045-7609, USA

M. Ghanta · B. Subramaniam (✉)
Department of Chemical and Petroleum Engineering, University of Kansas, Lawrence, KS 66045-7609, USA
e-mail: bsubramaniam@ku.edu

significant interest in green polyethylene from major companies such as Procter & Gamble (consumer goods manufacturer), Tetra Pak (packaging company) and Shiseido (cosmetic company) [7]. Dow Chemical Company and Braskem have announced plans to construct an integrated complex for the production of polyethylene based on sugarcane ethanol in Brazil [8]. While there is strong consumer interest in producing green polyethylene from biomass, the increased availability in the USA of relatively inexpensive ethane feedstock has significantly eroded the cost competitiveness of ethylene sourced from renewable feedstocks. However, in the longer term, bio-based feedstocks are the only sustainable option for producing chemicals.

The cracking of naphtha or of ethane to ethylene is highly energy intensive [9]. For ethylene production from corn via ethanol, the ethanol concentration in the effluent stream of the fermentation reactor dictates the energy intensity for ethanol enrichment and its subsequent dehydration to ethylene. Further, the type of fuel used for energy production influences the overall environmental impact. In this work, we perform a comparative environmental impact assessment (cradle-to-gate life-cycle analysis) to quantify the major contributors to the environmental impacts for ethylene production from naphtha, ethane and ethanol, employing natural gas as the energy source in each case. In addition, we also compare the environmental impacts when using other fossil fuels such as coal and oil as the energy sources. Where possible, we have also compared the GaBi® software predictions with reported plant emissions data in an attempt to establish the reliability of such predictions.

Methodology

Simulation

GaBi 4.4® software [10] is employed to perform comparative gate-to-gate, cradle-to-gate, and cradle-to-grave life-cycle assessments (LCA) for ethylene and energy production. The raw material and energy datasets provided by GaBi® are based on current technologies. The process simulation used in the GaBi® datasets incorporates process (heat, water and mass) integration and waste treatment technologies. Even though the GaBi® software is designed to perform environmental assessments and generate reports that conform to ISO 14040 [11] and ISO 14044 standards [12], the current analysis deviates from those rigorous standards in certain areas such as the definition of a functional unit, use of average market mix for representing diverse energy sources and the use of allocation. However, the environmental assessment *methodology* follows the

procedures generally adopted to ultimately develop ISO-compliant reports. Hence, the conclusions are unaltered by these deviations.

A USA-specific environmental assessment is performed by employing the US-specific life-cycle inventory (USLCI) and an embedded software tool known as tools for reduction and assessment of chemicals and other environmental impacts (TRACI) [10, 13]. The TRACI software, developed by the United States Environmental Protection Agency (USEPA), is designed based on the midpoint centric approach proposed by Intergovernmental Panel on Climate Change (IPCC). The TRACI methodology enables the generation of impact parameters that are USA specific. Empirical models developed by the US National Acid Precipitation Assessment Program and California Air Resource Board were utilized to estimate the acidification and smog formation potential. Human health cancer and non-cancer impact categories were estimated based on models developed using the USEPA Risk Assessment Guidance and USEPA's exposure factor handbook [14]. The potential effects of various production operations on environmental impact categories such as acidification, greenhouse gas emissions, ecotoxicity, human carcinogenic and non-carcinogenic effects, and eutrophication are estimated (see definitions in Supplementary Materials, Appendix A) [15].

Basis of estimations and common assumptions

The production basis for the estimated environmental impacts is assumed as *400,000 tonnes of ethylene/year* from each of the following sources: naphtha (petroleum crude), ethane (natural gas) and ethanol derived from corn (biomass). We chose this basis to facilitate comparison of the GaBi®-predicted emissions/impacts with those reported by the ExxonMobil Baytown ethylene cracker with a similar production capacity. It should be clear that even though we do not use a functional unit of 1-kg ethylene produced (as per ISO guidelines), our quantitative results may be suitably scaled to obtain environmental impacts for a functional unit of 1-kg ethylene produced. For each source, a proportional allocation method based on the energy content of the various products formed is employed to estimate the environmental impacts of ethylene production [10, 13]. We further assume that the electricity requirement for all the feedstocks is met with natural gas as fuel (later in this manuscript, we also assess the environmental impacts of using other fossil-based fuels).

Although the current US electricity generation capacities are similar for coal and natural gas [16], the majority (approximately 80 %) of the newer electricity generation capacity in the US uses natural gas [17]. Hence, natural gas is considered as the fuel source in this analysis. Given that valuable co-products are formed during the production of ethylene, the absolute environmental impact is estimated using a proportional energy allocation method, which is based on the energy content of the desired products relative to the energy content of all products and co-products formed with a particular feedstock. The allocation factor is estimated as the net calorific value of the desired product to the total calorific value of all products formed during the production of ethylene with each feedstock. While ISO guidelines suggest against allocation, this should be less of a concern in a comparative analysis if the same type of allocation is used for processes being compared. Further, our methodology allows us to predict the environmental impacts per capita for the various sources (i.e., per unit of ethylene feedstock source and per unit of energy source). From such predictions, it is possible to predict the impacts of using average mixtures of feed and energy sources for any region and time period, as required by ISO guidelines. "Weak point analyzer", a tool embedded in the GaBi® software was employed to perform a *dominance/contribution analysis* to identify the major environmental impact categories for producing ethylene. The common assumptions and boundaries for USA-based ethylene production from the three feedstocks are described in the following section.

System description

Ethylene from naphtha

Figure 1 shows the various processing steps, from crude oil recovery to crude oil transport to refinery processes, involved in producing ethylene from naphtha as the feedstock and natural gas as the energy source. For each of these steps, the various inputs and outputs considered when evaluating the overall environmental impacts are also shown. The schematic shown in Fig. 1 represents a cradle-to-gate life-cycle analysis. The environmental impact analysis further assumes the following: (a) a US crude oil mix dataset; (b) the naphtha obtained during the atmospheric distillation of crude oil has the following composition: C_3–C_4 (8 %), C_5 (22.4 %), C_6 (19.9 %), C_7 (18.2 %), C_8 (12.4 %), C_9 (11.5 %), C_{10}–C_{15} (8.6 %); and (c) the yield of ethylene from cracking naphtha is 30 % with the following co-products: H_2 and CH_4 (17 %), propylene (3 %), butadiene (2 %), C_4 olefins (1 %), pyrolysis gasoline (2 %) and benzene (1 %) [10, 18]. The fuel and power requirements for the steam cracking step are 20.1 and 0.3 GJ/tonne of ethylene, respectively [19]. A weighting factor of 0.058 (methodology shown in Supplementary Material, Section B) is utilized to estimate the environmental impacts associated with ethylene production from naphtha [20]. As shown in Fig. 1, the LCA analysis

incorporates the environmental impacts of producing the energy (from natural gas) required for the extraction of crude oil from reservoirs, transportation to a refinery in the USA and further processing to produce ethylene. The transportation involves the pumping of the crude oil from a Middle Eastern source to the nearest seaport via pipeline, subsequent shipping in a tanker to the USA (distance is assumed to be 8,000 km, typical of the distance from a Middle East destination), and delivery from the US port of entry to the refinery via pipeline.

Ethylene from ethane

Figure 2 shows the various processing steps, from natural gas recovery, its purification, transport to a refinery and ultimately the steam cracking of ethane to produce ethylene using energy sourced from natural gas. This cradle-to-gate environmental impact analysis assumes: (a) natural gas obtained from both conventional wells (65 %) and shale rock (35 %) with the following composition: methane (73 mol%); ethane (8 mol%); propane (5 mol%); and butane (3 mol%); carbon dioxide (5 mol%); oxygen (0.15 mol%); nitrogen (2 mol%); hydrogen sulfide (3 mol%) and traces of rare gases such as argon, helium, neon and xenon [21], (b) the recovered natural gas is processed to reduce the concentration of sulfur and moisture prior to pipeline transportation to the natural gas processing facility where it is fractionated into its individual components, and (c) the ethane fraction is cracked to produce ethylene (80 % selectivity or 56.4 % yield) along with coproducts (hydrogen, methane, propane, butane, propylene, acetylene, propadiene, vinylacetylene, propyne and butadiene) [10]. The fuel and power requirements for steam cracking of ethane are 13.7 and 0.2 MJ/kg of ethylene, respectively [19]. A weighting factor of 0.125 (rationale shown in Supplementary Materials, Section B) is utilized to estimate the environmental impacts associated with ethylene production from ethane.

Ethylene from ethanol

Figure 3 shows the various processes considered in this cradle-to-gate environmental impact analysis for ethylene production from corn-based ethanol. As shown in Fig. 3, the energy-intensive steps associated with ethanol production from corn include soil cultivation, planting, pesticide and fertilizer manufacture and its application, harvesting, transport to the refinery, fermentation, and distillation of ethanol to remove the water [22]. Approximately 308 million kilograms of pesticides and insecticides are used for corn production [23]. The energy requirement for the production of the active ingredient (assumed as glyphosphates,

Fig. 1 *Block diagram* describing the various processes included in the cradle-to-gate life-cycle assessment for the production of ethylene from naphtha sourced from petroleum crude

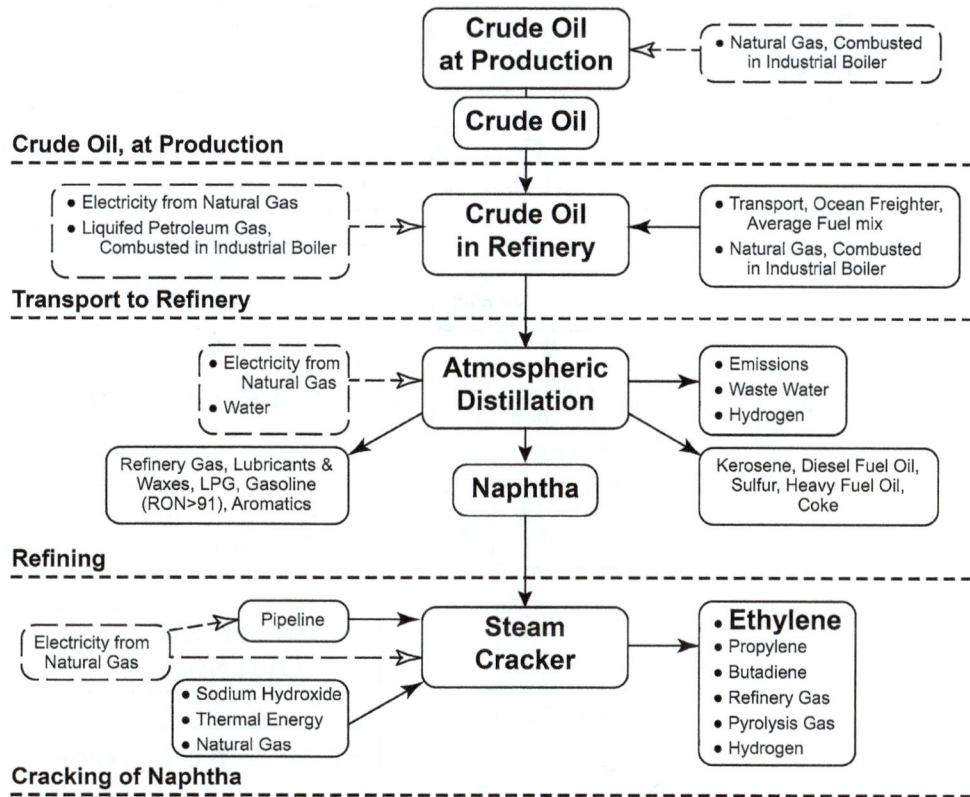

Fig. 2 *Block diagram* describing the various processes included in the cradle-to-gate life-cycle assessment for the production of ethylene from ethane sourced from natural gas recovered from both conventional wells and shale rock

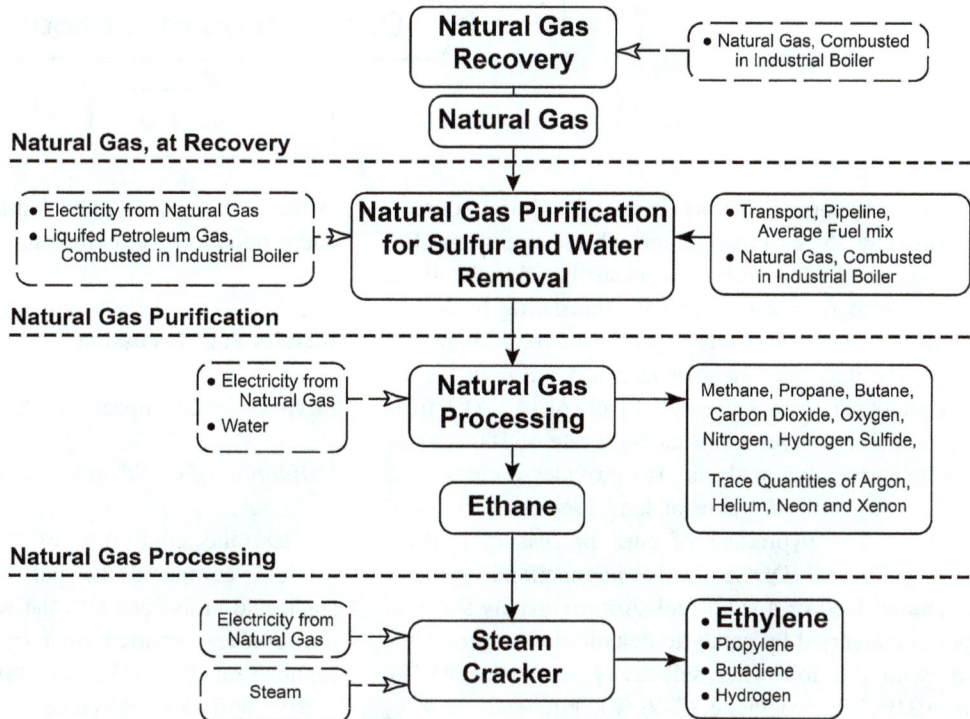

the newest pesticide extensively used for corn production), formulating the active ingredient into pesticide microgranules, packaging and transportation are approximately 457.6, 20, 2 and 1 MJ/kg, respectively [24]. The fertilizers used are urea, monoammonium phosphate, ammonium nitrate and NPK-15. The average values of nitrogen-, phosphate- and potash-based fertilizer consumed for corn production in the USA are 63.5, 27.2 and 35.8 kg/acre, respectively [25]. Approximately 73 % (217.9 million tonnes of CO_2 equivalent) of the overall US N_2O emissions (300.3 million

Fig. 3 *Block diagram* describing the various processes included in the cradle-to-gate life-cycle assessment for the production of ethylene from ethanol obtained from corn

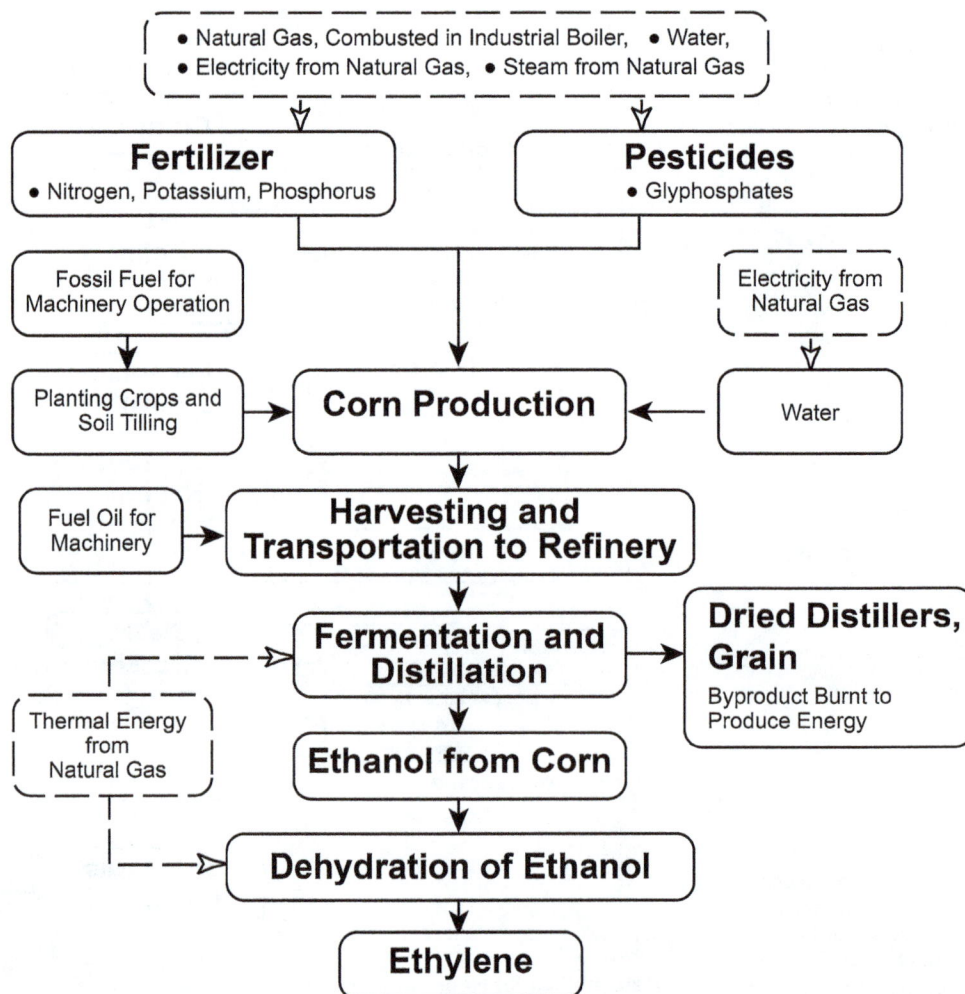

tonnes of CO_2 equivalent) are from agricultural sources. Approximately 75 % of the US agricultural emissions (165 million tonnes of CO_2 equivalent) is attributed to the direct emissions from fertilization of soil, translating to ~ 2.55 tonnes of CO_2 equivalent/acre of land used for corn growth [26, 27]. The data for ethanol sourcing from corn (in the USLCI database) assume an ethanol yield of 14.1 wt% from corn. Assuming an average production rate of 180 bushels of corn per acre, this analysis also provides a credit of 8 tonnes of CO_2 for every acre of land used for corn cultivation [10]. The byproduct of corn processing is dried distillers grain seed (DDGS), which has economic value as either animal feed or a solid fuel. Approximately 99 % of ethanol is converted by catalytic dehydration to produce a stream with the following selectivity: ethylene (96 %), ethane (0.05 %), propylene (0.06 %), butylenes (2.4 %) and acetaldehyde (0.2 %) [28]. The total energy required to dehydrate ethanol is 1.6 MJ/kg ethanol [29]. The net calorific values of DDGS and ethylene serve as the basis for allocating the environmental impact of ethylene production from ethanol. A weighting factor of 0.63 (methodology shown in Supplementary Materials, Section B) is utilized to

estimate the environmental impacts associated with ethylene production from ethanol.

Results and discussion

Environmental impacts of ethylene production

Validation of GaBi® predictions

To test the credibility of the computational approach, Table 1 compares the gate-to-gate emissions (fugitive, stack and emissions into the water stream) from an ethylene cracker reported by ExxonMobil (capacity: 400,000 tonnes/year) [30, 31] to the USEPA with those predicted by GaBi® software. Ethylene is sourced from naphtha [32–34]. A comparison of the *gate-to-gate* emissions shows that the GaBi® software reliably predicts the types of emissions and the qualitative trends. The quantitative predictions for a majority of impact categories (8 out of 15 with data for 5 categories not available in the public domain for comparison purposes) are of same order of magnitude as the

Table 1 Environmental impacts associated with ethylene production from naphtha with natural gas as the energy source: comparison of GaBi® predictions with toxic release inventory data reported by ExxonMobil (Baytown facility). The capacities for both are 400,000 tonnes of ethylene/year

Category	GaBi® gate-to-gate (millions)*	ExxonMobil (millions)	
		Released	Treated waste
Acidification (mol H$^+$ equivalent (eq.))	24.2	11.4	11.9
Eco-toxicity air (kg 2,4-dichlorophenoxyace eq.)	0.561	0.94	0.94
Ecotoxicity-surface soil (kg benzene eq.)	0.00167	n.a.	n.a.
Eco-toxicity water (kg 2,4-dichlorophenoxyace eq.)	9	8.1	56.7
Eutrophication (kg N-eq.)	0.019	0.015	0.016
Greenhouse gas (GHG) emissions (kg CO_2-eq.)	29.4	n.a.	n.a.
Human health cancer-air (kg benzene eq.)	0.018	0.078	1.09
Human health cancer-surface soil (kg benzene eq.)	6.61 (10^{-6})	2.2 (10^{-3})	0.08
Human health cancer water (kg benzene eq.)	0.012	0.029	1.5
Human health cancer air point source (kg benzene eq.)	0.186	n.a.	n.a.
Human health non-cancer air (kg toluene eq.)	36.4	13.9	27.8
Human health non-cancer surface soil (kg toluene eq.)	0.134	n.a.	n.a.
Human health non-cancer water (kg toluene eq.)	415	556	556
Ozone depletion potential (kg CFC-11 eq.)	2.49 (10^{-6})	n.a.	n.a.
Smog air (kg NO_x eq.)	4.08 (10^{-4})	2.4 (10^{-6})	1.8 (10^{-4})

* The results can be suitably scaled to obtain environmental impacts for a functional unit of 1 kg ethylene produced (for ISO-compliant reporting purposes) [31]

n.a. data not available at the toxic release inventory

Table 2 GaBi®-predicted cradle-to-gate environmental impacts associated with manufacturing 400,000 tonnes of ethylene from naphtha, ethane and ethanol using natural gas as energy source in all cases

Category	Naphtha (millions)	Ethane (millions)	Ethanol (millions)
Acidification (mol H$^+$ equivalent (eq.))	531.0	376	467.3
Eco-toxicity air (kg 2,4-dichlorophenoxyace eq.)	2.48	0.07	1.1
Ecotoxicity-surface soil (kg benzene eq.)	0	0	0.016
Eco-toxicity water (kg 2,4-dichlorophenoxyace eq.)	51	78	30
Eutrophication (kg N-eq.)	0.003	0	1.4
Greenhouse gas emissions (kg CO_2-eq.)	198	167	268
Human health cancer-air (kg benzene eq.)	0.24	0.11	0.32
Human health cancer-surface soil (kg benzene eq.)	0	0	0.74
Human health cancer water (kg benzene eq.)	0.6	0.26	1.4
Human health cancer air point source (kg benzene eq.)	3.5	1.8	2.0
Human health non-cancer air (kg toluene eq.)	1,130	20	300
Human health non-cancer surface soil (kg toluene eq.)	0	0	29,700
Human health non-cancer water (kg toluene eq.)	12,100	5,300	46,300
Ozone depletion potential (kg CFC-11 eq.)	27.4	0	47.6
Smog air (kg NO_x eq.)	5.9 (10^{-3})	1.4 (10^{-4})	3.8 (10^{-3})

reported emissions. This trend is similar to what we have reported elsewhere [35–37].

Cradle-to-gate life-cycle assessment

The environmental impacts of ethylene production from naphtha, ethane and ethanol as feedstocks, using natural gas as the energy source in all cases, are compared in Table 2. The cradle-to-gate environmental impacts (Table 2) for a majority of impact categories (11 of 15) are within an order of magnitude and thus their differences lie within prediction uncertainty. As expected, the predicted cradle-to-gate impacts are greater than the predicted gate-to-gate emissions (listed in Table 1) in most categories, from a few-fold to several orders of magnitude depending on the impact category. For the results listed in Table 2, a dominance analysis identified the following environmental impact categories to be noteworthy.

Greenhouse gas (GHG) emissions Ethylene production from ethanol involves highly energy-intensive steps in the overall life cycle, including H_2 production for ammonia fertilizer manufacture, the dehydration of ethanol (highly endothermic requiring 1.6 MJ/kg of ethylene) [38], and the separation of water from ethanol. Further, CO_2 is a byproduct in the steam reforming of CH_4, the dominant process for H_2 production [39]. The cumulative greenhouse gas emissions for the steam cracking of naphtha and ethane amount to 1,135 and 840 kg CO_2/tonne of ethylene, respectively. As shown in Table 2, natural gas burning to produce process energy is a major contributor to GHG emissions for ethylene production from all feedstocks. In the case of ethanol feedstock, CO_2 removal from the atmosphere by corn photosynthesis only partly offsets these emissions [40].

Acidification In general, this category is dominated by SO_X and NO_X emissions. Natural gas burning (for all ethylene sources) and ocean-based transportation of crude oil in ships powered by bunker-fuel (~ 15 %) are the major contributors [41, 42]. In contrast, transporting natural gas via pipeline accounts for approximately 3 % of the overall environmental impact.

Ecotoxicity—air In this category, the general causative factors are emissions of metals (copper, selenium and zinc), nonmetals (arsenic) and organic chemicals (such as polychlorinated biphenyls) into the atmosphere. The energy generated to process the crude oil (including extraction, refining and naphtha cracking) is the major contributor [43].

Ecotoxicity ground surface soil In general, contamination of soil by corn farming contributes to this impact category. Extensive use of chemical fertilizers and pesticides for corn production results in the contamination of soil by metals such as zinc, copper and nickel which constitute approximately 0.1 wt% of the fertilizer mass [44]. In 2009, the consumptions of nitrogen-, phosphate- and potash-based fertilizer were 4.8, 1.42 and 1.45 million nutrient tonnes, respectively, making corn production the most fertilizer intensive among all the crops grown in the USA [41]. Common agricultural practices such as conventional tilling (practice of turning or digging up soils) to prepare fields for seeding new corn remove organic residue from the top soil surface left by previous harvests or cover crops, further exacerbating the fertilizer requirement for cultivation. In comparison, soil pollution is negligible for ethylene production from crude oil and natural gas.

Ecotoxicity—Water In general, partitioning of metals (copper, nickel and chromium) and non-metals (arsenic)

into water reservoirs, lakes, and rivers contributes to this impact category. Leaching of the heavily fertilized top soil (during corn farming) by run-off of rain water or from irrigation is a major reason for this contamination. Inadequate rain and extensive irrigation during cultivation also adversely impact the local ecosystem due to the exhaustion of the water table and reduced water levels in water reservoirs, lakes and rivers. Water is also used in the fermentation of corn to ethanol (1.65 gallons of water/kg of ethylene produced). For comparison, 2–2.5 gallon of water is needed for the production of a kilogram of ethylene from naphtha [45].

Eutrophication Erosion of fertilized soil containing ammonia, nitrates and phosphates in corn farming and N_2O emissions are the main causative factors in eutrophication of fresh water [46]. In addition, wastewater from an ethanol processing facility has a high biological oxygen demand (BOD) value of 18,000–37,000 mg/L [47]. The direct emission caused by the microbial and chemical reduction of nitrates (biological denitrification and chemodenitrification), addition of mineral N-containing substrates (ammonium phosphate), animal manures, crop residues, nitrogen-fixing crops and sewage sludge to agricultural soils are the major sources of N_2O emissions [49, 50]. Approximately, 1.25 wt% of nitrogen present in the fertilizer is emitted into the atmosphere as N_2O. In comparison, the eutrophication potential for naphtha from crude oil is substantially lower (0.003 kg N eq. vs. 1.4 kg N eq. for corn ethanol), attributed to the NO_x emissions during ocean-based transportation of crude oil [48].

The overall environmental impact of ethylene production is thus similar for naphtha, ethane, and corn-ethanol feedstocks. It should be emphasized that ethylene sourced from cellulosic ethanol will have a different environmental impact. For all the ethylene sources, the environmental impact assessment identifies the energy production step (natural gas-based electricity) as the biggest contributor ($\sim$approximately 85 % of the overall environmental impact). In the following section, an analysis is performed to quantify and compare the environmental impact of producing energy from other fossil fuel sources (coal and fuel oil) as well.

Influence of energy source on environmental impacts

The foregoing LCA of ethylene production from various feedstocks assumes natural gas as the source of process energy. To determine the effect of the energy source, we also performed environmental impact assessments to quantify the impacts of generating process energy from various fossil fuels such as coal (hard coal, lignite), fuel oil (heavy fuel oil, light fuel oil) and natural gas. In each case, the cumulative environmental impacts of burning the fuels to produce 1,000 MJ of energy consider (a) extraction of the fuel from its source; (b) transportation of the extracted fuel to the power plant; and (c) production of 1,000 MJ energy at the power plant.

Impacts for coal

Coal is classified based on carbon, ash and inherent moisture content. Hard coal, also known as anthracite, is the best quality coal with a high carbon content and calorific value. Lignite, commonly known as brown coal, has a relatively lower energy content due to high inherent moisture and ash contents [51]. In the USA, lignite coal is primarily used for electricity production whereas hard coal is used for metal processing. Table 3 lists the GaBi® estimated impacts for the mining of coal alone and the overall *cradle-to-grave* impacts for producing energy from the mined coal. The difference in these impacts is attributed to

the emissions from a power generation facility. It must be noted that when producing energy by combustion of the fuel, the emissions beyond the power plant represent the "grave" for the fuel. As shown in Table 3, the predicted impacts for coal from an underground mine are greater than those for a surface mine in most categories except water pollution. However, the differences lie within the prediction uncertainty. Further, the Greenhouse Gas (GHG) emissions (global warming potential) from burning at the power plant [76.5 kg CO_2 Equivalent/1,000 MJ of energy] are similar to the actual emissions reported to the USEPA [78 kg CO_2 Equivalent/1,000 MJ of energy] by the Lawrence Energy Center (Lawrence, KS) [52]. This facility utilizes anthracite coal obtained from a surface mine [53, 54]. The emissions associated with burning fuel at the power plant contribute to 78 % of the overall environmental impact whereas energy usage during mining and transportation of the fuel contributes to approximately 17 and 5 %, respectively, of the overall impact.

Table 3 Predicted and actual impacts (italicized column) of producing 1,000 MJ of energy from coal

Category	Predicted impacts from coal mining		Predicted overall cradle-to-grave impacts for energy production		Predicted impacts of energy generation at power plant		TRI data for Lawrence Energy Center [51]
	Hard coal, mine	Lignite coal, surface mine	Hard coal	Lignite coal	Hard coal	Lignite coal	
Acidification (mol H$^+$ equivalent (eq.))	17.2	1.84	39.36	30.97	22.16	29.13	*7.35*
Eco-toxicity air (kg 2,4-dichlorophenoxyace eq.)	0.386	0.0189	0.457	0.518	0.071	0.4991	*0.00135*
Eco-toxicity surface soil (kg benzene eq.)	–	–	4.8 (10^{-5})	2.4 (10^{-5})	–	–	*N/A*
Eco-toxicity water (kg 2,4-dichlorophenoxyace eq.)	0.31 (10^{-4})	0.74 (10^{-4})	2.31 (10^{-4})	1.1 (10^{-4})	1.99 (10^{-4})	0.3 (10^{-4})	*1.5 (10^{-4})*
Eutrophication (kg N-eq.)	2.34 (10^{-3})	8.24 (10^{-4})	9.41 (10^{-3})	9.51 (10^{-3})	7.07 (10^{-3})	8.7 (10^{-3})	*11.4 (10^{-3})*
Greenhouse gas (GHG) emissions (kg CO_2-eq.)	20.8	7.72	97.32	100.31	76.5	92.6	*78*
Human health cancer-air (kg benzene eq.)	0.011	0.00323	0.099	0.0205	0.088	0.0172	*0.031*
Human health cancer-SS (kg benzene eq.)	–	–	1.5 (10^{-7})	0.5 (10^{-7})	–	–	*N/A*
Human health cancer water (kg benzene eq.)	0.000322	0.00126	0.011	0.002	0.01067	7.4 (10^{-4})	*N/A*
Human health cancer air point (kg benzene eq.)	0.123	0.014	0.227	0.184	0.104	0.17	*0.126*
Human health non-cancer air (kg toluene eq.)	24.099	5.39	49.9	32.44	25.80	27.05	*N/A*
Human health non-cancer GSS (kg toluene eq.)	–	–	0.0035	0.0014	–	–	*N/A*
Human health non-cancer water (kg toluene eq.)	7.59	16.6	9.117	29.8	1.52	13.44	*15*
Ozone depletion potential (kg CFC-11 eq.)	3.77 (10^{-11})	1.47 (10^{-11})	5 (10^{-8})	36 (10^{-8})	4.99 (10^{-8})	3.6 (10^{-7})	*N/A*
Smog air (kg NO$_x$ eq.)	5.35 (10^{-5})	1.91 (10^{-5})	19 (10^{-5})	21 (10^{-5})	13.6 (10^{-5})	19 (10^{-5})	*26 (10^{-5})*

N/A data not available from toxic release inventory

Impacts for natural gas

Natural gas, which is predominantly methane, has a low sulfur content and high specific energy (MJ/kg) compared to the other sources. Table 4 lists the environmental impacts of extracting and transporting natural gas from reservoirs, and of producing 1,000 MJ energy from natural gas at a power plant. Natural gas burning at the power plant contributes to 89 % of the overall impact whereas natural gas extraction and transportation contribute to 6 and 4 %, respectively, of the overall impact. As shown in Table 4, the predicted emissions in the various categories are of the same order of magnitude as those reported to the USEPA by the Astoria Generating Station (Queens, New York) [55, 56].

Impacts for fuel oil

Heavy fuel oil (Number 6, residual fuel oil, bunker fuel oil) mainly comprises residues from cracking and distillation units in the refinery. These fuels have higher mass density and high carbon/hydrogen ratios compared to light fuel oil (Number 3 fuel oil) [57]. Table 5 compares the predicted impacts associated with fuel oil production (high boiling fraction of crude oil) and emissions associated with the burning of both heavy and light fuel oils. The impacts of generating energy from heavy and light fuel oils are similar, with the differences being within prediction uncertainty. The impacts of producing 1,000 MJ energy at the

power plant account for approximately 93 % of the overall impact whereas oil extraction and oil transportation contribute to 3 and 4 % of the overall environmental impact, respectively.

Major adverse environmental impacts of various energy sources

For all energy sources considered in this work, the overall *cradle-to-grave* environmental impacts for energy production for a majority of environmental impact categories are within an order of magnitude with fuel burning at the power plant being the major contributor (78–93 % based on the energy source) to environmental pollution (Tables 3, 4, 5). The major sources of pollution are discussed in the following section.

Greenhouse gas (GHG) emissions The *cradle-to-grave* impacts estimated for coal (Table 3) and oil (Table 5) differ by approximately 10 %, which is within the prediction uncertainty. The *cradle-to-grave* carbon footprint for natural gas is lower by approximately 25 % (see Table 4) compared to either coal or oil. This is primarily attributed to the low carbon content and the higher calorific value of natural gas. This analysis however assumes that there is no contribution to GHG emissions by natural gas leakage.

Acidification potential of the various energy sources is dictated by the sulfur content and the associated SO_2 emissions during fuel burning. While NO_X emissions result

Table 4 Predicted and actual impacts (italicized column) of producing 1,000 MJ of energy from natural gas

Category	Predicted impacts from natural gas extraction	Predicted overall *cradle-to-grave* impacts for energy production	Predicted impacts of energy generation at plant	*Astoria generating station, TRI data* [53]
Acidification (mol H$^+$ eq.)	0.98	6.15	5.17	*N/A*
Eco-toxicity air (kg 2,4-dichlorophenoxyace eq.)	0.0031	0.013	9.9 (10^{-3})	*0.0016*
Eco-toxicity surface soil (kg benzene eq.)	–	–	–	*N/A*
Eco-toxicity water, (kg 2,4-dichlorophenoxyace eq.)	5.4 (10^{-4})	7.25 (10^{-4})	1.85 (10^{-4})	*5.4 (10^{-4})*
Eutrophication (kg N-eq.)	2.14 (10^{-3})	4.97 (10^{-3})	2.83 (10^{-3})	*N/A*
Greenhouse gas (GHG) emissions (kg CO$_2$-eq.)	7.6	74.86	67.26	*29.3*
Human health cancer-air (kg benzene eq.)	0.000681	0.0045	0.0038	*0.0017*
Human health cancer-GSS (kg benzene eq.)	–	–	–	*N/A*
Human health cancer water (kg benzene eq.)	0.0151	0.007	0.008	*0.007*
Human health cancer air point (kg benzene eq.)	0.00539	0.0474	0.0420	*N/A*
Human health non-cancer air (kg toluene eq.)	0.805	9.18	8.37	*19.9*
Human health non-cancer GSS (kg toluene eq.)	–	0.034	0.034	*N/A*
Human health non-cancer water (kg toluene eq.)	105.52	163	57.4	*39.7*
Ozone depletion potential (kg CFC-11 eq.)	1.18 (10^{-11})	63 (10^{-8})	6.29 (10^{-7})	*N/A*
Smog air (kg NO$_x$ eq.)	4.49 (10^{-6})	10 (10^{-5})	9.5 (10^{-5})	*N/A*

eq. equivalent, *N/A* data not available from toxic release inventory

Table 5 Predicted impacts of producing 1,000 MJ of energy from fuel oil

Category	Predicted impacts for fuel oil production (extraction and refining to obtain fuel oils)	Predicted cradle-to-grave impacts for energy production	
		Heavy fuel oil	Light fuel oil
Acidification (mol H$^+$ eq.)	1.23	24.18	9.58
Eco-toxicity air (kg 2,4-dichlorophenoxyace eq.)	0.004	0.356	0.054
Eco-toxicity surface soil (kg benzene eq.)	–	8.2 (10^{-4})	9 (10^{-4})
Eco-toxicity water (kg 2,4-dichlorophenoxyace eq.)	25.1 (10^{-4})	80.4 (10^{-4})	87.3 (10^{-4})
Eutrophication (kg N-eq.)	2.26 (10^{-3})	5.32 (10^{-3})	5.16 (10^{-3})
Greenhouse gas (GHG) emissions (kg CO$_2$-eq.)	4.83	95.38	89.54
Human health cancer-air (kg benzene eq.)	0.000854	0.0835	0.0017
Human health cancer-SS (kg benzene eq.)	–	28 (10^{-7})	30 (10^{-7})
Human health cancer water (kg benzene eq.)	0.00252	0.0031	0.0031
Human health cancer air point (kg benzene eq.)	0.0066	0.141	0.066
Human health non-cancer air (kg toluene eq.)	0.89	91.67	17.04
Human health non-cancer SS (kg toluene eq.)	–	0.0627	0.0679
Human health non-cancer water (kg toluene eq.)	60	60.95	63.65
Ozone depletion potential (kg CFC-11 eq.)	8.25 (10^{-10})	22 (10^{-8})	21 (10^{-8})
Smog air (kg NO$_x$ eq.)	5.59 (10^{-6})	11 (10^{-5})	11 (10^{-5})

eq. equivalent

from fuel burning, the actual amounts are relatively small. As shown in Table 3, the acidification potential of hard coal is higher than lignite coal by 21 % even though lignite contains more sulfur than hard coal. This is because of the fact that energy production from lignite requires state-of-the-art SO$_X$ and NO$_X$ abatement technologies to meet the stringent environmental regulations. The higher S content in heavy fuel oil results in higher acidification potential compared to light fuel oil (Table 5).

Ecotoxicity—air The metal emissions for coal and heavy fuel oil are greater than light fuel oil and natural gas by an order of magnitude (Tables 3, 4, 5). In coal, zinc is present in the sphalerite form that has a low melting point and hence is easily susceptible to vaporization resulting in metal emissions. Heavy metal emissions (such as chromium) in fuels depend on the properties and concentration of metals and the technologies used for combustion and post-combustion cleanup.

Human health cancer air Potential metal emissions for energy production from hard coal, lignite and heavy fuel oil are similar but an order of magnitude greater than those reported for natural gas and light fuel oil (Tables 3, 4, 5). Combustion of coal (anthracite and lignite) produces significant arsenic emissions, which have high toxicity and persistence [58]. Potential metal emissions for energy production from hard coal, lignite and heavy fuel oil are similar but an order of magnitude higher than that reported for natural gas and light fuel oil (Tables 3, 4, 5). Combustion of coal also produces mercury, nickel and chromium emissions [58]. In 2005, the amounts of SO$_X$ and

NO$_X$ emissions reported by the Lawrence Energy Center (KS) a coal fired power plant, are 0.066 (10^{-3}) and 0.0988 (10^{-3}) kg/MJ, respectively. In contrast, the natural gas fired power plant at Seminole (FL) reported SO$_X$ and NO$_X$ emissions of 0.158 (10^{-3}) and 0.02 (10^{-3}) kg/MJ in 2010, respectively [59]. In 2010, mercury emissions from coal-fired plants using state of the art mercury capture techniques are approximately 0.27 (10^{-9}) kg Hg/MJ of energy produced [60]. The mobility of arsenic in the atmosphere during mining, combustion and storage of coal is dependent on its mode of occurrence. Arsenic in hard coal and lignite is present in the pyrite organic phase. The storage facilities and waste material are major sources of arsenic mobilization. Clean coal technologies, employed to reduce sulfur content, are known to reduce arsenic concentration resulting in lower arsenic emissions during energy production from lignite [61].

The results from the foregoing analysis can be easily scaled to reflect *per capita* environmental impacts and can therefore be utilized to quantify the environmental impacts of energy production from various energy sources in general.

Situations that increase and reduce carbon footprint

For ethylene production from naphtha, the composition of the crude oil has a significant influence on the overall environmental impact. For example, increased sulfur and nitrogen contents in crude oil will adversely impact the process energy requirement and overall yield of the process. In the case of ethylene from ethane cracking, the inclusion of fugitive emissions during the handling of natural gas, such as methane (with a global warming potential of 25 times that of CO_2 on a weight basis), will significantly worsen the overall environmental impact. For ethylene from corn ethanol, the environmental impacts can be lower under the following scenarios: increased ethanol yields either due to the development of genetically modified corn or commercialization of technologies that can process both corn and corn stover (cellulose, hemi-cellulose and lignocellulose); and development of corn strains that require less fertilizer and water and also have a higher resistance to pests. Clearly, regardless of feedstock used in ethylene production, the production of energy from renewable sources and deployment of green energy technologies will significantly reduce the carbon footprint.

Conclusions

The cumulative emissions associated with the production of ethylene from naphtha, ethane and ethanol, with the process energy derived from natural gas, are similar and the differences lie within the prediction uncertainty of the cradle-to-gate life-cycle assessment. For ethylene produced from naphtha and ethane, the energy expended during the extraction and ocean-based transportation of fossil fuel sources (crude oil and natural gas) contributes significantly to adverse environmental impacts such as GHG emissions, acidification, and eco-toxicity (air and water). The eutrophication of water bodies is virtually negligible for these feedstocks. In the case of ethylene production from corn, the main contributor to adverse environmental impacts is the burning of natural gas to generate the energy needed for (a) producing the raw materials (including corn, fertilizers and pesticides), (b) endothermic dehydration of ethanol to ethylene, and (c) ethanol separation from water. The removal of carbon dioxide by plants due to photosynthesis only partly offsets the GHG emissions. Further, the leaching of the fertilized surface soil causes water pollution and eutrophication of the rivers and water bodies.

The cumulative cradle-to-grave environmental impacts for producing a given amount of energy from natural gas, coal and fuel oil are of the same order of magnitude for a majority of environmental impact categories. Energy sourced from natural gas has relatively lower global warming potential (by approximately 25 %) among all the energy sources due to its low sulfur and carbon contents. The predicted environmental impacts for energy production from coal and natural gas at power plants are similar to those reported by Lawrence Energy Center (coal based) and Astoria Generating Station (natural gas based). The *cradle-to-gate* analysis shows that in all cases, the fuel burning to produce energy at the power plant is by far the biggest contributor to the various adverse environmental impacts, ranging from approximately 78 to 93 % depending on the fuel. In other words, the choice of feedstock (naphtha, ethane or ethanol) used for the sourcing of ethylene does not significantly alter the overall environmental impact.

Acknowledgments This research was partly supported with funds from the following sources: National Science Foundation Accelerating Innovation Research Grant (IIP-1127765) and United States Department of Agriculture USDA/NIFA Award 2011-10006-30362.

References

1. Global Ethylene Oxide Market To Exceed 27 million tons by 2017, According to a New Report by Global Industry Analysts, Inc. (2011). http://www.prweb.com/releases/ethylene_oxide_EO/monoethylene_glycol/prweb8286140.htm
2. Weissermel K, Arpe H-J (2003) Industrial organic chemistry, 4th edn. Wiley-VCH, Washington DC
3. Zimmermann H, Walzl R (2009) Ethylene. In: Hawkins S, Russey WE, Pilkart-Muller M (eds) Ullmann's Encyclopedia of industrial chemistry. Wiley-VCH, New York, p 36
4. Petrochemicals (2012) Chemical engineering news. American Chemical Society, p 4
5. US firms detail investments (2011) Chemical engineering news. American Chemical Society, USA, p 56
6. Sasol announces feasibility study for Ethylene Cracker Complex (2012) SA business news. http://www.sa-businessnews.com/2011/12/07/sasol-announces-feasability-study-for-ethylene-cracker-complex/
7. Dow Studies Bio-Based Propylene Routes (2011) Chemical market reporter. http://www.icis.com/Articles/2011/02/28/9438198/dow-studies-bio-based-propylene-routes.html
8. Braskem to inaugurate sugar-to-ethylene plant on Friday (2010) Chemical market reporter; 1. http://www.icis.com/Articles/2010/09/23/9395849/braskem-to-inaugurate-sugar-to-ethylene-plant-on-friday.html
9. Choudhary VR, Uphade BS, Mulla ARS (1995) Coupling of endothermic thermal cracking with exothermic oxidative dehydrogenation of ethane to ethylene using a diluted SrO/La_2O_3 catalyst Angew. Chem Int Ed Engl 34:2
10. GaBi (2011) PE Solutions and Five Winds Corporation, USA
11. ISO 14040:2006- Environmental Management, Life Cycle Assessment, Principles and Framework (2006) ISO. http://www.iso.org/iso/catalogue_detail?csnumber=37456
12. ISO 14044:2006 Environmental Management, Life Cycle Assessment, Requirement and Guidelines (2006). http://www.iso.org/iso/catalogue_detail?csnumber=38498
13. Bare JC, Norris GA, Pennington DW, McKone T (2003) TRACI: the tool for the reduction and assessment of chemical and other environmental impacts. J Ind Ecol 6:49–78
14. Bare JC (2002) Developing a consistent decision-making framework by using the U.S. EPA's TRACI. AIChe J
15. Bare JC, Hofstetter P, Pennington DW, Udo de Haes HA (2000) Life cycle impact assessment workshop summary: midpoint

versus endpoints: the sacrifices and benefits. Int J Life Cycle Assess 5:319–326

16. Energy Do. Energy Information Administration (2010). http://www.eia.doe.gov/

17. Reisch MC (2012) Coal's enduring power. In: Chemical engineering news. The American Chemical Society. p 5

18. Canepa P (2011) Uncertainty in the GaBi datasets. GaBi, Seattle

19. Patel M (2003) Cumulative energy demand (CED) and cumulative CO(2) emissions for products of the organic chemical industry. Energy 28:721–740

20. Liptow C, Tillman A-M (2009) Comparative life cycle assessment of polyethylene based on sugarcane and crude oil. Department of Energy and Environment, Division of Environmental System Analysis, Chalmers University of Technology, Gothenburg

21. Natural Gas (2011). www.naturalgas.org

22. The Potential Impacts of Increased Corn Production for Ethanol in the Great Lakes-St. Lawrence River Region, 2007, Great Lakes Commission for the U.S. Army Corps of Engineers, p. 49

23. Donaldson D, Kiely T (2008) 2006–2007 pesticides industry sales and usage. In: Pesticides industry sales and usage. United States Environmental Protection Agency, Washington, DC

24. R. HZ. Energy Use and Efficiency in Pest Control, Including Pesticide Production, Use, and Management Options (2012) Farm energy home. http://www.extension.org/pages/62513/energy-use-and-efficiency-in-pest-control-including-pesticide-production-use-and-management-options

25. National Agricultural Statistics Service: Fertilizer consumption in the US (2011). http://www.nass.usda.gov/Data_and_Statistics/Pre-Defined_Queries/2010_Corn_Upland_Cotton_Fall_Potatoes/index.asp

26. Ogle SM, Del Grosso SJ, Adler PR, Parton WJ (2008) Soil nitrous oxide emissions with crop production for biofuel: implications for greenhouse gas mitigation. In: Biofuels, food & feed tradeoffs. Lifecycle Carbon Footprint of Biofuels Workshop, Miami Beach

27. Nitrous Oxide Emissions (2009) Emissions of greenhouse gas reports. US Energy Information Administration, Washington, DC

28. Ethylene from ethanol. In: Group CE (ed) Business area: biomass chemicals, 2002. Chematur Technologies AB, Karlskoga, Sweden, p 4

29. Dyck K, Ladisch MR (1979) Dehydration of ethanol: new approach gives positive energy balance. Science 205:898–900

30. Toxic Release Inventory-Exxon Mobil Chemical Company Olefins Plant (2012). http://iaspub.epa.gov/triexplorer/release_fac_profile?TRI=77522XXNCH3525D&year=2010&trilib=TRIQ1&FLD=&FLD=RELLBY&FLD=TSFDSP&OFFDISPD=&OTHDISPD=&ONDISPD=&OTHOFFD=

31. True WR. Global ethylene capacity continues advance in 2011 (2012) Processing news. http://www.ogj.com/articles/print/vol-110/issue-07/special-report-ethylene-report/global-ethylene-capacity.html

32. Fact Sheet of ExxonMobil Complex Baytown Facility (2012) ExxonMobil: Baytown, Texas. http://www.pdfport.com/view/542303-exxonmobil-complex-baytown-refinery-chemical-plant-5000.html

33. Graczyk M (2012) Exxon Mobil considering Baytown expansion. http://www.businessweek.com/ap/2012-06/D9V4F0MG0.htm

34. Robinson J (2012) Greenhouse gas prevention of dignificant deterioration permit application for ethylene expansion project. In: Company EMC (ed) United States Environmental Protection Agency, Baytown

35. Ghanta M (2012) Development of an economically viable H_2O_2-based, liquid-phase ethylene oxide technology: reactor engineering and catalyst development studies. In: Chemical and petroleum engineering. University of Kansas, Lawrence, p 238

36. Ghanta M, Ruddy T, Fahey D, Busch D, Subramaniam B (2013) Is the liquid-phase H_2O_2-based ethylene oxide process more economical and greener than the gas-phase O_2-based silver-catalyzed process? Ind Eng Chem Res 52:18–29

37. Ghanta M, Fahey DR, Busch DH, Subramaniam B (2013) Comparative economic and environmental assessments of H_2O_2-based and tertiary butyl hydroperoxide-based propylene oxide technologies. ACS Sustain Chem Eng 1:268–277

38. Morschbacker A (2009) Bio-ethanol based ethylene. J Macromol Sci Part C Polym Rev 49:5

39. Ahlgren S, Baky A, Bernesson S, Nordberg A, Noren O, Hansson PA (2008) Ammonium nitrate fertiliser production based on biomass—environmental effects from a life cycle perspective. Bioresour Technol 99:8034–8041

40. Patzek TW, Anti SM, Campos R, Ha KW, Lee J, Li B, Padnick J, Yee SA (2005) Ethanol from corn: clean renewable fuel for the future, or drain on our resources and pockets? Environ Dev Sustain 7:319–336

41. Huang W (2011) Fertilizer use and price. http://www.ers.usda.gov/Data/FertilizerUse/

42. Denton JE, Mazur L, Milanes C, Randles K, Salocks C (2004) Used oil in bunker fuel: a review of potential human health implications. Office of Environmental Health Hazard Assessment Integrated Risk Assessment Section (IRAS)

43. Crude Oil and Total Petroleum Imports (2011) ftp://eia.doe.gov/pub/oil_gas/petroleum/data_publications/company_level_imports/current/import.html

44. Background Report on Fertilizer Use, Contaminants and Regulations, 1999. National Program Chemicals Division, Office of Pollution Prevention and Toxics, Washington DC, p 395

45. Aden A (2008) Water usage for current and future ethanol production. National Renewable Energy laboratory, Denver, p 2

46. Dodds WK, Bouska WW, Eitzmann JL, Pilger TJ, Pitts KL, Riley AJ, Schloesser JT, Thornbrugh DJ (2009) Eutrophication of US freshwaters: analysis of potential economic damages. Environ Sci Technol 43:12–19

47. Pimentel D, Patzek TW (2008) Ethanol production: energy and economic issues related to U.S. and Brazilian sugarcane. In: Biofuels, solar and wind as renewable systems. Springer. Berlin, pp 357–371

48. Koroneos C, Dompros A, Roumbas G, Moussiopoulos N (2005) Life cycle assessment of Kerosene used in avitation. Int J LCA 10:8

49. Smith K, Bouwman L, Braatz B (1996) N2O: direct emissions from agricultural soils. In: Good practice guidance and uncertainty management in national greenhouse gas inventories. Intergovernmental Panel on Climate Change, pp 363–380

50. Nevison C (1996) Indirect N_2O emissions from agriculture. In: Good practice guidance and uncertainty management in national greenhouse gas inventories. Intergovernmental Panel on Climate Change, pp 381–397

51. Kambara S, Takarada T, Yamamoto Y, Kato K (1993) Relation between functional forms of coal nitrogen and formation of No(X) precursors during rapid pyrolysis. Energy Fuels 7:1013–1020

52. Greenhouse Gas Emissions-Lawrence Energy Center (2012). http://ghgdata.epa.gov/ghgp/main.do#/facilityDetail/?q=FacilityorLocation&st=KS&fid=521954&lowE=0&highE=23000000&&g1=1&g2=1&g3=1&g4=1&g5=1&g6=1&g7=1&s1=1&s2=0&s3=0&s4=0&s5=0&s6=0&s7=0&s8=0&s9=0&s301=1&s302=1&s303=1&s304=1&s305=1&s306=1&s401=1&s402=1&s403=1&s404=1&s701=1&s702=1&s703=1&s704=1&s705=1&s706=1&s707=1&s708=1&s709=1&s710=1&s711=1&ss=&so=0&ds=E

53. Toxic Release Inventory-Lawrence Energy Center (2012). http://iaspub.epa.gov/triexplorer/release_fac_profile?TRI=66044LWRNC1250N&year=2010&trilib=TRIQ1&FLD=&FLD=RELLBY&FLD=TSFDSP&OFFDISPD=&OTHDISPD=&ONDISPD=&OTHOFFD=

54. Coal Sourcing for Lawrence Energy Center (2000). http://www.
    westarenergy.com/wcm.nsf/content/lawrence
55. Toxic Release Inventory-Astoria Generating Station (2012).
    http://iaspub.epa.gov/triexplorer/release_fac_profile?TRI=11105
    STRGN18012&year=2010&trilib=TRIQ1&FLD=&FLD=RELL
    BY&FLD=TSFDSP&OFFDISPD=&OTHDISPD=&ONDISPD=
    &OTHOFFD=
56. Greenhouse Gas Emissions at Astoria Generating Station (2012).
    http://ghgdata.epa.gov/ghgp/main.do#/facilityDetail/?q=Facility
    orLocation&st=NY&fid=520539&lowE=0&highE=23000000&
    &g1=1&g2=1&g3=1&g4=1&g5=1&g6=1&g7=1&s1=1&s2=0&
    s3=0&s4=0&s5=0&s6=0&s7=0&s8=0&s9=0&s301=1&s302=1&
    s303=1&s304=1&s305=1&s306=1&s401=1&s402=1&s403=1&
    s404=1&s701=1&s702=1&s703=1&s704=1&s705=1&s706=1&
    s707=1&s708=1&s709=1&s710=1&s711=1&ss=&so=0&ds=E

57. Goldstein HL, Siegmund CW (1976) Influence of heavy fuel oil
    composition and boiler combustion conditions on particulate-
    emissions. Environ Sci Technol 10:1109–1114
58. Kolker A, Palmer CA, Bragg LJ, Bunnell JE (2005) Arsenic in
    coal. U.S. Geological Survey, Reston
59. Quarterly Emission Tracking Clean Air Markets and Air Radia-
    tion (2013) United States Environmental Protection Agency.
    http://www.epa.gov/airmarkt/quarterlytracking.html
60. Cleaner Power Plants (2012) Mercury and air toxics standards.
    http://www.epa.gov/airquality/powerplanttoxics/powerplants.html
61. Mercury in U.S. Coal-Abundance, Distribution, and Modes of
    Occurence, 2011. United States Geological Survey, Washington
    DC, p 5

18

Highly active and selective catalyst for synthetic natural gas (SNG) production

Yu Huang · Haoyi Chen · Jixin Su ·
Tiancun Xiao

Abstract A novel nickel-based methanation catalyst has been prepared and tested for CO and CO_2 hydrogenation to synthetic natural gas under various conditions. The catalysts before and after reaction have been characterized using XRD, Laser Raman and IR spectroscopes. It is showed that the novel catalyst can give high methane selectivity and yield at a very broad ranges of temperature and H_2/CO ratios, although the selectivity to methane increase with the H_2/CO ratio, which, however, increases with the temperature rising from 200 to 350 °C, then gradually decrease. The characterization results showed that the there is surface carbons forming over the catalyst surface, and the nickel crystalline structure changes during the reaction. Despite this, the catalyst still gives the same performance, which suggested that the catalyst can operate in a very broad range conditions.

Keywords Nickel catalyst preparation · Co-precipitation · Methanation · Synthetic natural gas

Introduction

Biomass development is predicted to grow the fastest of all renewable energy in the next decade, although it is already being utilized for a variety of purposes, including biofuels, biopower (electricity), biomaterials, biochemicals ("green chemistry"), and biopellets. The different categories of biomass are all interrelated, and of all the sources of renewable energy, biomass can be most honestly labeled "home-grown." Among these applications, Synthetic Natural Gas (SNG) production from biomass has attracted increasing attention in recent years, due to the rising price of natural gas, the wish for less dependency from natural gas imports and the opportunity of reducing green house gases. Although solid dry biomass has [1–5] been used directly (e.g., wood and straw) for ages, converting them into SNG has is a more preferential process.

The advantages SNG are, besides the relatively high efficiency of its production, the already existed infrastructure such as pipelines and the efficient end use technologies such as compressed natural gas cars, combined heat and power plants, or combined cycle plant. Therefore, in recent years there has been a dramatic increase interest in the manufacture of a clean, high BTU gas energy source which will meet pipeline standards by synthetic means biomass.

SNG normally is produced through the gasified products, e.g., H_2 and CO-containing gas stream. While a number of metallic species are known to be active and selective methanation catalysts including, inter alia, nickel, ruthenium, cobalt and iron, their application to the manufacture of high BTU or pipeline gas has been less than satisfactory for several reasons. First, methanation reactions are strong exothermic reaction, which easily heats up catalyst to 800 °C in an adiabatic reactor. However, these catalyst systems generally cannot stand very high temperature, the reactor temperature needs to be limited to temperatures below 400 °C to avoid sintering and deactivation of the catalyst, so the highly exothermic nature of the methanation reaction itself provides severe operational

Y. Huang · H. Chen · T. Xiao (✉)
Guangzhou Boxenergy Technology Ltd, Guangzhou Hi-Tech Development Zone, Guangzhou, People's Republic of China
e-mail: tiancun.xiao@chem.ox.ac.uk

J. Su
Environmental Engineering College, Shandong University, No. 27 Shanda Nanlu, Jinan, People's Republic of China

difficulties in controlling catalyst temperature when CO concentration of the feed gas is in the range required for methane-rich gas manufacture. Further, the methanation reaction itself is considered to be a combination of several reactions including the primary reaction and secondary reactions (2) and (3)

$$3H_2 + CO \rightarrow CH_4 + H_2O. \tag{1}$$

$$2H_2 + 2CO \rightarrow CH_4 + CO_2 \tag{2}$$

$$4H_2 + CO_2 \rightarrow CH_4 + 2H_2O \tag{3}$$

The thermodynamic calculation results for this process are shown in Table 1.

These thermodynamic equilibriums of the reaction 1 shows that the equilibrium yield of methane is adversely effected at high temperatures, i.e., above 900 K, as can be seen, the delta G at 900 K is +1,742 J/mol, which does not favor the methanation reaction. In fact, it can be seen that reaction (2) is a combination of reaction (1) and the water gas shift reaction (4).

$$CO + H_2O \rightarrow CO_2 + H_2 \tag{4}$$

So far, there have been 3 processes proposed for the SNG production, one is high temperature processes [6–10], which requires very highly stable catalyst and several steps for heat exchange, so as to cool down the reaction stream to improve the catalyst performance and avoid the catalyst deactivation. Another is to the low-temperature methanation [11–13], which proceeds in a reaction configuring with heat exchanger so as to remove the reaction heat quickly, and may have low requirements on the catalyst stability, but requires the catalyst to have high activity and selectivity. So far, the methanation catalyst is normally nickel-based system, which is prepared using impregnation method. The catalyst thus has high dispersion which can also be used in fluidized bed reaction and slurry reactor for SNG production, where the heat transfer is very fast, but attrition is a more important parameter [1, 14–18], also the separation of the catalyst particles from the reactor is also an issue.

Recently, a novel reactor system has been developed by Boxenergy Tech Ltd, which can take the reaction heat away in a fixed bed reactor, thus to maintain a constant low isothermal reaction temperature. In such case, the catalyst with low temperature activity and selectivity is more important. Overview of the previous publications, there are few reports on the low-temperature methanation catalyst for SNG production. It has shown that the organic matrix combustion method can give a Fischer–Tropsch catalyst with super high activity [19, 20], thus in this work, a novel method, e.g., organics assisted catalyst preparation method has been developed, and tested under various conditions, a very nickel based robust methanation catalyst with high activity and selectivity has been developed.

Experimental

Catalyst preparation

The catalyst has been prepared using organic induced partial combustion method. It may be an evolution of the chelating and organic combustion method [21, 22]. In short, a specific amount of 20.8 g $Ni(NO_3)_2 \cdot 6H_2O$, 1.06 g $ZrO(NO_3)_2$, and 5 g citric acid are mixed with 20 ml of water, heating to 50 °C while stirring for 30 min to form a transparent solution. 10 g of AlOOH (beomite) is then added to the solution, and statically placed under ambient conditions for 4 h, dried at 80 °C for 2 h and calcined in static air at 500 °C for 4 h to give a black powder. The nickel content in the black powder is analyzed using atom absorption, and the content is 30.1 %. The resultant powder is pelletized using a mechanical pelletizing machine. When using for catalyst, the sample is crushed and sieved into 60–100 mesh particles.

Catalyst testing

Every time, 0.1 g of the crushed $NiZrO_x/Al_2O_3$ catalyst is loaded in a quartz tube as the reactor. The reactor is

Table 1 The thermodynamic calculations of the methanation reaction 1

Reaction temperature (K)	Δ_H°(J/molCH$_4$)	Δ_G°(J/molCH$_4$)	Δ_{vol}(l)	Δ_S°(J/K)	Δ_{CP}(J/K)	K_{eq}
298.15	−250176.0	−150843.2	−7.3379E+01	−333.164	−4.642	2.6696E+26
300.00	−250184.6	−150226.8	−7.3834E+01	−333.192	−4.635	1.4327E+26
373.50	−250439.6	−125702.4	−9.1929E+01	−333.965	−1.580	3.7939E+17
600.00	−217924.0	−72548.9	−9.8469E+01	−242.292	−29.890	2.0691E+06
700.00	−220623.5	−48098.5	−1.1488E+02	−246.464	−24.193	3.8820E+03
800.00	−222782.8	−23298.6	−1.3129E+02	−249.355	−19.097	3.3205E+01
900.00	−224463.5	1742.5	−1.4770E+02	−251.340	−14.618	7.9226E−01

The above calculation is from: http://www.crct.polymtl.ca/fact/

inserted into a tubular furnace and connected to the gases. The catalyst is activated with 33.3 (vol) % H_2/N_2 flowing at 70 ml/min and heated at 2 °C/min to 350 °C and held at this temperature for 2 h. Then the reactor is cooled down to 200 °C in the flowing reduction atmosphere.

The gas is switched to a gas mixture of $N_2:H_2:CO = 60:30:10$ (volume) with a total following rate of 100 ml/min. when testing CO_2 hydrogenation, the mixture gas is set as $N_2:H_2:CO_2 = 50:40:10$. The other settings are all the same. The products flow through a cold trap where the produced vapour is condensed the flowing rate of the gas in and out were measured using a soap bubble meter, and the dried gaseous product concentration are analyzed using non-dispersive IR spectrometer.

The CO conversion and methane selectivity and yield are calculated as follows:

CO conversion: $X_{CO}(\%) = \dfrac{V_{CO.in} - V_{CO.out}}{V_{CO.in}} \times 100$

CH_4 selectivity: $S_{CH_4}(\%) = \dfrac{V_{CH_4.out}}{V_{CO.in} - V_{CO.out}} \times 100$

CH_4 yield: $Y_{CH_4}(\%) = \dfrac{X_{CO} S_{CH_4}}{100} = \dfrac{V_{CH_4.out}}{V_{CO.in}} \times 100$

where V_{CO}, V_{CH_4} in and out are the volume flowing rate under the ambient conditions calculated from the flowing rate and the content of CO and CH_4 in the inlet and outlet gas stream.

Catalyst characterization

XRD diffraction was carried out in a Philips PW1710 diffractometer equipped with Cu Kα radiation to detect the crystalline phase of nickel and alumina and coke after reaction. The morphology of the catalysts before and after reaction was observed in a JEOL-4000EX high-resolution electron microscope with an accelerating voltage of 400 kV. Raman spectra were recorded with a resolution of 2 cm^{-1} using a Yvon Jobin Labram spectrometer with an He$^+$ laser, running on a back scattered confocal arrangement.

Results and discussion

Normally when CO is hydrogenated, it is firstly converted into –CH_2– which either link together to form C_nH_m as liquid hydrocarbons. This is a well-known process for Fischer–Tropsch synthesis [23–26]. For SNG production, –CH_2– is expected to further hydrogenated into CH_4 rather than link to form the liquid. In such case, $C_2{}^+$ selectivity is more important. In SNG production, CH_4 selectivity as well as yield is more important [13, 15, 27, 28]. Therefore in the

following test results, only CH_4 selectivity and yield are presented under various conditions.

Figure 1 gives the temperature effect on the CH_4 selectivity and yield of CO hydrogenation test under different conditions. It is seen that the main product is methane, no $C_2{}^+$ hydrocarbons were detected, the methanation reaction starts up from 200 °C, however, significant CH_4 yield appear from 220 °C, which may be due to the low CO conversion, although the methane selectivity is significantly high. The CO selectivity reaches 87 % at 220 °C and almost unchanged from 220 °C to the reaction temperature to 400 °C, while the yield of methane decreases gradually with the methanation temperature. This can be explained by the strong exothermicity of the methanation reaction. The higher methanation temperature may lead to some steam reforming reaction, which is strong endothermic reaction. It is, therefore, inferred that the methanation reaction should be kept at temperatures from 260 to 300 °C. However, as showed before, once methanation starts, the generated heat is huge, thus how to remove the reaction heat away quickly is an issue when applying the low-temperature methanation process [29].

In contrast to the CO methanation, CO_2 hydrogenation to methane showed different trends. As shown in Fig. 2, the CO_2 hydrogenation starts up at 200 °C, although the methane yield is significantly lower, only about 1.4 % over the $NiZrO_x/Al_2O_3$ catalyst. When the reaction temperature is raised to 250 °C, methane selectivity increases to 95 %, while the CH_4 yield is about 35 %. When the methanation temperature rises to 280 °C, methane selectivity reaches about 98 %, while the CH_4 yield is about 62 %, suggesting that the catalyst prepared in this way gave a high CH_4 selectivity, although its activity for CO_2 conversion is lower than that for CO hydrogenation. The CH_4 yield reaches the maximum at 340 °C, while the CH_4 selectivity is about 98 %, suggesting that the suitable operation conditions for CO_2 methanation is 340 °C when pressure is 1 bar. Also the methanation temperatures above 340 °C have less effect on the CH_4 selectivity, and CO_2 conversion cannot reach more than 90 %.

CO_2 methanation performance with the temperature can be explained by the reaction process steps. Generally, CO_2 hydrogenation has two steps

$$CO_2 + H_2 \rightarrow H_2O + CO \tag{5}$$

$$CO + 3H_2 \rightarrow CH_4 + H_2O. \tag{6}$$

Reaction 5 is endothermic, and reaction 6 is strong exothermic. The overall reaction heat is less than CO methanation, therefore its CH_4 yield increase in a broader temperature range than the CO only methanation.

The stability of the catalyst for CO methanation with time at 250 °C and 1 bar and GHSV of 60,000 h^{-1} is

Fig. 1 Effect of temperature on the methanation reaction. Test conditions:
GHSV = 60,000 h^{-1},
N$_2$:H$_2$:CO = 6:3:1

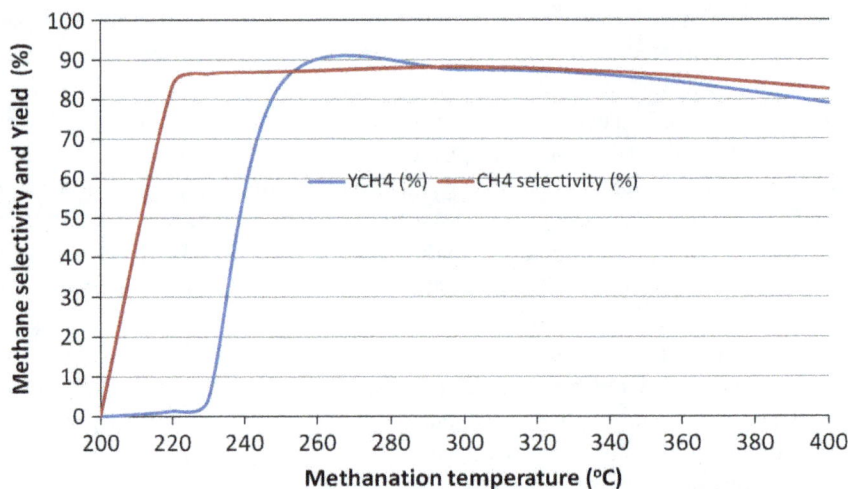

Fig. 2 Effect of reaction temperature on CO$_2$ methanation under the conditions of:
GHSV = 60,000 h^{-1}, P: 1 bar;
N$_2$:H$_2$:CO$_2$ = 5:4:1

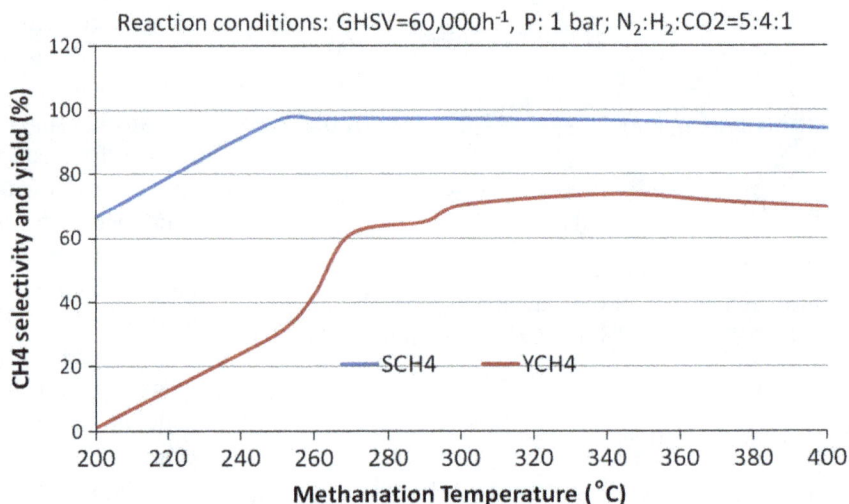

shown in Fig. 3. The CO conversion under this condition at the start of run is nearly 100 %, although the selectivity to methane is about 95 %, and the CH$_4$ yield is 92.2 %. The CH$_4$ selectivity drops to nearly 94 % after 3.5 h running, but changes to 95 %, which drops again after 8 h to 94 %. This change might result from the analysis errors, because the gas products were detected by non-dispersive IR. Thus, it can be inferred that the CH$_4$ selectivity in fact does not change much during the 600 min time on stream. However, the CH$_4$ yield as shown by the blue curve in Fig. 3 changes from 92.1 % at the start of the reaction to 89.8 %, which suggests that the CO conversion may drops during the reaction. From the decline trend it can be found that the yield drops faster in the first 260 min, but slower after 430 min. This can be explained by that the fresh catalyst in the initial reaction stage may be more sensitive to the reactants, the reactant can induce the catalyst surface change or leads to carbon deposition, which may cover some active site, thus leading to the activity drop. With the time on stream, the active site tends to go to steady state,

the carbon deposition rate is less or equivalent to the carbon hydrogenation rate, thus the yield tends to be stable. It is interesting to see that although the CH$_4$ yield drops slowly with the time on stream, which may be due to the loss of some active site or the nickel metal sintering, the methane selectivity over the supported nickel catalyst almost unchanged, suggesting that the catalyst active site change or nickel metal structure change does not alter the methanation reaction on the other active sites.

The effect of H$_2$/CO ratio on the products selectivity has also been studied over the NiZrO$_x$/Al$_2$O$_3$ catalyst. A methanation process normally involves in CO absorption over nickel site, which is then attacked by H$_2$ to give CH$_4$ and H$_2$O [30]. If there is not enough H$_2$ around the activated CO, it may convert into carbon and CO$_2$. So when H$_2$/CO ratio is 1.5, the selectivity to methane is about 80 %, and CO$_2$ about 20 %, no other hydrocarbons are produced. This suggests that the prepared nickel catalyst can work under low H$_2$/CO ratio gas mixture. With the H$_2$/CO ratios increasing, the CO$_2$ selectivity gradually drops,

Fig. 3 Stability test of CO methanation for SNG production under the conditions of: Methanation reaction conditions: $H_2:CO:N_2 = 30:10:60$, P: 1 bar; reaction temperature: 250 °C, GHSV: 60,000 h^{-1}

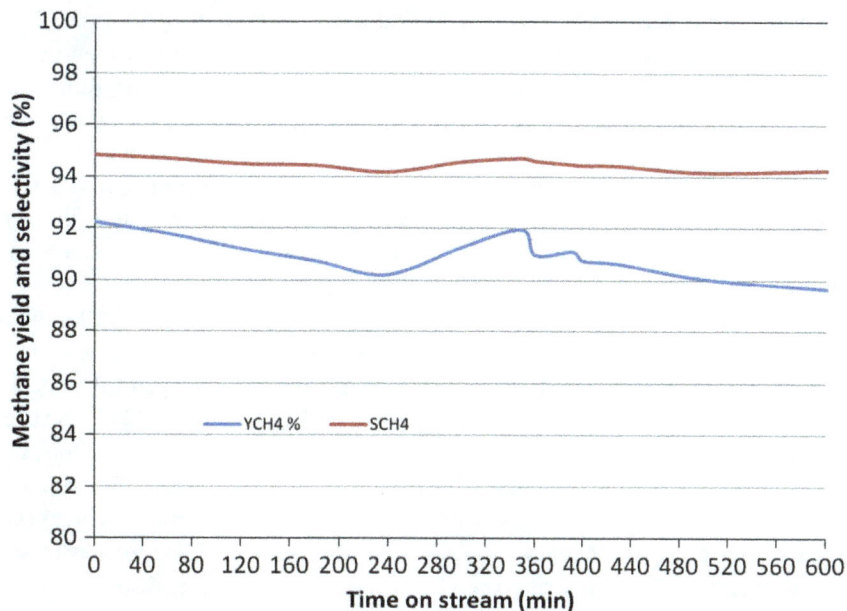

while CH_4 selectivity increases, this is in agreement with the thermodynamic prediction (Fig. 4). When H_2/CO ratio reaches the theoretical value for methanation, the CH_4 selectivity is about 84 %, while there is still some CO_2 generated, which may be due to competitive reaction between the Boudart reaction and methanation [23, 31].

Overview of the above test results shows that the NiZ-rO$_x$/Al$_2$O$_3$ catalyst prepared in this way has high methane selectivity, little C_2^+ side products are generated. The catalyst showed high activity at low temperature and can operate in a wide range.

The catalyst before and after catalytic test have been characterized using XRD, and the results are shown in

Fig. 5. For XRD pattern (the green curve in Fig. 5) of the as prepared catalyst only showed three broad diffraction peaks, which are assigned to the diffraction of gama-Al$_2$O$_3$ [32–34]. When the catalyst is tested for CO methanation in the mixture of $[(N_2 + CO_2)/H_2 = 4]$ at 350 °C for 5 h in stream, while the catalyst still had CH_4 selectivity more than 90 % and CO conversion more than 89 %, the XRD patterns of the alumina became sharper, compared to the fresh catalyst sample. This suggests that the catalyst may experience partly crystallization during the reaction. Also diffraction peaks at 38.7, 43.6, 58.5 and 62.5 appeared in this used sample. According to the literature, the peaks at 38.7 and 58.5 are due to the metallic nickel crystal with hcp

Fig. 4 Effcet of H_2/CO ratio on CO methanation. The test conditions are shown in the graph

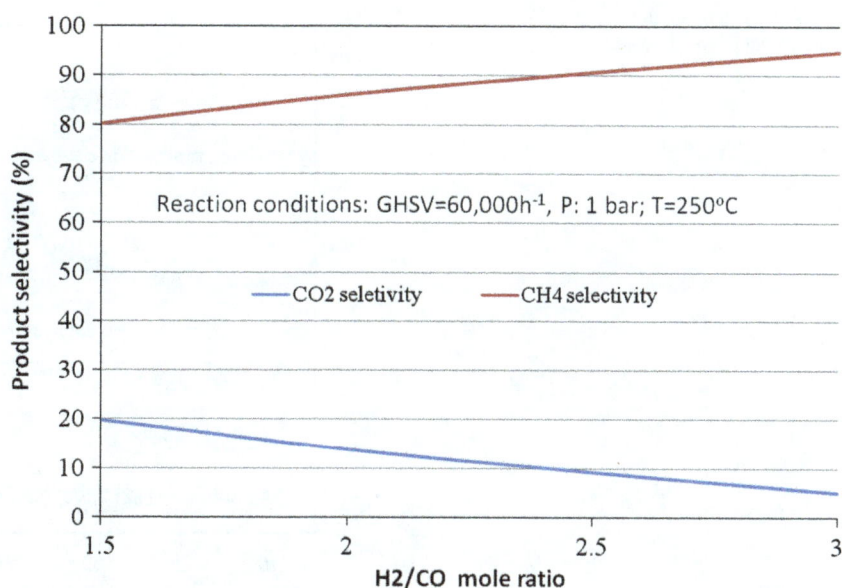

Reaction conditions: GHSV=60,000h^{-1}, P: 1 bar; T=250°C

structure, while the diffraction peaks at 43.6 and 62.5 are ascribed to the diffraction of metallic nickel with fcc structure. The peak of fcc form nickel is much stronger than that of hcp, suggesting that the main phase of nickel after CO_2 hydrogenation is fcc structure [35].

XRD patterns of the spent catalyst from the CO methanation reaction for 600 min is different from the catalyst unloaded from CO_2 methanation reaction. Probably the CO methanation reaction was controlled at 250 °C, the Al_2O_3 support diffraction peak is not as sharp as that from CO_2 system, suggesting that Al_2O_3 support still keeps most of its amorphous phases. Besides these peaks, there are small diffraction bands at 43.8°, 52.8° and 63.2° observed, which correspond to the diffraction of planes of 011, 200 and 220 of nickel crystal. Meanwhile, a small sharp diffraction peak at 27.6° is seen, which can be assigned to the deposited carbon [36, 37]. This suggests that there is carbon formed during the CO hydrogenation although the catalyst is still active.

The spent catalysts from different methanation reactions under various conditions have been measured using Laser Raman spectroscopy. In the fresh catalyst, almost no Raman band can be seen. The Raman spectrum of the spent catalyst after CO_2 methanation showed two distinct peaks at 1,404 and 1,882 cm^{-1}, and a shoulder peak at around 1,590 cm^{-1}. According to literatures, the Raman band at 1404 cm^{-1} can be assigned to the D band, and that at 1,590 cm^{-1} is due to the G band [38, 39]. The D band has arisen from structure defects or imperfection of graphite, whereas the G band is associated with a splitting of the E_{2g} stretching mode of graphite. In addition, a very weak D' band is present at ca. 1,600 cm^{-1} as a shoulder of the G band, which has stemmed from the dangling band of

disorder graphite. The broad bang at 1,882 cm^{-1} is unknown, and may need further study.

In the Raman spectrum of the spent catalyst from CO methanation, these peaks intensity increases significantly, suggesting that both D and G bands graphite increase significantly. This implies that over the spent catalyst from CO methanation more carbon is formed over the catalyst surface, which is in agreement with the XRD results (Fig. 6).

The TEM images of the nickel catalyst from CO methanation and CO_2 methanation are shown in Fig. 7. The image of the spent catalyst from CO_2 has smaller particles, the nickel particles size ranges from 20 to 60 nm. No whisker carbon is observed in the catalyst surface. The dark black fake may results from nickel particle or amorphous carbon as suggested by Laser Raman. However, in the TEM images of spent nickel from CO methanation, clearly the nickel particles and the surface carbon sizes are bigger, as suggested by XRD and Laser Raman. More carbon formed in the spent catalyst from CO methanation, However, no whisker carbon were seen over the catalyst, which might be the reason the catalyst is still active even after 600 min time on stream, because it is already pointed out that the whisker carbon normally account for the catalyst deactivation [36, 37].

Conclusion

Al_2O_3 supported Ni catalyst for CO_x methanation to produce synthetic natural gas has been prepared using organic decomposition method and tested under various conditions.

The catalyst prepared from this method catalyses CO methanation from 220 °C and CO_2 methanation starting

Fig. 5 XRD patterns of the methanation catalysts at various stages

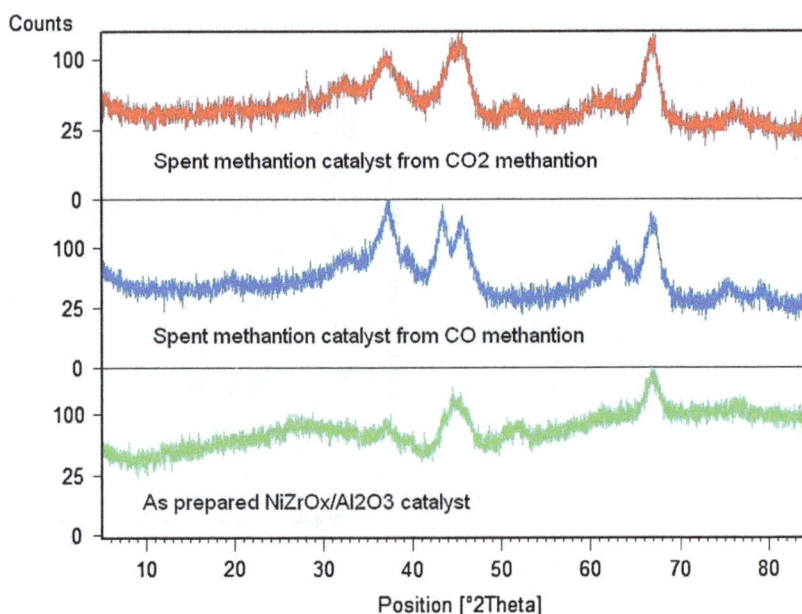

Fig. 6 Laser Raman spectra of the methanation catalyst at different stages

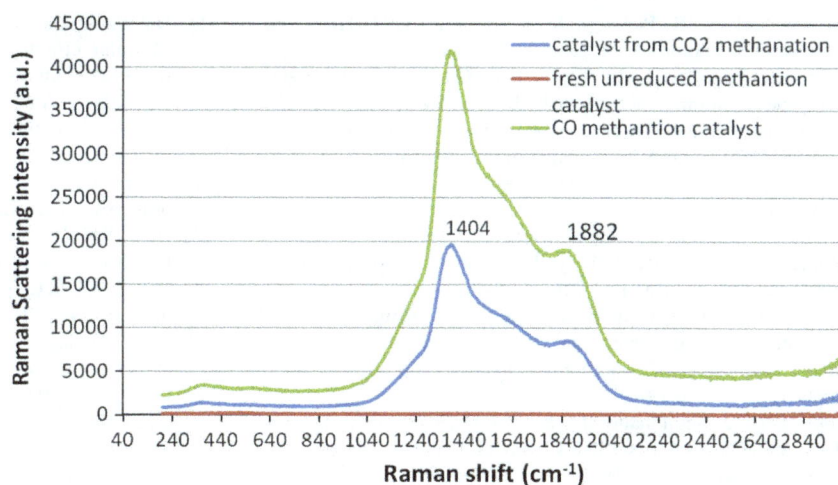

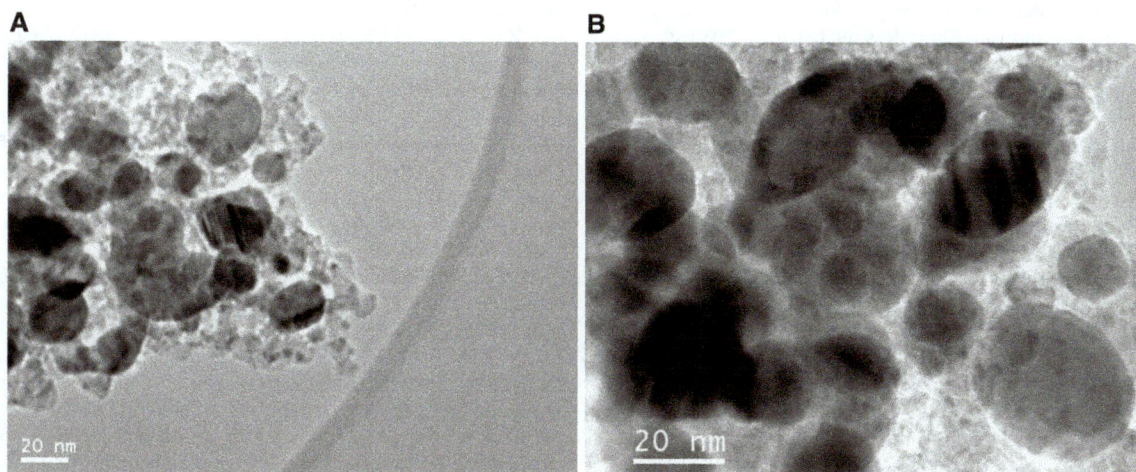

Fig. 7 TEM of the spent NiZrO$_x$/Al$_2$O$_3$ catalysts from **a** CO$_2$ methanation at 300 °C; **b** CO methanation at 250 °C

from 260 °C. It showed high CH$_4$ selectivity and yield, the catalyst can operate in a broad range of temperature.

The catalyst showed high stability and CH$_4$ selectivity for CO methanation, although more surface coke is detected over the catalyst surface.

The nickel crystallite becomes bigger and has different forms in the spent catalyst from CO$_2$ methanation reactions, while the one from CO methanation has different metallic form and more crystalline carbon.

Acknowledgments This work has been financially supported by Guangzhou Bairen (Hundred Talents) program. We would like to thank Dr. Zhenxing Liang for his discussion and professor Jixin Su for his activity test.

References

1. van der Meijden CM et al (2008) Production of Bio-CNG by gasification. In: Proceedings of 25th Annual International Pittsburgh Coal Conference, pp 133/1–133/8
2. Weiss AJ (1973) CRG [catalytic rich gas]-hydrogasification process for SNG [synthetic natural gas] production. Chem Eng Progr 69(5):84–90
3. Woodward C (1977) Methanation in substitute natural gas production. Energiespectrum 1(12):342–347
4. Wix C (2010) Process for the production of substitute natural gas. (Den.). Application, US, p 10
5. Lessard RR, Reitz RA (1981) Catalytic coal gasification: an emerging technology for SNG. Baytown Res. Dev. Div., Exxon Res. Eng. Co., Baytown, TX, USA, p 26
6. Hoehlein B et al (1984) Methane from synthesis gas and operation of high-temperature methanation. Nucl Eng Des 78(2):241–250
7. Luterbacher Jeremy S et al (2009) Hydrothermal gasification of waste biomass: process design and life cycle assessment. Environ Sci Technol 43(5):1578–1583

8. Rostrup-Nielsen JR, Pedersen K, Sehested J (2007) High temperature methanation. Appl Catal A 330:134–138

9. Udengaard NR, Olsen AN, Wix-Nielsen C (2006) High temperature methanation process-revisited. In: Proceedings of 25th annual international Pittsburgh coal conference, pp 25/1–25/5

10. Woodward C (1976) A high-temperature methanation catalyst for SNG applications. Am Chem Soc Div Fuel Chem Prep 21(4):22–29

11. Shinnar R, Fortuna G, Shapira D (1982) Thermodynamic and kinetic constraints of catalytic synthetic natural gas processes. Ind Eng Chem Process Des Dev 21(4):728–750

12. Yang Z et al (2010) Isothermal methanation for manufacture of substitute natural gas from coal or biomass. (Shanghai International Engineering Consulting Company, Peop. Rep. China). Application, CN, p 18

13. Kopyscinski J, Schildhauer TJ, Biollaz SMA (2010) Production of synthetic natural gas (SNG) from coal and dry biomass—a technology review from 1950 to 2009. Fuel 89(8):1763–1783

14. Bock HJ et al (1986) The Comflux-process for substitute natural gas (SNG). In: Proceedings of Eng. Found. Conf. Fluid, pp 489–496

15. Seemann MC, Schildhauer TJ, Biollaz SMA (2010) Fluidized bed methanation of wood-derived producer gas for the production of synthetic natural gas. Ind Eng Chem Res 49(15):7034–7038

16. van der Meijden CM et al (2009) Bioenergy II: scale-up of the MILENA biomass gasification process. Int J Chem React Eng 7

17. Blinn MB et al (1989) Advanced gasifier-desulfurizer process development for SNG (substitute natural gas) application. Final report August 1987–December 1988. KRW Energy Syst., Inc., Pittsburgh, PA, USA, p 210

18. Punwani DV, Arora JL, Tsaros CL (1978) SNG from peat by the PEATGAS Process. Inst. Gas Technol., Chicago, IL, USA, p 19

19. Xiao T, Chen H (2011) Methanation catalyst, its preparation process, and methanation reaction device having the same, Peop. Rep. China, p 11

20. Xiao T, Qian Y (2008) Promoted carbide-based Fischer-Tropsch catalyst, method for its preparation and uses thereof. Oxford Catalysts Limited, UK, p 35

21. Rezaei M et al (2006) Nanocrystalline zirconia as support for nickel catalyst in methane reforming with CO_2. Energy Fuels 20(3):923–929

22. Seo JG et al (2008) Preparation of Ni/Al_2O_3–ZrO_2 catalysts and their application to hydrogen production by steam reforming of LNG: effect of ZrO_2 content grafted on Al_2O_3. Catal Today 138(3–4):130–134

23. Boudart M, McDonald MA (1984) Structure sensitivity of hydrocarbon synthesis from carbon monoxide and hydrogen. J Phys Chem 88(11):2185–2195

24. Dalai AK, Davis BH (2008) Fischer-Tropsch synthesis: a review of water effects on the performances of unsupported and supported Co catalysts. Appl Catal A 348(1):1–15

25. Khodakov AY (2009) Fischer-Tropsch synthesis: relations between structure of cobalt catalysts and their catalytic performance. Catal Today 144(3–4):251–257

26. Stiegel GJ, Srivastava RD (1994) Natural gas conversion technologies. Chem Ind (London) 21:854–856

27. Bery RN (1973) SNG [substitute natural gas] from naphtha. Chem Eng (N. Y.) 80(13):90–91

28. Woodcock KE, Hill VL (1987) Coal gasification for synthetic natural gas (SNG) production. Energy (Oxford) 12(8–9):663–687

29. Tao P, Wang X (2007) Method for preparing synthetic natural gas (SNG) from coke oven gas (COG). (Southwest Research and Design Institute of Chemical Industry, Peop. Rep. China). Application, CN, p 9

30. Vogel F et al (2007) Synthetic natural gas from biomass by catalytic conversion in supercritical water. Green Chem 9(6):616–619

31. Ichikawa S, Poppa H, Boudart M (1985) Disproportionation of carbon monoxide on small particles of silica-supported palladium. J Catal 91(1):1–10

32. Hao Z et al (2009) Characterization of aerogel Ni/Al_2O_3 catalysts and investigation on their stability for CH_4-CO_2 reforming in a fluidized bed. Fuel Process Technol 90(1):113–121

33. Jun J et al (2008) Surface chemistry and catalytic activity of Ni/Al_2O_3 irradiated with high-energy electron beam. Appl Surf Sci 254(15):4557–4564

34. Xiao T-C et al (2004) Tungsten promoted Ni/Al_2O_3 catalysts for carbon dioxide reforming of methane to synthesis gas. Chem Res Chin Univ 20(4):470–477

35. Li H et al (2008) Fabrication and growth mechanism of Ni-filled carbon nanotubes by the catalytic method. J Alloys Compd 465(1–2):51–55

36. Koo KY et al (2008) Coke study on MgO-promoted Ni/Al_2O_3 catalyst in combined H_2O and CO_2 reforming of methane for gas to liquid (GTL) process. Appl Catal A 340(2):183–190

37. Martinez R et al (2004) CO_2 reforming of methane over coprecipitated Ni–Al catalysts modified with lanthanum. Appl Catal A 274(1–2):139–149

38. Reshetenko TV et al (2003) Catalytic filamentous carbon. Structural and textural properties. Carbon 41(8):1605–1615

39. Zhu X, Cheng D, Kuai P (2008) Catalytic decomposition of methane over Ni/Al_2O_3 catalysts: effect of plasma treatment on carbon formation. Energy Fuels 22(3):1480–1484

19

Functionalized regio-regular linear polyethylenes from the ROMP of 3-substituted cyclooctenes

**Henry Martinez · Jihua Zhang · Shingo Kobayashi ·
Yuewen Xu · Louis M. Pitet · Megan E. Matta ·
Marc A. Hillmyer**

Abstract We demonstrated that the ring-opening metathesis polymerization (ROMP) of 3-substituted cyclooctenes bearing polar substituents allows the synthesis of highly regio- and stereo-regular polymers. A series of polyalkenamers with 90–99 % head-to-tail/*trans* configuration were synthetized in good yields (33–87 % yield). Upon saturation of the backbone using diimide, these polymers represent a class of linear polyethylene derivatives where the polar side chain is located on every eighth carbon. The thermal properties of both the saturated and unsaturated polymers depend strongly on the size and polarity of the functional side groups. The results presented here demonstrate that the 3-substituted cyclooctenes can be used not only for the synthesis of precisely functionalized polyethylene derivatives, but also to participate in the ring-opening metathesis copolymerization with unfunctionalized cyclic olefins to generate copolymers with tunable properties.

Keywords ROMP · Polyethylene · Substituted polyolefin · Metathesis

H. Martinez · J. Zhang · S. Kobayashi · Y. Xu ·
L. M. Pitet · M. E. Matta · M. A. Hillmyer (✉)
Department of Chemistry, University of Minnesota, 207 Pleasant
St. SE, Minneapolis, MN 55455-0431, USA
e-mail: hillmyer@umn.edu

Introduction

Polyethylene (PE) is one of the most common plastics, the derivatives of which have found themselves useful in a variety of applications [1–3]. Incorporation of side chains, especially polar functional side chains, is desirable and often necessary to give the polymer desired properties and better performance. Due to the physical and chemical stability of PE, direct functionalization of PE is challenging [4]. Therefore, PE derivatives bearing functional side chains are typically produced by radical or coordination copolymerization of ethylene and functionalized olefin(s). These polymerizations, however, frequently suffer from dramatically different comonomer reaction rates, intolerance of catalyst to polar functional groups and lack of control of polymer microstructure. As a consequence, controlling the scope, content and distribution of functional side chains in the polymer is undermined. A number of methods have been developed to overcome these problems, and notable examples include development of compatible catalysts [5, 6] and metathesis polymerization of functionalized monomers. The latter method is interesting because of catalyst compatibility to polar functional groups and, more attractively, potential to precisely control the polymer structure. Wagener and coworkers have utilized acyclic diene metathesis polymerization (ADMET) [7] of functionalized α,ω-dienes to prepare linear PE derivatives bearing a broad range of polar functional groups [8–29]. Many of the ADMET-derived polymers feature precisely placed polar side chains when a symmetric monomer is used. As opposed to the condensation polymerization nature of ADMET, ring-opening metathesis polymerization (ROMP) is an addition polymerization which requires less demanding conditions and allows better control of the polymer molar mass [30]. The use of functionalized cyclic

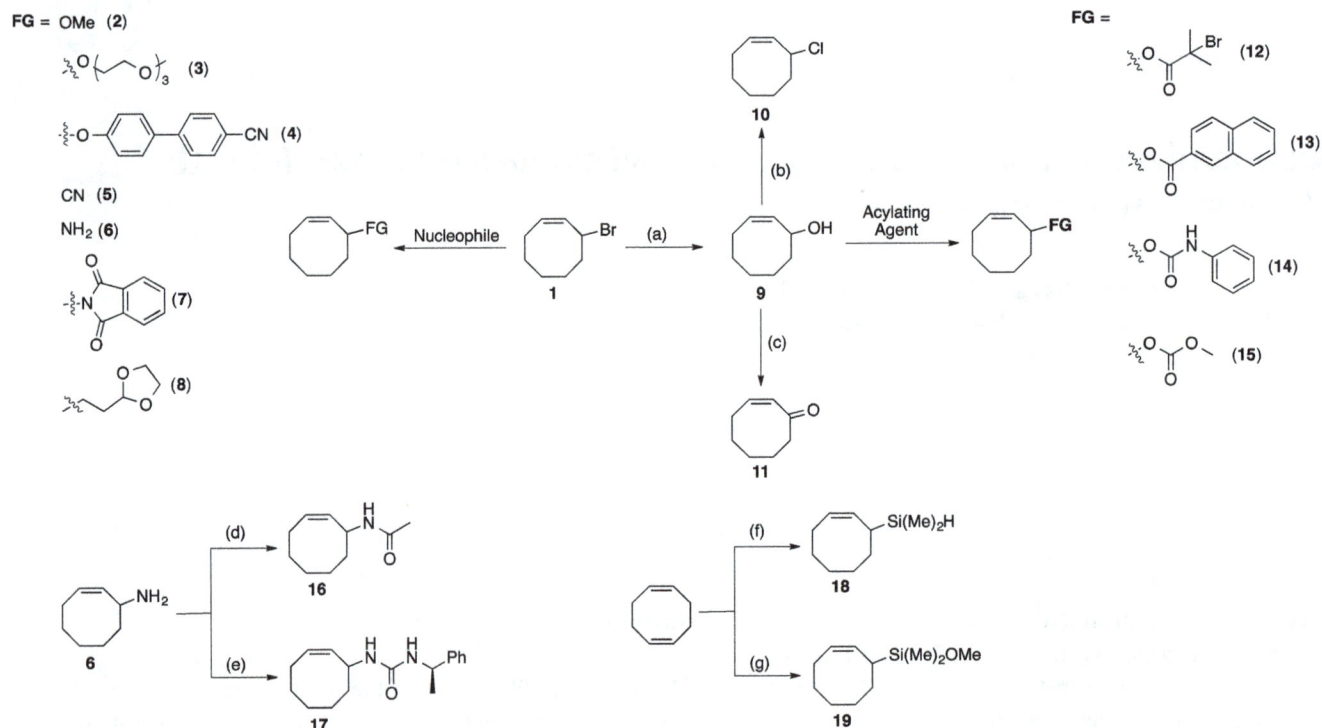

Scheme 1 Synthesis of various 3-substituted COEs. (*a*) Acetone/ water, NaHCO$_3$, reflux 1 h. (*b*) SOCl$_2$, Py. (*c*) Pyridinium chlorochromate, CH$_2$Cl$_2$. (*d*) Et$_3$N/AcCOCl, CH$_2$Cl$_2$. (*e*) DMAP, isocyanate. (*f*) 1. H$_2$PtCl$_6$.6H$_2$O, ClSiMe$_2$H, Cy, 2. LAH, THF. (*g*) 1. H$_2$PtCl$_6$.6H$_2$O, ClSiMe$_2$H, Cy, 2. MeOH, Py

olefins in ROMP has also been explored by many research groups to access to linear PE derivatives with diverse polar side chains, including boranes, thioethers, halides, nitrile, hydroxyl, carbonyl, esters, amides, urethanes and liquid crystal moieties [31–54]. However, excluding those in which the monomer is symmetrically substituted, only few examples resulted in regio-regular products [46, 48].

We recently published a series of reports [55–59] demonstrating that single substitution at the 3-position (i.e., the allylic position) of *cis*-cyclooctene (COE) was an effective way to synthesize regio- and stereo-regular polyalkenamers by ROMP using the well-defined Grubbs second (**G2**) or third (**G3**) generation catalysts. Such substituents we have explored included hydrocarbon groups (3**R**COE, R = methyl, ethyl, hexyl or phenyl) [55, 56] and the acetoxy group [57–59]. All the reported polyalkenamers possess high levels (>95 %) of head-to-tail (HT) repeating unit connectivity and *E* double bond configuration. Hydrogenation of these polymers afforded PE derivatives with precisely positioned branches, which exhibit distinct physical properties compared to their random counterparts. More importantly, these reports suggest that ROMP of any COE substituted with a sufficiently bulky group at the 3-position could proceed in a regio- and stereo-selective manner, which in combination with the exceptional functional group tolerance of Ru-based catalysts allows us to conveniently synthesize diversely functionalized precision PE derivatives.

In the present study, we explore the scope of polar functional groups that, when placed at the 3-position (3**FG**COE, Scheme 1), are amenable to controlled ROMP with Ru-based catalysts. The polymer regio-regularity [head-to-tail (HT), head-to-head (HH) and tail-to-tail (TT)] and stereo-regularity (*cis* and *trans*) were analyzed in this study. These polymers, after the backbone is saturated, represent a class of linear PE derivatives where the polar side chain is located on every eighth carbon. The precision microstructure further enables investigation of the effect of side chain size and polarity on the polymer properties.

Results and discussion

Monomer synthesis

3-bromocyclooctene (**1**) was chosen as the principal starting material because of its scalable and highly efficient preparation. Several synthetic pathways were explored as shown in Scheme 1. The bromo group exhibited a wide range of reactivity towards the hydroxyl group in alcohols and phenol derivatives (compounds **2–4**). Stirring compound **1** in anhydrous methanol at room temperature afforded 3-methoxy-COE (**2**) in 24 h; and elevated temperature and extended reaction time was required to obtain compound **3** from its respective alcohol. However, no reaction of **1** with more hindered alcohols (e.g., secondary alcohols) or their deprotonated forms was observed, even after months. The phenoxide anion derived from phenol 4′-hydroxy-4-biphenylcarbonitrile, which is a liquid crystal building block, reacts readily with **1** in DMF to give 3-substituted COE with an aryl ether group (compound **4**). Other types of nucleophiles, including cyanide, ammonia, phthalimide and a Grignard reagent, reacted with **1** to generate the corresponding 3**FG**COEs (compounds **5–7**). 3-amino-COE (**6**) was used to synthesize 3**FG**COE having amide (**16**) or urea (**17**) groups substituted at the 3-position (Scheme 1).

Treatment of 3-hydroxy-COE (**9**) [59] with thionyl chloride in the presence of pyridine gave 3-chloro-COE (**10**). The hydroxyl group in **9** was also oxidized to give 2-cyclooctenone (**11**). Furthermore, a variety of ester or urethane groups were introduced through the hydroxyl group by acylation or reaction with isocyanate (3**FG**COE **12–15**).

Two silicon-containing monomers (**18** and **19**) were synthetized from hydrosilylation of *cis*-1,5-cyclooctadiene followed by nucleophilic attack of hydride or methanol to replace the chloride atom. The hydrosilylation at the 3 position was expected if taken into account that the isomerization of cyclic dienes from non-conjugated to conjugated dienes has been previously reported using H_2PtCl_6 [60]. Based on NMR <5 % of 5-substituted COE monomer was obtained using this method.

Polymer synthesis and characterization

The compatibility of these 3**FG**COEs with **G2** and the capability of these functional groups to give regio- and stereo-selective polymerizations were tested in ROMP under various conditions (Scheme 2) and the results are summarized in Table 1. To have a fair basis to compare the reactivity of these monomers, all polymerizations were carried out at the same temperature (60 °C) with comparable initial monomer concentration (1–2 M). Some 3**FG**COEs with highly polar functional groups (**4**, **7**, **16** and **17**) were essentially insoluble in $CHCl_3$ at the prescribed concentration; a co-solvent was employed in these

cases to give a homogeneous solution. We attempted to use *cis*-4-octene as a chain transfer agent (CTA) to regulate the polymer molar mass as demonstrated in previous studies [55]. However, this effort only succeeded with some monomers while the others required higher catalyst loading to give polymers, suggesting these particular polar functional groups may inhibit the activity of **G2**.

Polymerization of ether-functionalized monomers **2** and **3** and carbonate monomer **15** under standard conditions resulted in polyalkenamers with the highest yields (87, 85 and 88 %, respectively) among all the monomers in this study. Polymerization and isolation of these polymers were straightforward. In contrast, polymerization of monomers **8** and **13** under standard conditions resulted in reaction mixtures containing considerable quantities of residual monomer, even after 24 h of reaction, suggesting a lower reactivity of such monomers under the chosen conditions. Initial attempts to polymerize monomer **11** yielded insoluble gels upon precipitation in methanol. However, the polymerization proceeded successfully in the presence of butylated hydroxytoluene (BHT). Additional BHT was added to a solution of the dissolved polymer immediately after precipitation to avoid subsequent cross-linking. Monomers bearing a cyclic imide (**7**), urethane (**14**) or aryl-cyanide group (**4**) were successfully polymerized, but required a relatively high catalyst loading.

Three N-containing monomers, **6**, **16** and **17**, as well as nitrile monomer **5**, did not generate any polymer using **G2** or the more reactive **G3**. In all cases, the reaction mixture turned a brown color within a few minutes, suggesting deactivation of the Ru catalyst. The failure of **5** or **6** was not surprising, as compounds containing nitrile groups or primary amines are known to deactivate the Ru catalysts [33, 36, 37, 61, 62]. However, the monomers bearing acetamide (**16**) and urea (**17**) functional groups did not polymerize either. Some studies suggested that amides and ureas were compatible with the Ru catalysts in both polymerizations [49–51] and small molecule reactions [63]. The reason for the inactivity in our trials is unknown.

The two Si-containing monomers, **18** and **19**, did not polymerize even at high catalyst loading. In both cases, the reaction mixture remained a yellow color even after 24 h, suggesting that the catalyst was active. Addition of COE after 24 h to both reaction mixtures yielded poly(COE) only in the solution that contained monomer **19**, which not

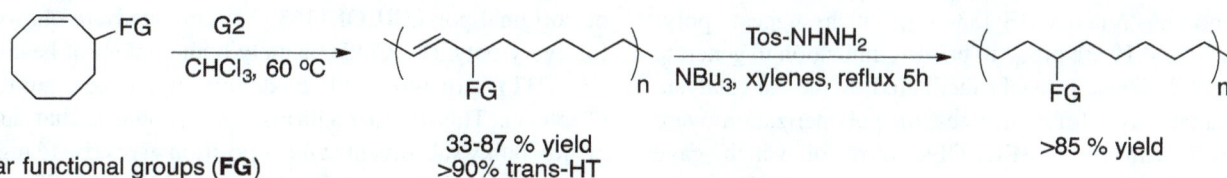

11 Polar functional groups (**FG**)

33-87 % yield
>90% trans-HT

>85 % yield

Scheme 2 Synthesis of linear polyethylene derivatives via ROMP followed by hydrogenation

Table 1 Polymers from ROMP of various **3FGCOEs** bearing polar functional groups using **G2**[a]

3FGCOE	[G2][a] (mol %)	[CTA][a] (mol %)	Conv.[b] (%)	Yield[c] (%)	M_n (kg/mol) Cal.[d]	SEC[e]	Đ[e]	E-HT[f] (%)	T_g^g (°C)
1[h]	0.20	N/A	–	69	–	–	–	–	–
2	0.04	1.0	>99	87	13.5	29.0	1.8	95	−54
3	0.10	1.0	>99	85	21.0	33.6	2.5	90	−69
4[i]	0.10	N/A	89	33	270	258	2.0	96	32
7[j]	0.10	N/A	98	71	250	240	1.4	>99	58
8	0.025	1.0	91	64	18.7	26.0	1.5	98	−34
10[h]	0.2	N/A	–	81	–	–	–	–	–
11[k]	0.025	1.0	92	73	12.1	27.6	1.7	>99	−26
13	0.025	1.0	66	42	18.1	15.3	2.0	>99	2
14	0.2	N/A	82	66	101	121	1.7	92	37
15	0.025	1.0	>99	88	18.0	38.3	1.9	97	−28

All polymerizations conducted at 60 °C with initial monomer concentration of 1–2 M in CHCl₃ if not otherwise stated

[a] Relative to monomer concentration, CTA = *cis*-4-octene

[b] Determined by ¹H NMR analysis of the reaction mixture

[c] Isolated yield

[d] M_n (cal) = MW (monomer) × Conv. × [M]₀/([G2] + [CTA])

[e] Determined by SEC in CHCl₃ at 30 °C versus polystyrene standards

[f] Estimated by ¹H NMR of the isolated polymer in CDCl₃

[g] Determined by DSC (2nd heating cycle) at 10 °C min⁻¹

[h] At room temperature

[i] 1:1 (v/v) CHCl₃:toluene

[j] 3:1 (v/v) CHCl₃:THF

[k] This polymer showed a T_m at 106 °C (ΔH = 64 J g⁻¹)

only suggested that the catalyst was active, but also that the SiMe₂OMe group in position 3 of the COE is bulky enough to inhibit the catalyst approach to the double bond. The fact that COE did not polymerize in the solution that contained monomer **18** might be due to oxidative addition of the Si–H group to the Ru center. Similar reactions have been used to perform hydrosilylation on terminal alkynes using Grubbs' first generation catalyst (**G1**) or **G2** catalyst [64, 65].

Both 3-hydroxy-COE (**9**) and 3-α-bromoisobutyrate-COE (**12**) [66] monomers quickly gave insoluble gels, indicating the formation of physically cross-linked networks possibly through strong hydrogen bonding (**9**) or covalent bonds (**12**) among polymer chains.

Polymers obtained from the halide-containing monomers (**1** and **10**) decomposed easily during the polymerization, generating dark insoluble rubbery materials. This result was not observed during the ROMP of 5-chloro or 5-bromo substituted COE [33–35], which suggests polymers of **1** and **10** are prone to halide elimination to generate conjugated dienes, possibly facilitated by the Ru catalyst.

Qualitatively, different rates of polymerization were observed with these 3FGCOEs, most of which gave polymers in high or quantitative conversion. However, significantly lower conversions were obtained with

3FGCOEs bearing aryl groups (**4** and **13**). This result is consistent with that reported on 3-phenyl COE, suggesting that bulky and rigid substituents result in lower polymerization rate. The polymer number average molar mass (M_n) determined by size exclusion chromatography (SEC) compared with polystyrene standards and the corresponding theoretical value were only in good agreement for the polymers containing aryl groups. This phenomenon can be attributed to the systematic difference of hydrodynamic radius between the polymer and the polystyrene standard, which is likely smaller in structurally similar polymers. The molar mass dispersity (Đ) ranged from 1.4 to 2.5, in accordance with the typical values for products from ROMP with extensive cross-metathesis.

The regio- and stereo-regularity of poly(3FGCOE) samples was characterized by ¹H and ¹³C NMR spectroscopy (see spectra in SI). Results comparable to those reported for our original poly(3RCOE) [55, 59] samples were observed for every poly(3FGCOE), namely high content of head-to-tail (HT) structure and *E* double bond configuration (Table 1). This result confirms our hypothesis that sufficiently bulky substituent at the 3-position gives rise to regio- and stereo-selective polymerization [55–59]. Thermal characterization using differential scanning calorimetry

Table 2 Characterization of hydrogenated polymers

	Yield[a] (%)	M_n^b (kg mol^{-1})	$Đ^b$	T_g^c (°C)	T_m^c (°C)	ΔH_m^c (J g^{-1})
PH2	80	36.8	1.6	−70	15[d]	25
PH3	85	34.1	1.9	−74	−5	25
PH7	90	38.3	4.1	37	–	–
PH8	92	24.9	1.6	−49	–	–
PH13	60	17.0	1.8	18	–	–
PH14	77	57.6	3.4	51	–	–
PH15	83	49.4	2.1	−32	49	0.4

[a] Isolated yield

[b] Determined by SEC in CHCl$_3$ at 30 °C versus polystyrene standards

[c] Determined by DSC (2nd heating cycle) at 10 °C min^{-1}

[d] Cold crystallization at −38 °C ($\Delta H = 25$ J g^{-1})

(DSC) suggested nearly all poly(3**FG**COE) are amorphous with glass transition temperatures (T_g) in the range of −69 to 58 °C (Table 1). The T_g of these polyalkenamers vary substantially (from −69 to 58 °C) and are clearly dependent on the functional group identity. Poly(3**FG**COE) with relatively small and less polar functional groups (e.g., **P2**) consistently exhibit lower T_g than those with relatively bulky and more polar functional groups (e.g., **P7**). Interestingly, **P11** showed a T_m at 106 °C ($\Delta H = 64$ J g^{-1}). When compared with the regio-irregular poly(cyclooct-4-enone) reported by Hillmyer et al. [33] in 1995 via ROMP using a Ru-based catalyst, **P11** has a higher T_g (−58 vs. −26 °C) and T_m (34 vs. 106 °C). This is consistent with the fact that at higher regio- and stereo-regularity both T_g and T_m increases [55, 59]. Similarly, the polymer microstructure has been shown to have profound influence on the thermal properties in precision polyethylenes prepared by ADMET [67].

Chemical hydrogenation of polyalkenamers using diimide has shown to be an effective way to generate saturated backbone [68]. We applied this method to hydrogenate poly(3**FG**COE), which resulted in a polymer equivalent to a copolymer of one part functionalized alpha olefin and three parts ethylene. With the exception of **P4** and **P11**, the corresponding hydrogenated polymers were obtained in decent yields (Table 2). **P4** has poor solubility in xylenes, while **P11** cross-linked while heating the solution of xylenes (even in the presence of large amounts of BHT) resulting in an insoluble gel. NMR analysis revealed the presence of the expected functional group in the expected content in the hydrogenated polymers (see SI), suggesting the reaction conditions are compatible with these functional groups. The clean and well-defined ^{13}C NMR spectra further support the regio-regular structure in the hydrogenated polymers as well as in their unsaturated precursors. With the exception of **P7** and **P14**, SEC in

CHCl$_3$ versus polystyrene standard showed slight increase of molar mass and comparable dispersity for all hydrogenated polymers. Hydrogenated polymers of **P7** and **P14** (**PH7** and **PH14**, respectively) exhibited a large decrease in apparent molar mass and much broader dispersity compared to the unsaturated precursors. Considering that no noticeable change of functional group was observed in NMR spectra, we assume that the change of molar mass and dispersity may be due solely to conformational changes (i.e., hydrodynamic radius) of the hydrogenated polymers in CHCl$_3$. However, we cannot rule out some level of chain scission.

Table 2 summarizes the thermal characterization using DSC of the hydrogenated polymers. **PH7**, **PH8**, **PH13** and **PH14**, where the functional group is bulky, remained amorphous. The T_g of the saturated polymers range from −74 to 51 °C, and dependent on the functional group. Similarly to the unsaturated polymers, small substituents exhibit a much lower T_g than those with relatively bulky substituents. In addition to a glass transition, both **PH2** and **PH3** displayed well-defined melting transition transitions and **PH15** showed a small but unambiguous melting peak, indicating that these polymers are semi-crystalline. **PH3** with the larger substituent (PEG) showed a lower melting point than that in **PH2** or **PH15** with the smaller substituents (OMe and OCOMe, respectively). The only report of a 5-methoxy-COE polymerized via ROMP with a Ru-based catalyst does not report thermal properties for such polymer [69]. We previously reported that the saturated polymer of 3-methyl-COE (poly(3MeCOE)) exhibited a T_g at −59 °C, a $T_m = 2$ °C and $\Delta H_m = 30$ J g^{-1} [55]. When the T_m of poly(3MeCOE) is compared with that of **PH2** and **PH15**, the more polar polymers (**PH2** and **PH15**) show a higher T_m despite being larger groups; however, the ΔH_m was lower in both cases. The more polar **PH3**, despite having the largest of all substituents, shows a T_m with equal enthalpy of melting to that of **PH2**. These results indicate that the polarity of the substituent and that of the polymer has a strong influence in the final thermal properties of the polymer. The polarity appears to have greater effect on crystallization than does the size; this could possibly be due to the polar functional group strengthening inter-chain interaction and thus favoring crystallization.

Conclusions

In summary, we demonstrated an efficient pathway to rapidly expand the scope of 3-substituted COE bearing diverse polar functional groups (3**FG**COEs). Many of these 3**FG**COEs allow ROMP using the Grubbs Ru-based catalyst (**G2**) to afford functionalized polyalkenamers, which possess HT regio-regularity and *E*-stereo-regularity. Both poly(3**FG**COE) and the saturated poly(3**FG**COE) exhibit thermal properties dependent on the size and polarity of the functional groups. This methodology, complementary to the ADMET approach, can be utilized to prepare PE derivatives with diverse polar side chains, including but not limited to those described in this study, which are difficult to be synthesized through radical or coordination copolymerization. Some examples demonstrated here further enable the incorporation of functional groups that are problematic with typical ROMP catalysts and conditions. For instance, PH**7** can be properly treated to release free amine side or PH**8** can be use to generate free aldehyde groups. It is envisioned that 3**FG**COEs would be used not only to synthesize precision functionalized PE derivatives but also to participate in the ring-opening metathesis copolymerization with unfunctionalized cyclic olefins. These advances are a step forward to new graft copolymers [66, 70] and complex supramolecular structures with PE backbone.

Acknowledgments We thank the Abu Dhabi-Minnesota Institute for Research Excellence (ADMIRE) and Dow Chemical Company for support.

References

1. Peacock A (2000) Handbook of polyethylene: structures: properties, and applications, 1st edn. CRC Press, FL
2. Fink JK (2010) Handbook of engineering and specialty thermoplastics: polyolefins and styrenics. Scrivener Publishing LLC, USA
3. Cordeiro CF, Petrocelli FP (2004) Encyclopedia of polymer science and technology. Wiley-VCH, Weinheim
4. Boaen NK, Hillmyer MA (2005) Post-polymerization functionalization of polyolefins. Chem Soc Rev 34:267–275
5. Boffa LS, Novak BM (2000) Copolymerization of polar monomers with olefins using transition-metal complexes. Chem Rev 100:1479–1494
6. Nakamura A, Ito S, Nozaki K (2009) Coordination–insertion copolymerization of fundamental polar monomers. Chem Rev 109:5215–5244
7. Baughman TW, Wagener KB (2005) Recent advances in ADMET polymerization. Adv Polym Sci 176:1–42
8. Valenti DJ, Wagener KB (1998) Direct synthesis of well-defined alcohol-functionalized polymers via acyclic diene metathesis (ADMET) polymerization. Macromolecules 31:2764–2773
9. Watson MD, Wagener KB (2000) Ethylene/vinyl acetate copolymers via acyclic diene metathesis polymerization. Examining the effect of "long" precise ethylene run lengths. Macromolecules 33:5411–5417
10. Watson MD, Wagener KB (2000) Functionalized polyethylene via acyclic diene metathesis polymerization: effect of precise placement of functional groups. Macromolecules 33:8963–8970
11. Hopkins TE, Pawlow JH, Koren DL, Deters KS, Solivan SM, Davis JA, Gomez FJ, Wagener KB (2001) Chiral polyolefins bearing amino acids. Macromolecules 34:7920–7922
12. Hopkins TE, Wagener KB (2003) Amino acid and dipeptide functionalized polyolefins. Macromolecules 36:2206–2214
13. Baughman TW, van der Aa E, Lehman SE, Wagener KB (2005) Circumventing the reactivity ratio dilemma: synthesis of ethylene-co-methyl vinyl ether copolymer. Macromolecules 38:2550–2551
14. Baughman TW, van der Aa E, Wagener KB (2006) Linear ethylene–vinyl ether copolymers: synthesis and thermal characterization. Macromolecules 39:7015–7021
15. Boz E, Wagener KB, Ghosal A, Fu RQ, Alamo RG (2006) Synthesis and crystallization of precision ADMET polyolefins containing halogens. Macromolecules 39:4437–4447
16. Baughman TW, Chan CD, Winey KI, Wagener KB (2007) Synthesis and morphology of well-defined poly(ethylene-co-acrylic acid) copolymers. Macromolecules 40:6564–6571
17. Berda EB, Lande RE, Wagener KB (2007) Precisely defined amphiphilic graft copolymers. Macromolecules 40:8547–8552
18. Boz E, Wagener KB (2007) Progress in the development of well-defined ethylene–vinyl halide polymers. Polym Rev 47:511–541
19. Berda EB, Wagener KB (2008) Inducing pendant group interactions in precision polyolefins: synthesis and thermal behavior. Macromolecules 41:5116–5122
20. Boz E, Ghiviriga I, Nemeth AJ, Jeon K, Alamo RG, Wagener KB (2008) Random, defect-free ethylene/vinyl halide model copolymers via condensation polymerization. Macromolecules 41:25–30
21. Opper KL, Fassbender B, Brunklaus G, Spiess HW, Wagener KB (2009) Polyethylene functionalized with precisely spaced phosphonic acid groups. Macromolecules 42:4407–4409
22. Opper KL, Wagener KB (2009) Precision sulfonic acid ester copolymers. Macromol Rapid Commun 30:915–919
23. Mei JG, Aitken BS, Graham KR, Wagener KB, Reynolds JR (2010) Regioregular electroactive polyolefins with precisely sequenced π-conjugated chromophores. Macromolecules 43:5909–5913
24. Aitken BS, Lee M, Hunley MT, Gibson HW, Wagener KB (2010) Synthesis of precision ionic polyolefins derived from ionic liquids. Macromolecules 43:1699–1701
25. Opper KL, Markova D, Klapper M, Mullen K, Wagener KB (2010) Precision phosphonic acid functionalized polyolefin architectures. Macromolecules 43:3690–3698
26. Aitken BS, Wieruszewski PM, Graham KR, Reynolds JR, Wagener KB (2012) Control of charge-carrier mobility via in-chain spacer length variation in sequenced triarylamine functionalized polyolefins. ACS Macro Lett 1:324–327
27. Aitken BS, Wieruszewski PM, Graham KR, Reynolds JR, Wagener KB (2012) Perfectly regioregular electroactive polyolefins: impact of inter-chromophore distance on PLED EQE. Macromolecules 45:705–712
28. Aitken BS, Buitrago CF, Heffley JD, Lee M, Gibson HW, Winey KI, Wagener KB (2012) Precision ionomers: synthesis and thermal/mechanical characterization. Macromolecules 45:681–687
29. Leonard JK, Wei YY, Wagener KB (2012) Synthesis and thermal characterization of precision poly(ethylene-co-vinyl amine) copolymers. Macromolecules 45:671–680
30. Bielawski CW, Grubbs RH (2007) Living ring-opening metathesis polymerization. Prog Polym Sci 32:1–29
31. Ramakrishnan S, Chung TC (1990) Poly(5-hydroxyoctenylene) and its derivatives: synthesis via metathesis polymerization of an organoborane monomer. Macromolecules 23:4519–4524
32. Couturier JL, Tanaka K, Leconte M, Basset JM, Ollivier J (1993) Metathesis of sulfur-containing olefins with a metallacyclic

aryloxo(chloro)neopentylidenetungsten complex. Angew Chem Int Ed 32:112–115

33. Hillmyer MA, Laredo WR, Grubbs RH (1995) Ring-opening metathesis polymerization of functionalized cyclooctenes by a ruthenium-based metathesis catalyst. Macromolecules 28:6311–6316

34. Yang HL, Islam M, Budde C, Rowan SJ (2003) Ring-opening metathesis polymerization as a route to controlled copolymers of ethylene and polar monomers: synthesis of ethylene–vinyl chloride-like copolymers. J Polym Sci Part A Polym Chem 41:2107–2116

35. Xu GJ, Wang DR, Buchmeiser MR (2012) Functional polyolefins: poly(ethylene)-graft-poly(tert-butyl acrylate) via atom transfer radical polymerization from a polybrominated alkane. Macromol Rapid Commun 33:75–79

36. Schneider MF, Gantner C, Obrecht W, Nuyken O (2010) Ring-opening metathesis polymerization of nitrile substituted cis-cyclooctenes. Macromol Rapid Commun 31:1731–1735

37. Schneider MF, Gantner C, Obrecht W, Nuyken O (2011) Ring-opening metathesis polymerization of cis-cyanocyclooct-4-ene: search for active catalysts, variation of monomer to catalyst ratios and monomer concentrations. J Polym Sci Part A Polym Chem 49:879–885

38. Scherman OA, Kim HM, Grubbs RH (2002) Synthesis of well-defined poly((vinyl alcohol)2-alt-methylene) via ring-opening metathesis polymerization. Macromolecules 35:5366–5371

39. Scherman OA, Walker R, Grubbs RH (2005) Synthesis and characterization of stereoregular ethylene–vinyl alcohol copolymers made by ring-opening metathesis polymerization. Macromolecules 38:9009–9014

40. Hejl A, Scherman OA, Grubbs RH (2005) Ring-opening metathesis polymerization of functionalized low-strain monomers with ruthenium-based catalysts. Macromolecules 38:7214–7218

41. Maughon BR, Grubbs RH (1996) Synthesis and controlled cross-linking of polymers derived from ring-opening metathesis polymerization (ROMP). Macromolecules 29:5765–5769

42. Breitenkamp K, Simeone J, Jin E, Emrick T (2002) Novel amphiphilic graft copolymers prepared by ring-opening metathesis polymerization of poly(ethylene glycol)-substituted cyclooctene macromonomers. Macromolecules 35:9249–9252

43. Breitenkamp K, Emrick T (2003) Novel polymer capsules from amphiphilic graft copolymers and cross-metathesis. J Am Chem Soc 125:12070–12071

44. Lehman SE, Wagener KB, Baugh LS, Rucker SP, Schulz DN, Varma-Nair M, Berluche E (2007) Linear copolymers of ethylene and polar vinyl monomers via olefin metathesis–hydrogenation: synthesis, characterization, and comparison to branched analogues. Macromolecules 40:2643–2656

45. Han HJ, Chen FX, Yu JH, Dang JY, Ma Z, Zhang YQ, Xie MR (2007) Ring-opening metathesis polymerization of functionalized cyclooctene by a ruthenium-based catalyst in ionic liquid. J Polym Sci Part A Polym Chem 45:3986–3993

46. Song AR, Parker KA, Sampson NS (2009) Synthesis of copolymers by alternating ROMP (AROMP). J Am Chem Soc 131:3444–3445

47. Kratz K, Breitenkamp K, Hule R, Pochan D, Emrick T (2009) PC-polyolefins: synthesis and assembly behavior in water. Macromolecules 42:3227–3229

48. Song AR, Lee JC, Parker KA, Sampson NS (2010) Scope of the ring-opening metathesis polymerization (ROMP) reaction of 1-substituted cyclobutenes. J Am Chem Soc 132:10513–10520

49. Breitenkamp RB, Ou Z, Breitenkamp K, Muthukumar M, Emrick T (2007) Synthesis and characterization of polyolefin-graft-oligopeptide polyelectrolytes. Macromolecules 40:7617–7624

50. Revanur R, McCloskey B, Breitenkamp K, Freeman BD, Emrick T (2007) Reactive amphiphilic graft copolymer coatings applied to poly(vinylidene fluoride) ultrafiltration membranes. Macromolecules 40:3624–3630

51. Kobayashi S, Kim H, Macosko CW, Hillmyer MA (2013) Functionalized linear low-density polyethylene by ring-opening metathesis polymerization. Polym Chem 4:1193

52. Shi HC, Shi DA, Yin LG, Luan SF, Zhao J, Yin JH (2010) Synthesis of amphiphilic polycyclooctene-graft-poly(ethylene glycol) copolymers by ring-opening metathesis polymerization. React Funct Polym 70:449–455

53. Winkler B, Rehab A, Ungerank M, Stelzer F (1997) A novel side-chain liquid crystal polymer of 5-substituted cis-cyclooctene via ring-opening metathesis polymerization. Macromol Chem Phys 198:1417–1425

54. Xia Y, Verduzco R, Grubbs RH, Kornfield JA (2008) Well-defined liquid crystal gels from telechelic polymers. J Am Chem Soc 130:1735–1740

55. Kobayashi S, Pitet LM, Hillmyer MA (2011) Regio- and stereoselective ring-opening metathesis polymerization of 3-substituted cyclooctenes. J Am Chem Soc 133:5794–5797

56. Martinez H, Miró P, Charbonneau P, Hillmyer MA, Cramer CJ (2012) Selectivity in ring-opening metathesis polymerization of Z-cyclooctenes catalyzed by a second-generation Grubbs catalyst. ACS Catal 2:2547–2556

57. Zhang J, Matta ME, Hillmyer MA (2012) Synthesis of sequence-specific vinyl copolymers by regioselective ROMP of multiply substituted cyclooctenes. ACS Macro Lett 1:1383–1387

58. Pitet LM, Zhang J, Hillmyer MA (2013) Sequential ROMP of cyclooctenes as a route to linear polyethylene block copolymers. Dalton Trans 42:9079–9088

59. Zhang J, Matta ME, Martinez H, Hillmyer MA (2013) Precision vinyl acetate/ethylene (VAE) Copolymers by ROMP of acetoxy-substituted cyclic alkenes. Macromolecules 46:2535–2543

60. Benkeser RA, Mozdzen EC, Muench WC, Roche RT, Siklosi MP (1979) Organic chemistry of dichlorosilane. Additions to conjugated and unconjugated diene systems followed by intramolecular cyclizations. J Org Chem 44:1370–1376

61. Slugovc C (2004) The ring opening metathesis polymerisation toolbox. Macromol Rapid Commun 25:1283–1297

62. Leitgeb A, Wappel J, Slugovc C (2010) The ROMP toolbox upgraded. Polymer 51:2927–2946

63. Formentin P, Gimeno N, Steinke JHG, Vilar R (2005) Reactivity of Grubbs' catalysts with urea- and amide-substituted olefins. Metathesis and isomerization. J Org Chem 70:8235–8238

64. Menozzi C, Dalko PI, Cossy J (2005) Hydrosilylation of terminal alkynes with alkylidene ruthenium complexes and silanes. J Org Chem 70:10717–10719

65. Polshettiwar V, Varma RS (2008) Olefin ring closing metathesis and hydrosilylation reaction in aqueous medium by Grubbs second generation ruthenium catalyst. J Org Chem 73:7417–7419

66. Xu Y, Thurber CM, Lodge TP, Hillmyer MA (2012) Synthesis and remarkable efficacy of model polyethylene-graft-poly(methyl methacrylate) copolymers as compatibilizers in polyethylene/poly(methyl methacrylate) blends. Macromolecules 45:9604–9610

67. Rojas G, Berda EB, Wagener KB (2008) Precision polyolefin structure: modeling polyethylene containing alkyl branches. Polymer 49:2985–2995

68. Hahn SF (1992) An improved method for the diimide hydrogenation of butadiene and isoprene containing polymers. J Polym Sci Part A Polym Chem 30:397–408

69. Demonceau A, Stumpf AW, Saive E, Noels AF (1997) Novel ruthenium-based catalyst systems for the ring-opening metathesis polymerization of low-strain cyclic olefins. Macromolecules 30:3127

70. Chung TC (2002) Synthesis of functional polyolefin copolymers with graft and block structures. Prog Polym Sci 27:39–85

20

Comparing Pt/SrTiO$_3$ to Rh/SrTiO$_3$ for hydrogen photocatalytic production from ethanol

A. K. Wahab · T. Odedairo · J. Labis ·
M. Hedhili · A. Delavar · H. Idriss

Abstract Photocatalytic hydrogen production from ethanol as an example of biofuel is studied over 0.5 wt% Rh/SrTiO$_3$ and 0.5 wt% Pt/SrTiO$_3$ perovskite materials. The rate of hydrogen production, r_{H2}, over Pt/SrTiO$_3$ is found to be far higher than that observed over Rh/SrTiO$_3$ (4×10^{-6} mol of H$_2$ g$_{catal.}^{-1}$ min^{-1} (1.1×10^{-6} mol of H$_2$ m$_{catal.}^{-2}$ min^{-1}) compared to 0.7×10^{-6} mol of H$_2$ g$_{catal.}^{-1}$ min^{-1} (5.5×10^{-8} mol of H$_2$ m$_{catal.}^{-2}$ min^{-1}), respectively, under UV excitation with a flux equivalent to that from the sun light (ca. 1 mW cm^{-2}). Analyses of the XPS Rh3d and XPS Pt4f indicate that Rh is mainly present in its ionic form (Rh^{3+}) while Pt is mainly present in its metallic form (Pt^0). A fraction of the non-metallic state of Rh in the catalyst persisted even after argon ion sputtering. The tendency of Rh to be oxidized compared to Pt might be the reason behind the lower activity of the former compared to the later. On the contrary, a larger amount of methane are formed on the Rh containing catalyst compared to that observed on the Pt containing catalyst due to the capacity of Rh to break the carbon–carbon bond of the organic compound.

Keywords Ethanol-photoreaction · XPS Rh3d ·
XPS Pt4f · Hydrogen production · Perovskite materials ·
Band gap · SrTiO$_3$ · Carbon–carbon bond dissociation

Introduction

Photo-catalytic production of hydrogen from renewables is poised to be one of its main sources in the future once successful catalytic materials are made possible. The reaction requires the presence of a semiconductor with bang gap energy within the energy of the solar radiation, a conductor such as a noble metal to accept electrons from the conduction band in addition to hydrogen-containing compounds [1]. Ultimately, the desired compound for hydrogen production is water. Next to water are alcohols and, in particular, ethanol because it is provided from biomass and is therefore renewable [2].

SrTiO$_3$ is stable in water as well as in presence of organic compounds; more importantly it endures corrosion under UV excitation and unlike other non oxygen containing compounds is already oxygen terminated and therefore cannot be over oxidized. It has two band gaps: one indirect at 3.25 eV, similar to anatase the perovskite structure with TiO$_2$, and one direct at 3.75 eV [3]. The indirect band gap is between the upper valence band, mainly composed of O2p, and the empty states Ti3d (t$_{2g}$). The direct band gap is between O2p and Ti3d (e$_g$) levels [4] (Fig. 1).

Numerous works have addressed the photo-catalytic activity of SrTiO$_3$ alone, with other transition metals as well as doped with other ions [5–11]. Results differ strongly from one study to the other due to difference in crystallinity, metal dispersion, effect of dopant and reaction conditions. In a recent work [6], the reactivity of SrTiO$_3$ with different particle sizes was tested for hydrogen

A. K. Wahab · T. Odedairo · H. Idriss (✉)
SABIC: T&I Riyadh and CRI-KAUST, Riyadh, Saudi Arabia
e-mail: idrissh@sabic.com

J. Labis
KAIN, Riyadh, Saudi Arabia

M. Hedhili · A. Delavar
KAUST, Thuwal, Saudi Arabia

Fig. 1 Top view (**a**), side view (**b**) and perspective view (**c**) of SrTiO$_3$. Small black spheres (Sr^{2+}), large gray spheres (O^{2-}), smaller yellow spheres in **c** are those of Ti^{4+} cations. Also indicated in **d** are the electronic transitions between O2p and Ti3d levels (redrawn from Ref. [4])

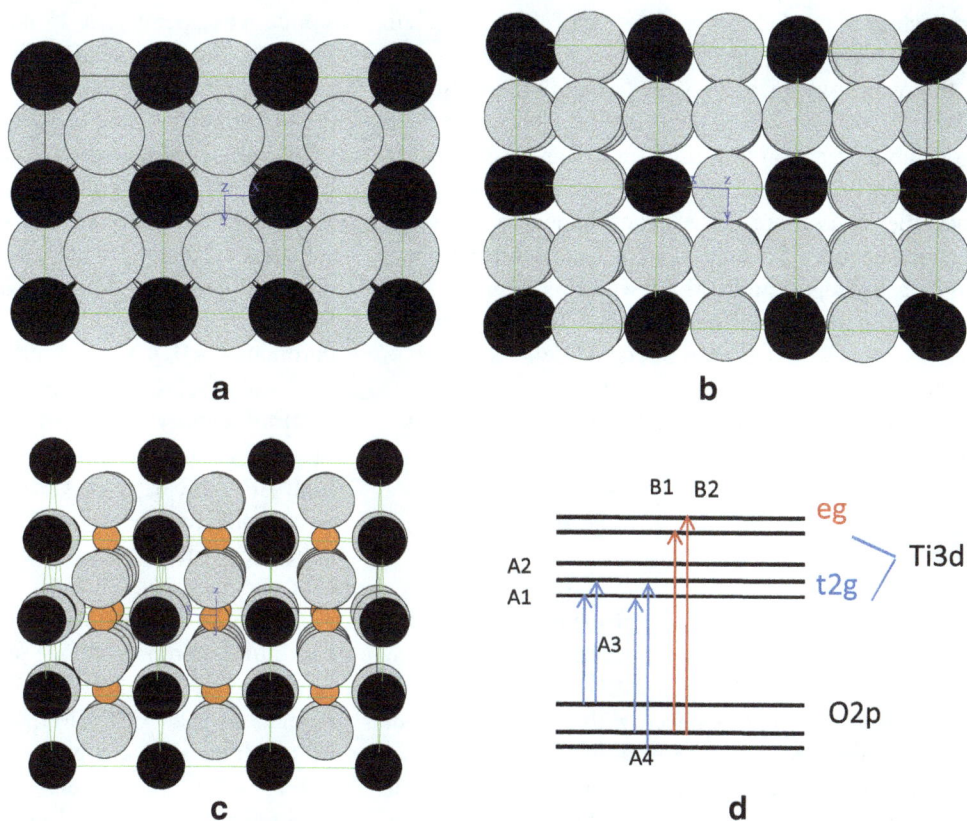

production from water. It was found that bulk material (particles dimension >100 nm) was more active than 30 nm size particles which in turn were more active than 6.5 nm size particles (producing 28, 19.4 and 3 μmol of H$_2$ g$_{catal.}^{-1}$ h^{-1}, respectively [at 26.3 mW cm^{-2} with λ in the (250–380 nm)]. Reasons for this decrease are attributed to an increase in the water oxidation overpotential for the smaller particles and reduced light absorption due to quantum size effect. In another work [7], probing the anisotropy of the reactivity of SrTiO$_3$ microcrystals indicated that both reduced and oxidized products are formed preferentially on {100} surfaces. This anisotropy was explained as being due to differences in the electronic band structure. Because direct optical transitions for charge carriers having momentum vectors in the <100> direction overlap well with the spectral distribution of the absorbed photons, more photogenerated carriers are moving toward {100} surfaces than other surfaces and, as a result, the {100} surfaces are more active. Other work [8] has addressed the photocatalytic water splitting activity for hydrogen production over the mesoporous-assembled SrTiO$_3$ nanocrystal-based photocatalysts with various hole scavengers: methanol, ethanol, 2-propanol, D-glucose, and Na$_2$SO$_3$. Pristine mesoporous-assembled SrTiO$_3$ photocatalysts exhibited higher photocatalytic activity in hydrogen production than the non-mesoporous-assembled commercial photocatalysts or commercial SrTiO$_3$. These results

indicate that the mesoporous assembly of nanocrystals with high pore uniformity plays a significant role, affecting the photocatalytic hydrogen production activity.

Moreover, it was seen that the Pt co-catalyst enhances the visible light harvesting ability of the mesoporous assembly with an optimum Pt loading of 0.5 wt% on the mesoporous-assembled SrTiO$_3$ photocatalyst providing the highest photocatalytic activity, with hydrogen production rate of 276 μmol h^{-1} g$_{catal.}$ and a quantum efficiency of 1.9 % under UV light irradiation. Other works indicated that Rh (1 %)-doped SrTiO$_3$ photocatalyst loaded with a Pt co-catalyst (0.1 wt%) gave 5.2 % of the quantum yield at 420 nm for the H$_2$ evolution reaction from a methanol solution (10 vol%) [9]. Another work [10] focused on doping SrTiO$_3$/TiO$_2$ with N ions (using hexamethylene-tetramine) and optimized photocatalytic activity of hydrogen production (average hydrogen production rate = 5.1 mmol g$_{catal.}^{-1}$ h^{-1} with 2 wt% loaded Pt) under visible light was seen although little is known about the catalyst stability and comparison with the non-doped semiconductor material. A recent computational study (using the hybrid DFT method) of doping SrTiO$_3$ with metal cations coupled with experimental study showed that co-doping Cr and La ions had considerable enhancement effect on hydrogen production of water/methanol solution [11]. The reason invoked is that doping with La ions raised the Fermi level of Cr ions and stabilizes its oxidation state of Cr^{3+} which is needed for the red shift of the band gap to extend light absorption into the visible region.

In this work, we have conducted a study of Rh/SrTiO$_3$ and Pt/SrTiO$_3$ materials in addition to monitoring their photo-catalytic reaction to further understand the extent of their activity. In particular, we attempt to answer which of the two metals is more active when added to the semiconductor support. We find that Rh has a weaker effect than Pt on hydrogen production (which is similar to the case where these metals are deposited on TiO$_2$ anatase [12]). Although we have not conducted detailed study of the particle size effect of the semiconductor, we do not find considerable difference in the reaction rates upon changing its morphology.

Experimental

SrTiO$_3$ was prepared by the sol–gel method where TiCl$_4$ was added to a strontium-nitrate solution in stoichiometric amounts. After the addition of TiCl$_4$ to the strontium-nitrate solution, the pH was raised with sodium hydroxide to a value between 8 and 9 where strontium hydroxide and titanium hydroxide precipitated. The precipitate was left to stand for about 12 h at room temperature to ensure completion of the reaction after which it was filtered and washed with de-ionized water until neutral pH (~ 7). The resulting material was then dried in an oven at 100 °C for a period of at least 12 h. Next the material was calcined at 800 °C. X-ray diffraction techniques were used to indicate formation of SrTiO$_3$ (Fig. S1). Several SrTiO$_3$ materials were tested in addition to a commercial SrTiO$_3$ (Sigma-Aldrich). Rh metals were impregnated from a solution containing RhCl$_3$ in 1 N HCl. The resulting catalyst differed from the initial SrTiO$_3$ as its particle size was far smaller and its BET surface area higher. Pt metals were impregnated the same manner (from a PtCl$_4$ precursor). Other techniques used to study the material included XPS, XRD, TEM, and UV–vis. Photoreaction was conducted in a batch reactor (100–250 mL) containing 10–25 mg of materials under stirring conditions with a UV lamp flux of about 1 mW cm^{-2}. Analysis was conducted using GC-TCD equipped with a Porapack Q with a N$_2$ carrier gas isothermally at 50 °C (N$_2$ flow rate = 20 mL min^{-1}). At these conditions, the products were eluted from the column with the following order: hydrogen, oxygen, methane, carbon dioxide, ethylene, ethane, propylene, propane, acetaldehyde, followed by ethanol. The powder XRD patterns of the samples were recorded on a Philips X'pert-MPD X-ray powder diffractometer. A 2θ interval between 10° and 90° was used with a step size of 0.010° and a step time of 0.5 s. The diffractometer was equipped with a Ni-filtered Cu K$_\alpha$ radiation source ($\lambda = 1.5418$ Å). The X-ray

source was operated at 45 mA and 40 kV. Sample preparation for the X-ray analysis involved gentle grinding of the solid into a fine powder and packing of approximately 0.1–0.3 g of the sample into an aluminum sample holder with light compression to make it flat and tight. XRD patterns of the samples were recorded with the X'pert HighScore Plus software and saved in XRDML text format for further manipulation and processing. X-Ray Photoelectron spectroscopy was conducted using a Thermo Scientific ESCALAB 250 Xi, equipped with a monochromated AlK$_\alpha$ X-ray source, Ultra Violet He lamp for UPS, ion scattering spectroscopy (ISS), and reflected electron energy loss spectroscopy (REELS). The base pressure of the chamber was typically in the low 10^{-10} to high 10^{-11} mbar range. Charge neutralization was used for all samples (compensating shifts of ~ 1 eV). Spectra were calibrated with respect to C1s at 284.7 eV. The Rh3d, Pt4f, O1s, Sr3d, Ti2p, C1s and valence band binding energy regions were scanned for all materials. Typical acquisition conditions were as follows: pass energy = 20 eV and scan rate = 0.1 eV per 200 ms. Ar ion bombardment was performed with an EX06 ion gun at 2 kV beam energy and 10 mA emission current; sample current was typically 0.9–1.0 µA. The sputtered area of 900 × 900 µm^2 was larger than the analyzed area: 600 × 600 µm^2. Self-supported oxide disks of approximately 0.5 cm diameter were loaded into the chamber for analysis. Data acquisition and treatment were done using the Avantage software.

Results

Figure 2 presents a TEM image of the Rh/SrTiO$_3$ catalyst. Rh particles with mean size of 3–4 nm are seen. SrTiO$_3$ is composed of small crystallites with sizes between 10 and 20 nm. The BET surface area of the material is found to be ca. 13 m^2 g$_{catal.}^{-1}$. Figure S2 presents high resolution image of an Rh particle; the spots in the FT image unambiguously identify it by its crystalline structure. Figure 3 presents TEM of two types of Pt/SrTiO$_3$: one where Pt was deposited on commercial SrTiO$_3$ and the other on SrTiO$_3$ that was prepared by the sol–gel method followed by annealing at 800 °C. The commercial SrTiO$_3$ is composed of particles with dimension larger than 1 µm while the one prepared by the sol–gel method is made of typical perovskite (cubic) particles with a mean size of about 50 nm; Pt particles cannot be seen at this resolution.

XP spectra Ti2p, Sr3d, O1s, and Rh3d (or Pt4f) were collected among other lines; here we focus on the main lines relevant to this study. Figure 4 presents XPS Rh3d and Ti2p of the as-prepared catalyst (a, b) and Ar ions sputtered (c, d). The presence of Rh^{3+} (mainly as Rh$_2$O$_3$) and Rh metal can be seen with binding energy = 307.0 and

308.8 eV, respectively, in line with reference elements [13]. To further confirm the presence of Rh metal as well to probe the extent of its reduction, the same area was Ar ions sputtered for 2 min (2 kV, 10 mA; sample current 10 mA). Clearly, the signal from Rh metal increased compared to that of Rh^{3+} due to reduction of Rh ions under bombardment. However, only about 50 % of Rh ions are reduced to metallic Rh. Our objective was not to reduce all Rh ions but to observe for the effects of mild ion sputtering on the reduction of the noble metal. Associated with Rh ions reduction is a reduction of Ti^{4+} ions to Ti^{3+} and Ti^{2+} ions (b, d) as has been seen in numerous work for Ar ions reduction of TiO_2 [14, 15].

Figure 5 presents the Sr3d, Ti2p and Pt4f of the as-prepared 0.5 wt% Pt/SrTiO$_3$ catalyst. Sr3d is typical of the doublet ($3d_{5/2}$, $3d_{3/2}$) of Sr^{2+} ions (spin orbit splitting close to 2 eV), the narrow lines of the $Ti2p_{3/2,1/2}$ and their binding energy positions indicate the absence of Ti^{3+} while the $Pt4f_{7/2,5/2}$ is that of metallic Pt (binding energy of $Pt_{7/2}$

at 72.5 eV with a spin orbit splitting of 3.4 eV). The presence of Pt metal in the as-prepared catalyst is in sharp contrast to that of the Rh/SrTiO$_3$ where a large fraction of Rh is found to be in the form or Rh^{3+} ions. This is typical of Pt deposited on reducible metal oxides where it is often largely present in its metallic form [16, 17].

Figure 6 presents results of the hydrogen production using photons of 3.3 eV in a batch reactor containing 0.5 wt% Pt/SrTiO$_3$ and 0.5 wt% Rh/SrTiO$_3$. Hydrogen production is seen together with methane. The production rate is comparable to that observed on Au/TiO$_2$ anatase previously and far higher than that of Au/TiO$_2$ rutile [18]. The ratio Rh/Ti is found equal to 0.07 while that of Pt/Ti is equal to 0.16. Two main observations are clear from Fig. 6. First Pt/SrTiO$_3$ is about six times higher than Rh/SrTiO$_3$ based on weight and about 20 times higher based on area. Even if we account for difference in the M to Ti^{4+} ratios, the rate of hydrogen production is still far higher for the Pt-containing catalysts compared to that of the Rh one. Second, Rh/SrTiO$_3$ produces large amounts of methane when compared to Pt/SrTiO$_3$. The larger amounts of methane can be rationalized by the fact that Rh is more active in breaking the C–C bond compared to Pt [19–21]. In this case, the ratio H$_2$ to CH$_4$ is equal to 2 for Rh/SrTiO$_3$ while it is equal to 8 in the case of Pt/SrTiO$_3$.

The reaction occurs due to electron excitation from the valence band (VB) O2p to the conduction band (CB) Ti3d upon UV illumination, as presented in Fig. 1. Alcohols are known as hole scavenges [22–24]. Ethanol is oxidized to acetaldehyde via two electron injection into the VB [25] with a time scale in the nanosecond range [26]. Mechanistically ethanol is first dissociatively adsorbed on the surface forming ethoxides (according to Eq. 2) and then upon two electron injections (through α-oxy radical [18, 25]) acetaldehyde is formed. The hydrogen ions released through this process are reduced to one hydrogen molecule [27, 28]. Reactions 1–4 represent these steps.

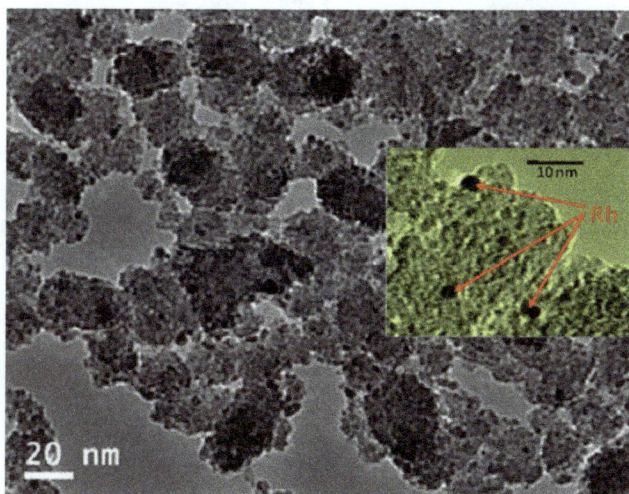

Fig. 2 TEM of 0.5 wt% Rh/SrTiO$_3$. Rh can be seen as dark particles in the main figure as well as the inset

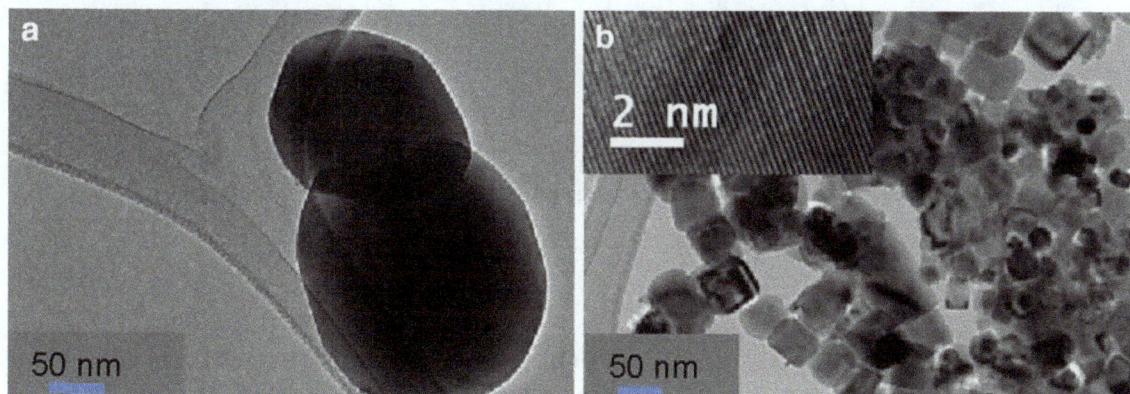

Fig. 3 TEM of 0.5 wt% Pt/SrTiO$_3$. **a** Commercial SrTiO$_3$ and **b** SrTiO$_3$ prepared by the sol–gel method; the *inset* in **b** presents the [100] cubic structure of SrTiO$_3$ evidence of a high degree of crystallinity

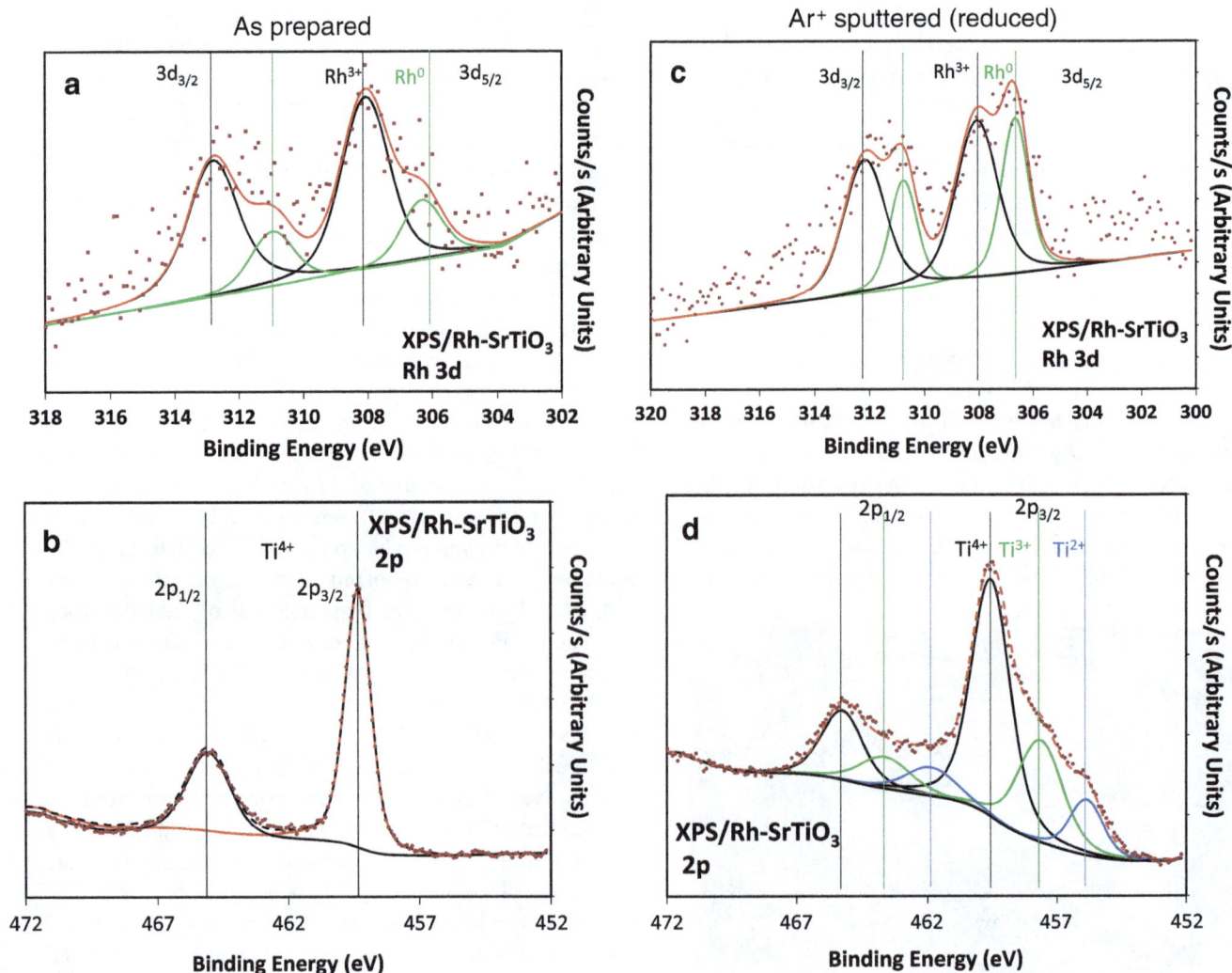

Fig. 4 XPS Rh 3d and Ti2p of 0.5wt% Rh/SrTiO₃. **a, b** As-prepared and **c, d** after Ar ions sputtering

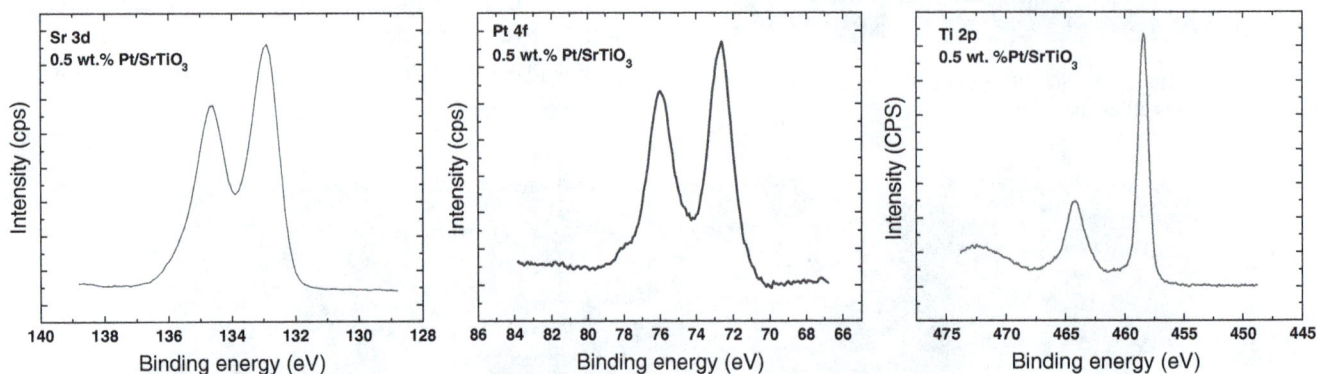

Fig. 5 XPS Sr3d, Ti2p and Pt 4f of the as-prepared 0.5 wt% Pt/SrTiO₃. The Sr to Ti atomic ratio is about 1

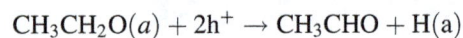

$$SrTiO_3 + UV \rightarrow e^- + h^+ \tag{1}$$

$$CH_3CH_2OH \rightarrow CH_3CH_2O(a) + H(a) \tag{2}$$

$$CH_3CH_2O(a) + 2h^+ \rightarrow CH_3CHO + H(a) \tag{3}$$

$$2H(a) + 2e^- \rightarrow H_2 \tag{4}$$

(a) adsorbed

The reaction over Pt/SrTiO₃ is similar to that over Au/TiO₂ where the main products are H₂ and acetaldehyde [1, 18], with traces of methane. However, over Rh/SrTiO₃ the reaction proceeds to CH₄. Therefore, a large fraction of CH₃CHO farther reacts with CH₄ and CO.

Fig. 6 Hydrogen and methane photo-production from ethanol over 0.5 wt% Rh/SrTiO$_3$ and 0.5 wt% Pt/SrTiO$_3$. BET surface area of Rh/SrTiO$_3$ = 13 m^2 g^{-1} while that of Pt/SrTiO$_3$ is 3.5 m^2 g^{-1}. The *numbers* inside the figure indicate the slope (the rate in moles of hydrogen per g$_{catal.}$ per minute)

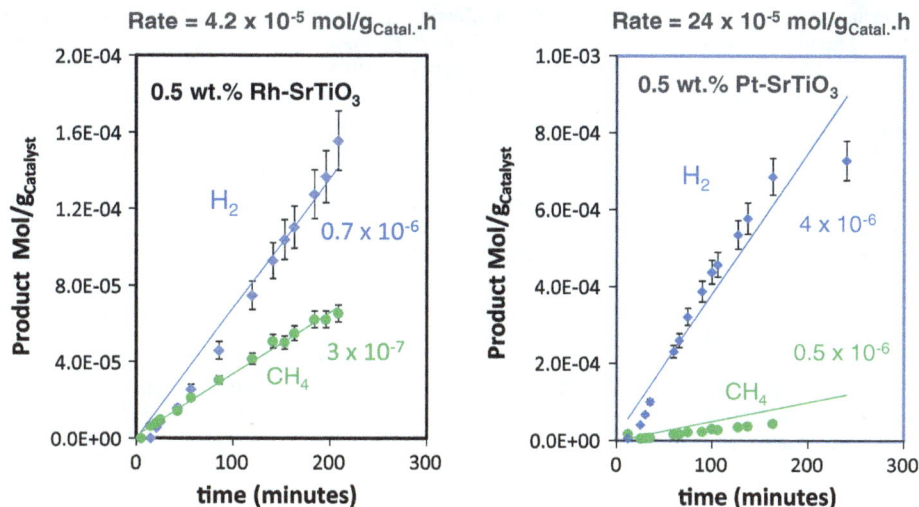

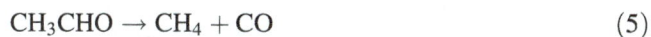

$$CH_3CHO \rightarrow CH_4 + CO \tag{5}$$

As presented in Fig. 6, CH$_4$ is formed. Therefore, one should except if the reaction of Eq. 5 is complete, equal amounts of H$_2$ and CH$_4$ would be formed. We have conducted numerous runs and have found that in all cases CH$_4$ concentration was lower than that of H$_2$. Recent DFT studies on SrTiO$_3$ indicated the possibility of a pathway for ethanol decomposition involving CH$_3$CO radical (with the release of three hydrogen ions) [29]. In this case, CH$_3$CO may react with surface oxygen making acetate species or may split to CH$_3$ radical and CO; CH$_3$ radicals may react with OH radicals giving methanol, the latter would easily decompose to CO and hydrogen [30]. This route may explain the higher ratio H$_2$ to CH$_4$ observed. To extract all hydrogen atoms from ethanol, the formation of CH$_4$ is not desired and at present work is in progress to find a method to more efficiently break the carbon–carbon bond before obtaining acetaldehyde at 300 K under photon irradiation in the presence of water providing additional hydrogen and oxygen ions to complete the reforming reaction, as in high temperature ethanol steam reforming [31].

Conclusions

The rate of photo-catalytic reaction of ethanol to hydrogen is found to be far higher on Pt/SrTiO$_3$ compared to Rh/SrTiO$_3$. The most likely reason for the higher activity of the Pt containing catalysts is the ease by which metallic Pt is formed compared to Rh (where a large fraction of the as-prepared material is in Rh^{3+}). Ar ions sputtering Rh/SrTiO$_3$ considerably reduced Ti in SrTiO$_3$, but only reduced about half of Rh ions to metallic Rh. Another important difference is noticed between the two catalytic systems and it is related to reaction selectivity. Rh/SrTiO$_3$

produces large amounts of CH$_4$ compared to Pt/SrTiO$_3$; this is most likely related to the capacity of Rh to break the carbon–carbon bond compared to that of Pt.

References

1. Connelly KA, Idriss H (2012) The photoreaction of TiO$_2$ and Au/TiO$_2$ single crystal and powder surfaces with organic adsorbates. Emphasis on hydrogen production from renewables. Green Chem 14:260–280
2. Scott M, Idriss H (2009) Heterogeneous catalysis for hydrogen production. In: Anstis P, Crabtree RH (eds) Handbook of green chemistry—green catalysis, vol 1, chap. 10, ISBN-10: 3-527-31577-2
3. Yamada Y, Kanemitsu Y (2010) Band-to-band photoluminescence in SrTiO$_3$. Phys Rev B 82:121103 (R) 1–4
4. van Benthem K, Elsässer C, French R (2001) Bulk electronic structure of SrTiO$_3$: experiment and theory. J Appl Phys 90:6156–6164
5. Kudo A, Miseki Y (2009) Heterogeneous photocatalyst materials for water splitting. Chem Soc Rev 38:25–278
6. Townsend TK, Browning ND, Osterloh FE (2012) Nanoscale strontium titanate photocatalysts for overall water splitting. ACS Nano 6:7420–7426
7. Giocondia JL, Salvador PA, Rohrer GS (2007) The origin of photochemical anisotropy in SrTiO$_3$. Topics Catal. 44:529–533
8. Puangpetch T, Sreethawong T, Yoshikawa S, Chavadej S (2009) Hydrogen production from photocatalytic water splitting over mesoporous-assembled SrTiO$_3$ nanocrystal-based photocatalysts. J Mol Catal A Chem 312:97–106
9. Konta R, Ishii T, Kato H, Kudo A (2004) Photocatalytic activities of noble metal ion doped SrTiO$_3$ under visible light irradiation. J Phys Chem B 108:8992–8995
10. Jian-Hui Y, Yi-Rong Z, You-Gen T, Shu-Qin Z (2009) Nitrogen-doped SrTiO$_3$/TiO$_2$ composite photocatalysts for hydrogen production under visible light irradiation. J Alloy Compd 472:429–433

11. Reunchan P, Ouyang S, Umezawa N, Xu H, Zhang Y, Ye J (2013) Theoretical design of highly active SrTiO$_3$-based photocatalysts by a codoping scheme towards solar energy utilization for hydrogen production. J. Mater Chem A 1:4221–4227

12. Yang YZ, Chang CH, Idriss H (2006) Photo-catalytic production of hydrogen from ethanol over M/TiO$_2$ catalysts (M = Pd, Pt or Rh). Appl Cat B Environ 67:217–222

13. Briggs D, Seah MP (eds) Practical surface analysis, 2nd edn, vol 1. Wiley, Chichester

14. Idriss H, Barteau MA (1994) Characterization of TiO$_2$ surfaces active for novel organic syntheses. Catal Lett 26:123–139

15. Idriss H, Barteau MA (1994) Carbon-carbon bond formation on metal oxides: from single crystals toward catalysis. Langmuir 10:3693–3700

16. Yee A, Morrison S, Idriss H (2000) A Study of ethanol reactions over Pt/CeO$_2$ by temperature-programmed desorption and in situ FT-IR spectroscopy: evidence of benzene formation. J Catal 191:30–45

17. Chen H-W, Ku Y, Kuo Y-L (2007) Effect of Pt/TiO$_2$ characteristics on temporal behavior of o-cresol decomposition by visible light-induced photocatalysis. Water Res 41:2069–2078

18. Murdoch M, Waterhouse GIN, Nadeem MA, Metson JB, Keane MA, Howe RF, Llorca J, Idriss H (2011) The effect of gold loading and particle size on photocatalytic hydrogen production from ethanol over Au/TiO$_2$ nanoparticles. Nat Chem 3:489–492

19. Idriss H (2004) Ethanol reactions over the surfaces of noble metal/cerium oxide catalysts. Platinum Metals Rev. 48:105–115

20. Idriss H, Scott M, Llorca J, Chan S.C, Chiu W, Sheng PY, Yee A, Blackford MA, Pas SJ, Hill AJ, Alamgir FM, Rettew R, Petersburg C, Senanayake SD, Barteau MA (2008) A phenomenological study of the metal–oxide interface: the role of catalysis in hydrogen production from renewable resources. Chem Sus Chem 1:905–910

21. Chen H-L, Liu S-H, Ho J-J (2006) Theoretical calculation of the dehydrogenation of ethanol on a Rh/CeO$_2$ (111) surface. J Phys Chem B 110:14816–14823

22. Muir JMR, Choi YM, Idriss H (2012) DFT study of ethanol on TiO$_2$ (110) rutile surface. Phys Chem Chem Phys 14: 11910–11919

23. Yu Z, Chuang SSC (2007) In situ IR study of adsorbed species and photogenerated electrons during photocatalytic oxidation of ethanol on TiO$_2$. J Catal 246:118–126

24. Wu G, Chen T, Zong X, Yan H, Ma G, Wang X, Xu Q, Wang D, Lei Z, Li C (2008) Suppressing CO formation by anion adsorption and Pt deposition on TiO$_2$ in H$_2$ production from photocatalytic reforming of methanol. J Catal 253:225–227

25. Mller BR, Majoni S, Memming R, Meissner D (1997) Particle size and surface chemistry in photoelectrochemical reactions at semiconductor particles. J. Phys. Chem. B 101:2501–2507

26. Hoffmann MR, Martin ST, Choi W, Bahneman DW (1995) Environmental applications of semiconductor photocatalysis. Chem Rev 95:69–96

27. Nadeem MA, Murdoch M, Waterhouse GIN, Metson JB, Keane MA, Llorca J, Idriss H (2010) Photoreaction of ethanol on Au/TiO$_2$ anatase. Comparing the micro to nano particle size activities of the support for hydrogen production. J Photochem Photobiol A Chem 216:250–255

28. Bowker M, Davies PR, Al-Mazroai LS (2009) Photocatalytic reforming of glycerol over gold and palladium as an alternative fuel source. Catal Lett 128:253–255

29. Adeagbo WA, Fischer G, Hergert W (2011) First-principles investigations of electronic and magnetic properties of SrTiO3 (001) surfaces with adsorbed ethanol and acetone molecules. Phys Rev B 83:195428-1–195428-8

30. Bowker M, James D, Stone P, Bennett R, Perkins N, Millard L, Greaves J, Dickinson A (2003) Catalysis at the metal-support interface: exemplified by the photocatalytic reforming of methanol on Pd/TiO$_2$. J Catal 217:427–433

31. Scott M, Geoffrey M, Chiu W, Blackford MA, Idriss H (2008) Hydrogen production from ethanol over Rh-Pd/CeO$_2$ catalysts. Topics Catal. 51:39–48

21

The feedstock curve: novel fuel resources, environmental conservation, the force of economics and the renewed east–west power struggle

Oliver R. Inderwildi · Fabian Siegrist ·
Robert Duane Dickson · Andrew J. Hagan

Abstract Rapid technological advancements can make previously uneconomic resources and/or feedstock available within significantly reduced timeframes. This can and will further transform the global energy landscape and moreover, will impact the mix of feedstock we use for energy provision and material production—the so-called Feedstock Curve. Herein, three current examples are assessed to illustrate that this restructuring has by far wider reaching implications: Firstly, we examine how unconventional resources—mainly produced using fractured cracking techniques—have restructured the US energy landscape, are now fueling the US economic recovery and will impact the geopolitical balance. Secondly, we assess how unconventional resources could impact European energy security, the Crimean crisis and redirect global cash flows. Thirdly, we analyse the potential impact of so-called methane hydrates deposited off the shores of Japan on the energy transition of the Island nation and how they might impact its trade deficit and long-term economic outlook. Last but not least, we will present arguments that unconventional resources, when regulated properly, may be a blessing for the environment. With these examples, this think piece and concept note will illustrate the interconnectedness of economics, politics, environmental conservation and technology.

Keywords Energy security · Foreign policy · International security · Resources · Global change

The contribution reflects the views of the author, not the institution.

O. R. Inderwildi (✉)
Smith School, University of Oxford, South Parks Road,
Oxford OX1 3QY, UK
e-mail: oliver.inderwildi@smithschool.ox.ac.uk

F. Siegrist · R. D. Dickson
Deloitte LLP, 30 Rockefeller Plaza, New York City, NY
10111, USA

A. J. Hagan
World Economic Forum, 91–93 route de la Capite,
1223 Cologny, Geneva, Switzerland

Introduction

Not long ago, anyone who had mentioned that the United States would likely become energy independent or would have said that they could consider exporting fossil resources would have received roaring laughter. The energy market in North America has, however, undergone a full transformation and neither energy independence nor the US as a fuel exporting country is an unrealistic assumption [1]. Novel feedstock such as shale gas and tight oil are impacting the resource mix or as we coin it—the *Feedstock Curve*. Why Feedstock Curve? Because these novel fossil resources do not only impact the energy sector but also the petrochemical industries as they are their critical feedstock.

Main fossil fuel resources used to date were so-called conventional resources, meanwhile substantial amounts of unconventional resources are adding to the feedstock curve. For a detailed discussion of resources grades, we refer to a previous publication [2]. Conventional reserves are the most accessible and least technically challenging to bring into production; they require relatively little capital and energy investments for production and hence are the most profitable resources. Most of the remaining conventional oil and gas reserves are located in Russia and the Middle East. In contrast, unconventional reserves are not readily recovered because they are deposited within rock formations, such as shale gas and tight oil, or are combined with sand as in case of oil sands. Technology such as fractured cracking (fracking) of rock formations or extraction of oil from oil sands is required to produce the resource; however, the technology requires significant upfront investments while production requires significant supplementary energy [3]. Consequently, unconventionals have a considerably higher production cost, deliver less net energy [4] and cause more greenhouse gas emissions. Yet, significant amounts of these resources are located in the Western World!

One of the authors has previously reported on the decline of conventional reserves; unconventional reserves are likely to mitigate shortcomings in conventional production but are only economic at sustained, high resource prices. Especially in North America, this novel feedstock has had and will remain to have a significant impact on macroeconomics, trade balances and energy security and, consequently, the North American economic outlook as well as competitiveness. We hereafter analyse illustrative cases in order to draw a picture of the potential impact of conventional resources.

Analysis

North America and the energy transformation

In order to illustrate the importance of unconventional resources for supply security, we will hereafter assess the impact and the prospects for the United States, the country most impacted by the transition of the Feedstock Curve. We base our assessment on estimates of the US Government's Energy Information Administration (EIA) [5], the Organisation for Economic Cooperation and Development (OECD) as well as the Census Bureau [6].

Unconventional Gas: For decades the United States was dependent on imports of natural gas, historically in gaseous form and more recently, approximately since 1985, also in form of liquefied natural gas (LNG). While diversified supply of natural gas and oil from abroad ensured US energy security, it obviously came with cash transfers abroad and consequently detrimentally affected the trade balance [7]. At the height of import dependence, the United States imported well over four trillion cubic feet (tcf); however, imports have dropped below 3 tcf, while domestic production of natural gas has risen to more than 8 tcf [8] per annum. The enabler of this shift in supply is a combination of technological advancements such as horizontal drilling, fractured cracking of shale rock and advanced chemicals that allow cracking of shale under relatively benign conditions [9]. This enabler had the mentioned transformative impact of the American feedstock curve and is likely to continue to have a significant impact. Figure 1 depicts estimates by the EIA that conclude that domestic gas production will continue to rise—albeit at

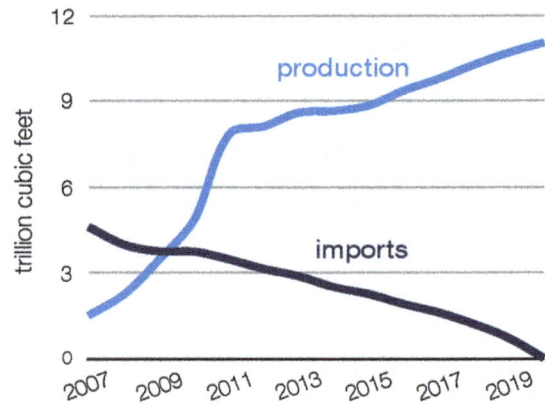

Fig. 1 US production and imports of natural gas in tcf; data taken from EIA (footnote)

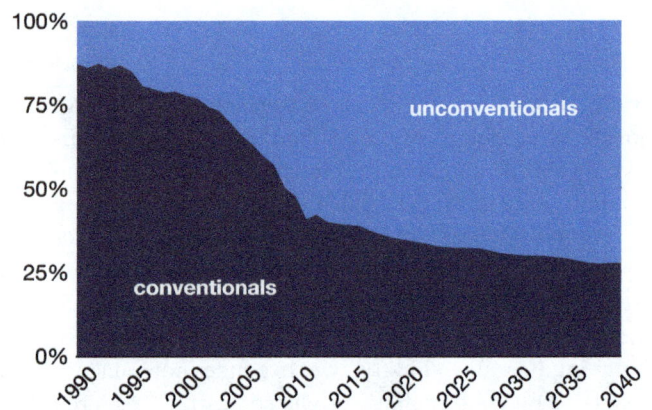

Fig. 2 Forecasted natural gas consumption by type (conventional vs. unconventional); data from the EIA [11]

a slower rate—and imports will reduce to negligible amounts before 2020. This assumptions is supported by the fact that in the US degasification terminals for LNG are being refitted to be liquefaction units as the shale gas boom is continuing to affect the feedstock so drastically that surplus of gas rather than shortages are expected. For instance, Cheniere Energy decided to entirely alter the purpose of their LNG terminals and refit them to liquefaction rather than regasification units [10], a decision well supported by data displayed in Fig. 1.

What has spurred this drastic restructuring of the US situation? Technological advancements, for instance fracking and horizontal drilling, allow unconventional gas reserves to be released from soft rock such as shale and sandstone, vide supra. Already today, more than 50 % of the US' natural gas production stem from unconventional wells, such as shale and tight gas wells, by 2040 more than 75 % will come from unconventionals [11]. Figure 2 illustrates how unconventional (domestic) gas forces conventional (imported) gas out of the feedstock curve.

The increased shale gas production has a beneficial effect as it is expected that the long-term volume of natural

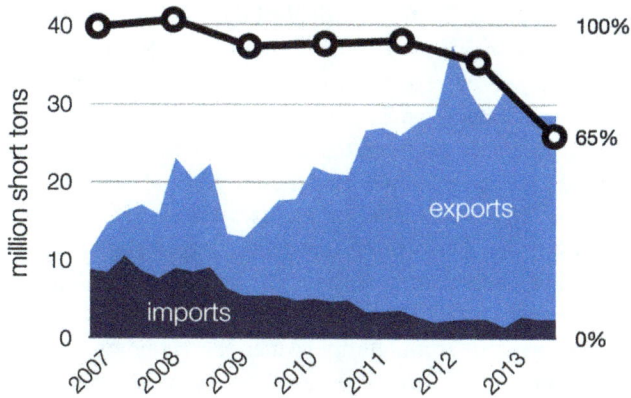

Fig. 3 US coal production, imports and exports

gas liquids (NGLs). These are the more important feedstock for the (petro-)chemical industry [12] as they have many values beyond their ability to produce heat, i.e. the calorific value and are hence very valuable for the chemical industries [13]. The production of NGLs is, however, uncertain as it will vary considerably with the so-called wetness of the natural gas produced. In 2013, Troner estimated that NGL production could amount to 1.5–1.8 million barrels per day (mbd) [14].

While this is excellent news for the recovery of the American economy, many claim that it is bad news for the environment as production of unconventional reserves has significant environmental externalities as it can impact water supply as well as purity (1) [15] and is likely to be more emission intense than conventional natural gas (2) [16]. While both issues are valid concerns when utilising these resources, they are not insurmountable: Firstly, water impact can be mitigated by technological improvement which can be driven by regulation and hence, this should be relatively straightforward to solve. Secondly, while more emission intense, unconventionals can crowd out coal which will lead to an overall reduction of greenhouse-gas emissions. An approximation by the International Energy Agency (IEA) suggests that the greenhouse-gas emissions caused by unconventional gas are half those of coal (calculated in $gCO_{2(eq)}/kWh$) [17]. Consequently, overall emissions are likely to be reduced significantly if unconventional gas crowds out coal. Indeed, the EIA assessed a sustained decline in US greenhouse gas emission since 2010 while the national economy is growing at several % annually [18].

But is this happening? Is the reduction really due to a crowding out of coal? In order to assess whether is crowding out is indeed happening, we look at the imports and exports of coal from and to the US as well as domestic coal production after 2007, when the shale gas boom really took off (Fig. 3). From this graph it can be seen that coal imports have fallen to negligible levels from 2007 onwards,

while exports have almost tripled during the same period. Overall production on the other hand declined by one-third! Figure 3 depicts the exports and imports of coal to/from the United States as well as domestic production.

From this graph it can be clearly seen that coal is leaving the feedstock curve in the US, most likely crowded out by cheap shale gas. A global shale gas boom could therefore indeed lead to a crowding out of most pollution energy sources such as coal. However, utilising these resources will not be as straightforward in other regions of the world owing to differences in legal frameworks (see below) and geological circumstances (see "Shale gas and the rest of the world").

Due to different property rights, utilising shale gas will be much more challenging in other parts of the world, e.g. the European Union. For instance, in the United States resources beneath your land are considered your property which gives a private or legal person a clear incentive to produce these or sell the rights for production [19]. In the EU, resources are a common good and hence belong to the government which leads to a clear incentive to oppose techniques such as fracking below your property as they could pose a risk for both real estate and value of premises. However, the IEA estimates that the EU has more than 600 tcf of unconventional gas reserves, which could act as a terrific back-up fuel to balance out intermittency of renewables and/or assisting the transition away from nuclear power. However, analogue to the negative public perception of nuclear power, unconventional resources encounter scepticism from the environmentally conscious European public. Recent happenings in Eastern Europe, however, could drastically alter public perception, see "Europe, the Loss of Resources, and the Renewed Power Struggle".

Unconventional Oil: Even more important for the US economy is—of course—the most precious fossil fuel resource, oil. American oil production has peaked in the 1960s [20] and endeavours to increase energy security using bio-based or synthetic fuels have essentially failed [21, 22].

In case of oil, the "unconventional revolution" transformation has not yet taken place, but it is likely that so-called tight oil will play a crucial role in the continued transformation of the feedstock curve. Tight oil is—analogue to shale gas—conventional oil that is trapped in soft stone formations and can be produced using fracking techniques. Also here, technology transformed uneconomical and inaccessible resources into viable feedstock within years. Tight oil is pushing current US oil production past 10 mbd reducing the need for imports from 60 % in 2005 to 40 % in 2012 (Fig. 4). Forecasts by the EIA assume that by 2016 the need for imports will be reduced to 25 %. Especially with crude oil prices sustained over $100 [23], this will have very positive effects on the US trade

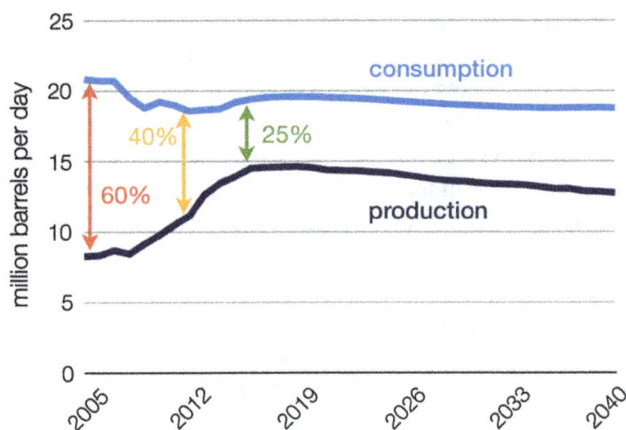

Fig. 4 US oil production and consumption, a gap that widened over decades is now narrowing again

balance and hence economic outlook, vide infra. Moreover, this has wide ranging political implications as most of these imports could come from neighbouring Canada and Mexico both friendly to the US and dependent on revenues from resource exports [24]. At present US military presence not only provides stability to the Middle East but US forces are also key in policing critical trade routes. Will the need for this presence and the associated cost persist or will the US reconsider? This issue will be considered in a later section of this think piece.

Resource imports, trade balances and macroeconomics

For more than a decade, the US has borrowed money, mainly from China, in order to transfer it to resource-exporting states to satisfy its domestic resource hunger. A significant proportion of this resource demand is due to a notoriously inefficient car fleet. Will this affect the long-term economic outlook?

There is no short answer to this very important question, nor is there one correct answer. Many academics and policymakers have expressed grave concerns about the widening US trade deficit and the consequently increasing foreign debt, however, there is a lot of disagreement about the severity of the problem and the potential consequences [25]. The United States have had trade deficits for most years in the last 3 decades, which led to negative balance of payments and subsequently to a very significant foreign debt. This works as long as there are sufficient foreign investors (mainly governments) that lend the US money, but should these sources run dry, the US might pay dearly [26]. Credit ratings of the US Government have already been reduced by most rating agencies, initially by Standard & Poor's in late 2011 [27]. A heavy debt burden can hamper economic growth as governments have less financial resources for stimulating innovation and economic

growth [28]. Hence, the foreign debt of the US creates risk and uncertainty for its long-term economic outlook as it's likely to hamper or redirect foreign direct investments. However, the Council on Diplomatic Relations estimates that recently roughly 50 % of the US trade deficit was due to *just* oil [29]. The narrowing of the differential between consumption and production of oil as well as gas, vide supra, will hence reduce the trade deficit, ease balance of payment issues and consequently will affect foreign debt and credit rating of the United States [30]. All this will thus free up resources to stimulate the American economy and hence, technological advancements in unconventional resource production not only transformed the feedstock curve and hence the energy landscape, but are also likely to fuel US economic recovery. One of the authors has previously outlined that advanced economies are indeed able to translate resource wealth into economic growth and innovation without suffering from the so-called resources or carbon curse [31]. The US were already able to become the worldwide leading economy based on conventional resource wealth post WWII and are likely to transform this newly gained, unconventional resource wealth in a similar manner. Innovation in fuel production has therefore clearly had an impact on the energy landscape, domestic greenhouse-gas emissions and the long-term economic outlook. Will this success be transferable to other areas of the world and what are the political consequences of this transition? These issues will be explored in the second part of this think piece.

Shale gas and the rest of the world

Unconventional fossil fuel reserves are not an American phenomenon, many regions of the world have significant reserves of shale gas and tight oil [32]. Figure 5 gives an overview over shale gas reserves as an example of abundant unconventionals: In South America reserves are mainly located in Argentina and Brazil, the two largest countries on this continent, while in North America the reserves are spread out fairly evenly over Mexico, the US and Canada. In Asia, shale gas reserves are centered in China and in Australia within Australasia.

The Chinese shale gas reserves are significant, approximately twice the size the reserves the US have at hand [33]. Since energy security is very high on the Chinese political agenda and air pollution could be reduced by replacing coal with clean-burning natural gas, utilization of unconventional gas is a no-brainer for China. However, the Chinese resources have proven to be more difficult to produce than US reserves which has delayed production. Ultimately, however, China will tap into these resources as energy security and economic concerns will be major driving forces; for these reasons China even acquired

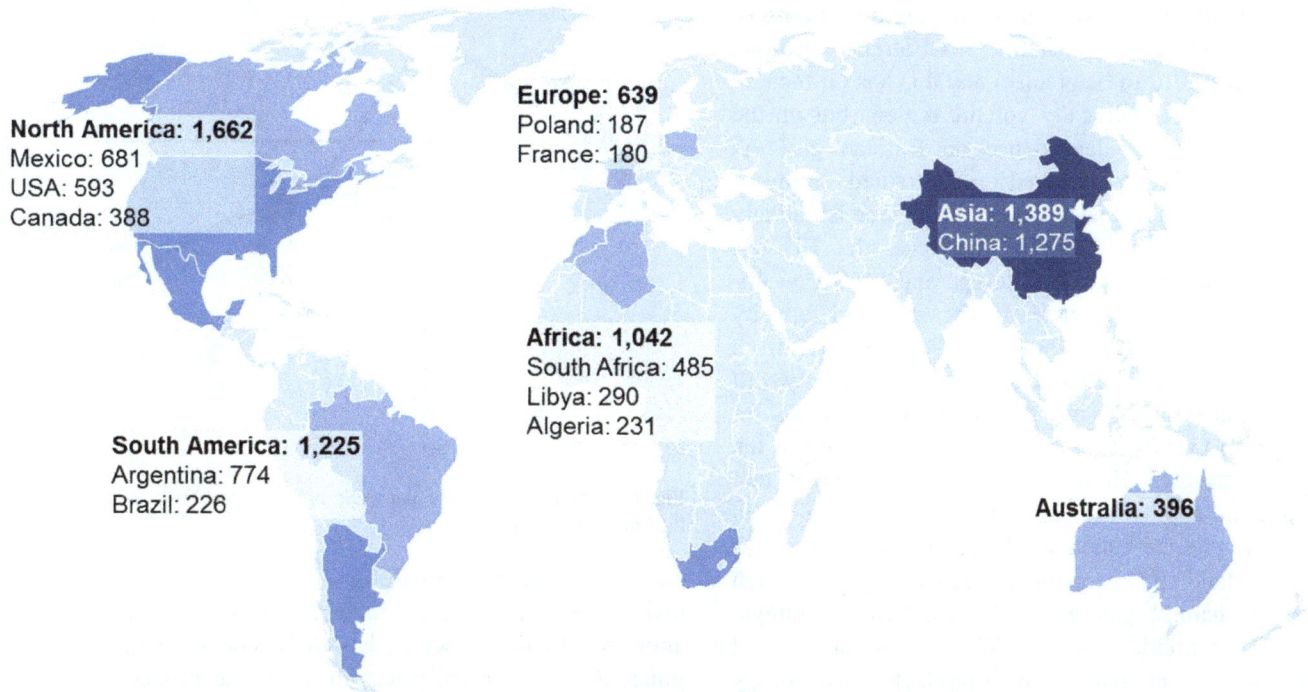

Fig. 5 Estimation of technically recoverable global shale gas resources (in tcf) Source: US EIA, World Shale Gas [38]

companies with specific know-how in shale gas production [34].

Also Europe has shale-gas reserves albeit significantly smaller than the larger continents, and at the same time the countries with the greatest endowments, France and Poland, are both reluctant to produce—in fact France has even banned exploration of shale gas [35]. A critical obstacle preventing deployment on shale gas in Europe is public perception of the environmental risks posed by unconventionals production. However, many European countries have stringent targets for decarbonisation and, analogue to the US, gas can crowd out coal thereby reducing carbon emissions [36]. Moreover, European shale gas causes fewer emissions than natural gas from outside the EU, imported via LNG terminals or pipelines, mainly because of emissions from liquefaction and regasification or long-distance gas transport [37]. These are two very strong arguments for shale gas utilization, another strong argument is to come! A drawback of shale gas is that the potentially cheap source of energy, might pose a threat to the significant investments renewable energy requires (crowding out). On the other hand natural gas, due to its flexibility, can supplement the transformation of the energy landscape. Gas-fired power plants are very flexible as they can easily be powered up and down and could therefore provide a suitable backup needed to cope with intermittent power supply from solar and wind sources. It will be important to smartly incentivize the energy transition in order to balance renewables and gas to gain maximum benefits for both the climate and the economy.

Europe, the loss of resources, and the renewed power struggle

The current situation in the Ukraine and the consequently rising tension between the east and west could, however, draw public attention away from environmental issues towards energy security concerns. At the moment, Europe (especially Germany) is dependent on Russian gas: roughly a third of the gas used in the OECD Europe is imported from Russia [39] approximately the amount Norway, the biggest domestic producer, provides. The de facto annexation of the Crimean by the Russian Federation will deplete the resource base Western Europe has easy access to, detrimentally affecting its already weak energy security and hence the Crimean question can be seen as the biggest security crisis Europe had to face since the end of the Cold War. The incorporation of the territory adjoining the Black Sea into the Russian Federation will annex the rights to resources located offshore in the Black Sea [40] and hence these resources are moved into Russian dominated territory, away from the EU. There are two straightforward solutions to the crisis: increasing domestic production of natural gas or diversifying imports [41]. During the first Ukrainian gas crisis in 2009, the EU has taken important steps and increased the number of interconnectors for both electricity and natural gas in order to increase its energy security and mitigate reliance on volatile regions in Eastern Europe. However, in the meantime Nord Stream has gone online [42] (in 2012 to be precise) and this has on the one hand circumvented politically volatile regions but on the other hand increased dependence on Russian gas! As always, energy security is based on diversified supply of different energy resources, however, this dogma is unlikely to hold right now in Europe. Therefore, the EU has expanded the capacity for gas storage and LNG regasification since 2005; for instance, the regasification capacity will raise almost four times between 2005 and 2015 from 2.5 tcf to 9.2 tcf per annum. This additional capacity will reduce reliance on Russian gas if sufficient LNG, for instance from the US, is on the market. Very recently, the speaker of the US House, John Boehmer, stated that expediting the approval to export natural gas to the EU in order to mitigate the Union's reliance on Russian gas was one way to stand up to geopolitical aggression [43]. In such a scenario, natural gas would be used as a strategic weapon—a geopolitical weapon. Were this to happen and effective, unconventionals and their production technology would have directly geopolitical impact as they are likely to reduce Russian influence in Western Europe! The EU has meanwhile welcomed such measures, which would intensify transatlantic relations [44, 45].

Asia and energy scarcity

In Asia, the energy situation has not undergone the radical transformations North America has encountered and consequently this area is still coping with relative energy scarcity [46]. As a consequence, energy as well as resource prices are comparatively high; in the case of natural gas for instance, the Asian price can be almost an order of magnitude higher than in North America, a phenomenon referred to as Asian premium [47]. In order to assess the Asian situation, we will herein assess the most extreme case—Japan. The island nation is resource-poor and has hence based its energy security policy on nuclear power in order to be relatively independent of resource imports (roughly 30 % of its electricity production had stemmed from nuclear power). Unfortunately, Japan is also the most earthquake-prone country in the world and the devastation of a nuclear power plant in Fukushima in 2011 has had a devastating impact on Japan and the perception of nuclear power. After the disaster in Fukushima, the authorities suspended nuclear power production pending a thorough and comprehensive review of reactor security. It is anticipated that some reactors will go online again after being cleared by the authorities; however, it is unlikely that former levels of nuclear power production will be reached in the foreseeable future. The restructuring of the Japanese energy landscape in the aftermath of the Fukushima

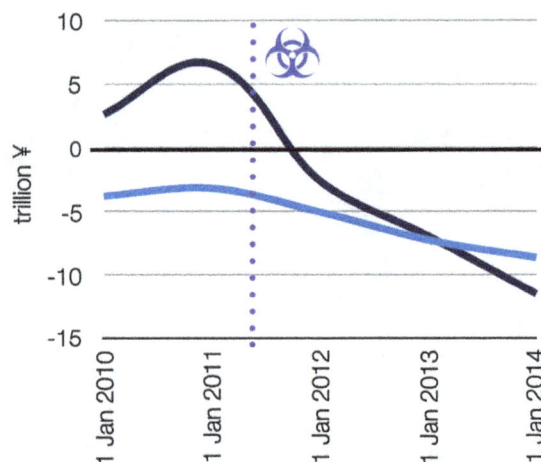

Fig. 6 Natural gas imports and trade balance, data from the OECD [48] and the EIA [49]

disaster was neither straightforward nor cheap. One of the main power (electricity) sources to replace nuclear generation is natural gas, which has to be bought at the mentioned Asian premium prices critically Japan's economic competitiveness. In Fig. 6, cost of natural gas imports into Japan as well as the Japanese trade balance since 2010 are depicted; from this illustration it can be seen that resource imports after the Fukushima accident in March 2011 increased, and hence cash flows abroad. Coupled to these cash flows out of Japan, the trade balance has for the first time in decades shifted into the negative! As in case of the US, the trade deficit directly affects balance of payments, foreign debt and thus Japan's long-term economic outlook.

In addition to this detrimental macroeconomic impact, particularly industries that use natural gas as a feedstock and are also energy-intensive are affected by imports of pricy resources and rising energy prices as costs of two input factors increase in a coupled manner. This additional cost entering production prices in a twofold manner directly affects competitiveness of companies located in Japan compared to companies located in areas with lower feedstock prices, such as those in the United States. Even though Japan does not possess significant amounts of shale gas, it has reasonable amounts of so-called methane hydrates. Methane hydrates also known as frozen methane or methane clathrate is methane trapped within a frozen water crystal and is deposited in significant amounts on the ground of deep waters off the Japanese shores. This unconventional resource could potentially be a game changer for Japan just as shale has changed the North American energy landscape. Figure 7 maps the confirmed, potential and possible methane hydrate reserves off Japan's shores: the Eastern Nankai trough contains confirmed reserves of 500 billion cubic meters and it is estimated that more than a trillion cubic meters, enough to satisfy Japan's natural gas needs for more than a decade. If Japan wants to

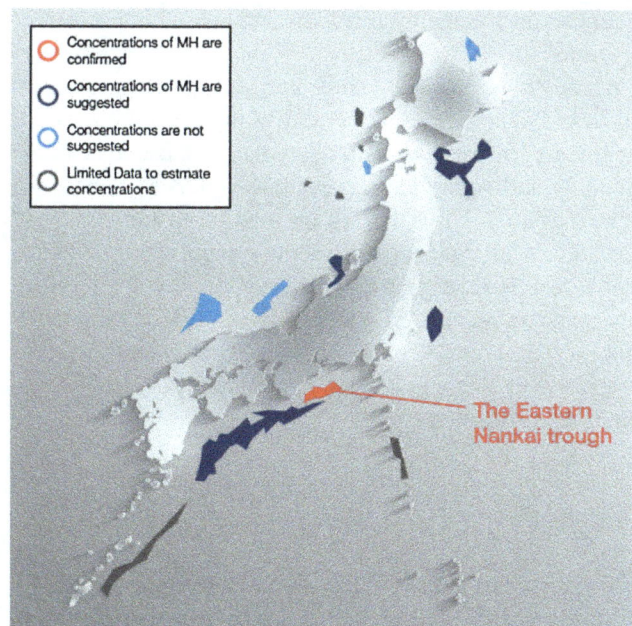

Fig. 7 Confirmed, potential and possible methane hydrate (MH) reserves

transition out of nuclear power, these unconventional gas resources could be the perfect transition fuel. Moreover, it is suggested that the actual reserves are five times as large and could thus have a long-term impact on Japan's energy security and economic outlook [50].

If Japan could utilise this unconventional resource, the macroeconomic implications would be as strong as in the US: the current trade deficit could be shifted back to a trade surplus, competitiveness with other regions of the world could be restored through declining energy and feedstock prices and last but not least, energy security could be increased which would likely mitigate if not resolve tensions with other Asian states—yet another geopolitical impact of the restructuring of the feedstock curve. There are, however, concerns about feasibility of these projects; for instance, Japan's food security is largely based on maritime supply and hence any environmental damage in relation to off-shore production of methane hydrates could directly affect food security. Nevertheless, also here unconventional feedstock and the transformation of the feedstock curve could have significant political and economic implications.

Discussion

Technological advancements in fossil fuel production have—within a few years—altered the mix of resources we utilize, the so-called feedstock curve. The transformation of the feedstock curve has in turn had a drastic impact on

the macroeconomic outlook of, for instance, the United States and could potentially affect other countries, like Japan, in a similar manner. The United States, once the biggest producer of petroleum, became highly dependent on fuel imports, but has undergone an energy transformation and is meanwhile independent of natural gas imports and is likely to become relatively independent of oil imports. Hence, the technological advancements that enabled the production of unconventional resources, have had significant macroeconomic impact as they are likely to reduce the enormous trade deficit of the US and hence will improve the country's macroeconomic outlook. What implications does this have for geopolitics? As an example, the United States is policing the seven global oil choke-points [51] and provides stability to the Middle East through diplomatic channels and military presence. Will this continue when the US only has to import relatively minor amounts of petroleum which are likely to come from Canada, Mexico and Venezuela? Will they continue to sustain a costly presence in the Middle East when Saudi oil supply is not critical for economic survivals anymore? On first glance this would seem that this would not make sense when most oil is domestically produced or imported from adjacent countries, but the US will still want to keep oil prices at least relatively low and since oil is a globally traded commodity, undisrupted trade will be key to ensure equilibrium prices [52]. Therefore, the US is unlikely to abandon the security arena for oil, but it is debatable if they will continue to be involved in the current manner. Import dependent countries such as China will be expected to contribute their fair share to naval security. These issues have to be carefully as well as holistically dealt by energy analysts.

Russia's renewed imperial ambitions are fuelled by fossil resources: almost a billion dollars is transferred to Russia daily for fossil fuel exports and Putin's confidence is growing with every dollar. Especially Europe is dependent on these exports and Russia is sure to have a tight grip on Europe, the current happenings in Eastern Europe are clear signs of this. Through ample gas production in the US and the potential of EU shale gas production, the West could strike back by intensifying transatlantic bonds and gaining independence of imports from the East. After all Russia is as dependent on foreign currency as we are dependent on fossil fuel imports and when cash flows change directions from East to West, many in Russia will be mighty disappointed. Moreover, the increasing energy independence is providing the US with a confidence boost, coined 'barrels of confidence' by Lee and Lahn [53].

Looking at the issue from the resource-rich countries' point of view is interesting as well: the US shale gas boom has kept gas prices in North America relatively low

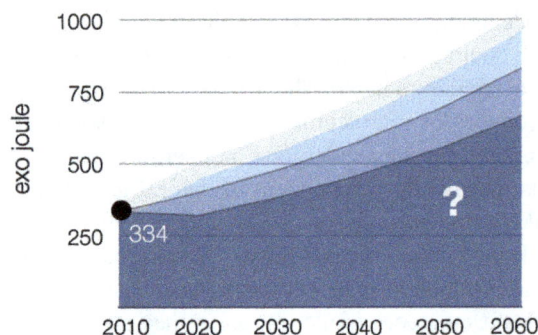

Fig. 8 Global energy demand and uncertainty around the feedstock utilised

compared to Europe and Asia, simultaneously reducing the amount of LNG imports needed. In a similar fashion, tight oil coming online will affect—most likely stabilize—oil prices. For countries relying on resource exports for economic stability, i.e. countries suffering from the resource curse, the shifting feedstock curve should be the final wake-up call: For how long will resource exports manage to finance their political systems? Should the resource wealth be immediately invested in innovation and technology to set these countries on course for a post-resource area? Will unconventional production just be a temporary phenomenon as carbon emission and climate change concerns will impact political agendas so that the resources have to be abandoned? These are critical question governmental strategists have to tackle.

Assessing how things will develop over the coming years is already tricky, but how will all this pan out in the long term? It is tough to forecast how this situation will develop as political, economic and technological issues interact in a highly complex manner. The uncertainty of the three factors listed above is so large that it is hard to predict what will power us in a decade's time. Will unconventionals really crowd out renewable energy or will unconventional gas power the transition to a renewable energy system? Will emerging technologies in battery and grid storage make gas backup for renewables unnecessary? Another tough call to make as technological developments can disrupt complete industries within a few years. Moreover, emerging technologies can—within years—impact the amount of energy we use and hence there is not only significant uncertainty around the resources we will utilise but also the amount of energy we demand to ensure a smooth functioning of the globalised economy. The only thing that is certain is that in 2010 we utilised 334 exo joule of nuclear, fossil and renewable energy (Fig. 8), how energy demand will evolve or which resource will be in demand is *highly uncertain*. Moreover, will cheap unconventional resources crowd out renewables or high-emissions resource out of the feedstock curve? Also this is

highly uncertain and therefore the environmental impact of unconventionals.

Resources have always had a tremendous impact on geopolitics, but resource accessibility by technological advancements has made this mix so volatile that tide turns very quickly and with wide ranging consequences. This is why the feedstock curve has to be monitored closely in collaboration with businesses, public officials and independent researchers to gain a clear view of how it affects international economics as well as geopolitics. Due to multiple, often coupled, uncertainties, predictions only make sense over a few years and consequently have to be continuously reassessed.

Conclusions

Unconventional resources will continue to restructure the energy landscape as well as the feedstock curve and they will be critical for the recovery of the global economy. Economic and political factors are driving the exploration of unconventional fuel resources which can provide energy security, while financial considerations push technological frontiers in resource production.

The provision of energy security has risen to the top of the political agenda due to renewed geopolitical tensions in Eastern Europe and the far East Asia as well as economic necessity due to large trade deficits of advanced economies which could lead to a long-term instability of the globalized economy.

Production of unconventionals could indeed have a detrimental effect on the environment by increasing greenhouse-gas emissions, impacting water and crowding out investment in renewables. When carefully regulated, impact on water could be minimized while greenhouse gas emissions would be reduced when unconventional gas crowds out coal and provides efficient backup for renewable power. Implementation of smart policies will be critical to ensure that these resources are not only a winner for politics and economics but also for the environment.

Moreover, we have to extend our assessments: at the moment, we carefully forecast energy demand and assess potential resources that could be used to supply this demand. Nevertheless, the ICT revolution has made technological progress both fast and volatile so that we have to use 'intelligence' and assess lab-stage technologies with regards to their potential impact on both energy supply and demand.

Resources and the energy security they provide have often not only impacted the economic success of countries but also the geopolitical balance of power, the renewed power struggle in Eastern Europe is clearly a sign of this trend. It has to be closely monitored how the technological redistribution of fossil resources will impact issues such as tensions between Japan and China, the reoccurring conflict in the Middle East as well as the renewed power struggle between East and West.

Collaborative endevours including acedemics, businesses and governmental agencies will be critical for monitioring the feedstock curve and its potential impact on the globalized economy.

References

1. http://online.wsj.com/news/articles/SB10001424127887324767004578489130300876450
2. Owen NA, Inderwildi OR, King DA (2010) The status of conventional world oil reserves—hype or cause for concern? Energy Policy 38(8):4743–4749
3. Hall CAS, Lambert JG, Balogh SB (2014) EROI of different fuels and the implications for society. Energy Policy 64:141–152
4. Hall CAS, Klitgaard KA (2012) Energy and the wealth of nations. Springer, Berlin
5. US Governments, Energy Information Administration, Washington D.C., USA
6. US Department of Commerce, US Census Bureau, https://www.census.gov/, Washington D.C., USA
7. US Department of Commerce, US Census Bureau, https://www.census.gov/, Washington D.C., USA
8. http://www.eia.gov/dnav/ng/hist/n9100us2a.htm
9. http://environment.yale.edu/envy/stories/fracking-outpaces-science-on-its-impact
10. http://www.ft.com/intl/cms/s/0/a5053c50-8d2b-11e1-9798-00144feab49a.html#axzz2x4YiJGsh
11. http://www.eia.gov/naturalgas/
12. Speight JG (2010) The refinery of the future. William Andrew, London
13. Hashiguchi BG et al (2014) Science 343(6176):1232–1237
14. Troner A (2013) Natural gas liquids in the shale revolution. Baker Institute, Rive University http://bakerinstitute.org/files/3961/download/
15. Brantley SL et al (2013) Water resource impacts during unconventional shale gas development: The Pennsylvania experience. Int J Coal Geol. http://dx.doi.org/10.1016/j.coal.2013.12.017
16. International Energy Agency (2013) Are we entering the golden age of shale gas. Paris, France. http://www.worldenergyoutlook.org/media/weowebsite/2011/WEO2011_GoldenAgeofGasReport.pdf
17. http://www.bbc.co.uk/news/mobile/business-12245633
18. http://www.eia.gov/environment/emissions/carbon/
19. Stevens P (2012) The shale gas revolution. Chatham House http://www.chathamhouse.org/sites/default/files/public/Research/Energy,%20Environment%20and%20Development/bp0812_stevens.pdf
20. Inderwildi OR, King DA (2012) Energy, transport & the environment. Springer, Berlin
21. Inderwildi OR, Jenkins SJ, King DA (2008) J Phys Chem C 112(5):1305–1307
22. Inderwildi OR, King DA (2009) Energy Environ Sci 2(4):343–346
23. http://www.bloomberg.com/energy/
24. http://www.eia.gov/dnav/pet/pet_move_impcus_a2_nus_ep00_im0_mbbl_m.htm
25. Scott RE (2013) http://www.epi.org/publication/trade-deficits-consequences-policy-implications/
26. http://www.frbsf.org/education/publications/doctor-econ/2007/june/trade-deficit-exchange-rate
27. http://www.reuters.com/article/2011/08/06/us-usa-debt-downgrade-idUSTRE7746VF20110806
28. Checherita-Westphal C, Rother P (2012) Eur Econ Rev 56(7):1392–1405
29. Diplomatic Council on Energy Security http://www.secureenergy.org/sites/default/files/DCES-Oil-and-the-Trade-Deficit.pdf
30. International Monetary Fund, World Economic Outlook (2013) http://www.imf.org/external/pubs/ft/weo/2013/01/pdf/text.pdf
31. Friedrichs J, Inderwildi O (2013) The carbon curse: are fuel rich countries doomed to high CO_2 intensities? Energy Policy.
32. http://www.iea-etsap.org/web/E-TechDS/PDF/P02-Uncon%20oil&gas-GS-gct.pdf
33. World Economic Forum (2013) Energy harnessing: new solutions for sustainability and growing demand
34. http://www.ibtimes.com/us-shale-gas-boom-attracting-investments-chinese-energy-companies-1126087
35. http://www.theguardian.com/environment/2013/oct/11/france-fracking-ban-shale-gas
36. http://www.politico.com/story/2014/03/wh-methane-strategy-next-step-in-climate-plan-105178.html?hp=r13
37. AEA Technology (2012) Climate impact of potential shale gas production in the EU
38. http://www.eia.gov/analysis/studies/worldshalegas/pdf/fullreport.pdf
39. http://www.nytimes.com/2014/03/24/business/international/weaning-europe-from-russian-gas.html?_r=0
40. http://www.independent.co.uk/voices/comment/oil-and-gas-could-explain-putins-costly-attempt-to-control-the-crimea-9193464.html
41. http://www.bloomberg.com/news/2014-03-20/eu-readies-natural-gas-plan-to-cut-reliance-on-russia-in-months.html
42. http://www.naturalgaseurope.com/russia-plans-massive-expansion-of-nord-stream-pipelines-
43. http://www.ft.com/intl/cms/s/0/e2cf61ba-a489-11e3-b915-00144feab7de.html#axzz2x4YiJGsh
44. http://www.ft.com/intl/cms/s/0/e539fc24-b37c-11e3-b891-00144feabdc0.html#axzz2x4YiJGsh
45. http://www.ft.com/intl/cms/s/0/67308bb8-97f2-11e3-8c0e-00144feab7de.html#axzz2x4YiJGsh
46. World Economic Forum (2013) Japanese Energy Transition., Tokyo Japan, http://www3.weforum.org/docs/IP/2014/CH/Japan_Energy_Report_2014.pdf
47. Ken Koyama IEEJ (2012) A Japanese perspective on the international energy landscape http://eneken.ieej.or.jp/data/4252.pdf
48. OECD, Paris, France; http://stats.oecd.org/index.aspx?queryid=166
49. http://www.eia.gov/countries/country-data.cfm?fips=ja
50. http://www.ft.com/intl/cms/s/2/8925cbb4-7157-11e3-8f92-00144feabdc0.html#axzz2xlXj751E
51. http://www.eia.gov/countries/regions-topics.cfm?fips=wotc&trk=p3
52. World Economic Forum (2013) The Japanese energy transitions
53. Lee B, Lahn G (2014) Barrels of confidence. Prospect

22

Dry reforming of methane over ZrO_2-supported Co–Mo carbide catalyst

X. Du · L. J. France · V. L. Kuznetsov ·
T. Xiao · P. P. Edwards · Hamid AlMegren ·
Abdulaziz Bagabas

Abstract The process of dry reforming of methane has the potential to be an effective route for CO_2 utilization via syn-gas production. In the present study, ZrO_2-supported Co–Mo bimetallic carbide catalysts were prepared via a co-precipitation method through a combined reduction and carburization procedure employing a CH_4/H_2 (20/80 %) mixture. All of the as-synthesized materials were tested at 850°, under atmospheric pressure and a CO_2:CH_4 ratio of 1. The importance of the ZrO_2 support became immediately apparent when it exhibited a higher conversion than the corresponding low-surface-area bulk Mo_2C catalyst, which we attribute to lewis acid and base active sites on the surface of ZrO_2. From catalytic tests and pre-and post-reaction X-ray diffraction (XRD) patterns, we observed that different dispersions of the monometallic carbides, caused by varying the pre-heating temperatures on ZrO_2, did not significantly affect conversion or yield. In contrast, incorporation of cobalt atoms into the Mo_2C lattice significantly enhanced the conversion, yield and stability of the catalysts. Post-reaction XRD patterns indicated that the bimetallic carbide had enhanced the resistance to the oxidation effect that is known to deactivate Mo_2C catalysts. In addition, increasing the Co loading in the mixed metal carbides was seen to enhance the resistance of the catalyst to the reverse water gas shift reaction, leading to improved stability of the H_2 yields.

Keywords Dry reforming of methane · Synthesis gas · Bimetallic carbide · Zirconia support

Introduction

The dry reforming of methane $(DRM, CO_2 + CH_4 \rightarrow 2CO + 2H_2, \Delta H^o_{298} = +247$ $KJmol^{-1})$ is an effective method to utilize CO_2, via a reaction with CH_4 to produce a mixture of CO and H_2 known as "synthesis gas (syn-gas)" [2]. This reaction is, however, highly endothermic and generally high temperatures are required both for significant levels of conversion as well as for reducing side reactions. The DRM process provides several advantages over steam reforming of methane $(SRM, H_2O + CH_4 \rightarrow CO + 3H_2, \Delta H^o_{298} = +206$ $KJmol^{-1})$, and perhaps the most important one is the production of syn-gas with a low H_2/CO ratio, which is suitable for use in forming higher level alcohols [10] (in a stoichiometric reaction, the H_2/CO ratio of DRM production is 1:1 while SRM has a product ratio of 3:1). Additionally, of course, DRM does not require the use of water to produce syn-gas.

Metal carbides are a relatively new family of catalysts for DRM. York et al. [24] studied DRM reactions over β-Mo_2C and WC with a direct comparison with noble metal catalysts. Importantly, the results showed that β-Mo_2C has an activity for DRM comparable to certain noble metals, while the price of molybdenum is much cheaper than noble metals. Furthermore, compared to Ni-based catalysts (arguably the earliest-found material to have a high activity

X. Du · L. J. France · V. L. Kuznetsov · T. Xiao · P. P. Edwards (✉)
Inorganic Chemistry Laboratory, Department of Chemistry, KACST-Oxford Petrochemical Research Centre (KOPRC), University of Oxford, Oxford OX1 3QR, UK
e-mail: peter.edwards@chem.ox.ac.uk

X. Du
e-mail: xian.du@chem.ox.ac.uk

H. AlMegren · A. Bagabas
Petrochemicals Research Institute (PRI), King Abdulaziz City of Science and Technology (KACST), P.O. Box 6086, Riyadh 11442, Saudi Arabia

in DRM), Mo_2C, has higher stability due to its enhanced resistance to coking.

The dispersion of catalyst onto a support surface is another critical factor which can significantly influence the catalytic activity of materials [1]. The supported catalyst can play an important role in promoting reaction, which has been proven by many researchers after making a direct comparison between bulk and supported materials on a variety of supports, such as TiO_2, Al_2O_3, SiO_2, and ZrO_2 [3, 12, 19].

In recent years, therefore, supported transition metal carbides have gained increased prominence since the support can improve both the efficiency and the stability of carbides in the DRM reaction. Systematic studies on Mo_2C loaded on different supports have been carried out by Brungs et al. [7] and Darujati et al. [11]. Their results revealed that during a long-term DRM test, when Mo_2C is supported by ZrO_2 and γ-Al_2O_3, the catalytic activity and stability were higher than that for materials loaded on other oxide based supports. The advantage of γ-Al_2O_3 appears to be its significantly higher surface area compared to other kinds of support. But the advantage of ZrO_2 is its amphoteric nature [8]. The Lewis acid sites enhance the dispersion of metal due to the preference of metal atoms to reside at Lewis acid sites on the support [6], whereas the Lewis base sites can enhance the adsorption of CO_2 on the support in conjunction with Lewis acid sites [4]. Moreover, on ZrO_2, it is proposed that CO_2 activation takes place at the interface between carbide and support, which leads to a low oxidation effect on carbides [17].

To-date, the activity and stability of catalysts during DRM reaction have been improved only by a small improvement. Moreover, the product selectivity of catalysts (i.e. specifically the H_2/CO ratio) appeared to differ from a supported catalyst because they have the different activities on CO_2 decomposition and CH_4 cracking and/or the existence of side reactions such as reverse Water–Gas shift (RWGS, $CO_2 + H_2 \rightarrow CO + H_2 O$, $\Delta H^o_{298} = +41$ KJmol^{-1}). Hence, one of the major initiatives for catalyst improvement stem from the need to improve selectivity, by reducing side reactions. Additionally the need to balance the two DRM half reactions, CH_4 cracking and the reverse boudouard reaction, is essential to avoid carbon deposition.

In seeking the high activities (including both the conversion of reactants and the yields of products) and high stabilities, here we prepared the ZrO_2-supported Mo/Co–Mo carbide catalysts for the dry reforming of methane. The incorporation of cobalt into the bimetallic system is because of its high activity in CH_4 decomposition [8] combined with relatively high resistance to coking, when compared to nickel metal [16]. Co–Mo bimetallic carbides have earlier been used for the DRM as an alternative to

expensive noble metal [18], and such Co–Mo bimetallic carbides were systematically prepared and characterized by Xiao et al. [22]. The prepared catalysts were tested under 850 °C, atmospheric pressure and CO_2:CH_4 1:1 conditions for 4 h, and the results not only exhibit the noticeable activity of ZrO_2 support itself but also show the high performance of the novel ZrO_2-supported Co–Mo bimetallic carbide catalyst both in activity and the resistance to deactivation.

Experimental

Catalyst preparation

Zirconia support (Alfa-Aesor, 90 m^2/g) was ground to 64–125 μm and pre-heated at 400 °C (marked as −400), the same temperature as the slurry calcination process. For the purpose of comparison, the same ZrO_2 with the same particle size was pre-heated at 750 °C (marked as −750), the same temperature as the oxide carburization process.

The required amount of ammonium molybdate tetrahydrate (98 %, Sigma Aldrich) and zirconia support were added to distilled water before stirring to generate slurry. The slurry was stirred and heated at 100 °C to remove water prior to calcination in a muffle furnace for 10 h at 400 °C to which the ramping rate was set at 10 °C/min. After the calcination process, the sample of MoO_3/ZrO_2 was obtained.

To obtain the carbide materials, 3.0 g of prepared oxides were carburized in a quartz tube under a gas flow of CH_4/H_2 (8 and 32 ml/min respectively) for 3 h at 750 °C, and both the ramping and cooling rate were 3 °C/min. After carburization, the carbide samples were passivated in static air for 48 h before unloading. The carbide samples were labelled with a percentage of Mo_2C loading in terms of their oxide precursors. All the synthesized catalysts were labeled as X %-Mo-Y (X is the loading of MoO_3 in wt% and Y denotes the pre-heating temperature).

Besides the supported Mo_2C catalysts, the pure ZrO_2 support and the bulk Mo_2C carbide were also prepared from zirconia catalyst support (Alfa-Aesor, 43815, 90 m^2/g) and ammonium molybdate tetrahydrate (98 %, Sigma Aldrich-A7302), respectively via the same method of calcination and carburization processes.

With the aim of preparing Co–Mo bimetallic carbide samples, the required amount of ammonium molybdate tetrahydrate and cobalt(II) nitrate hexahydrate (98 %, Sigma Aldrich) were dissolved in water using the same method as above and the percentage in weight of the bimetallic oxides were 10 and 15 wt% while the Co/Mo ratio was kept at 0.4/0.6 which has been employed previously and found to possess the highest stability in pyridine hydrodenitrogenation reactions [22].

Catalyst activity test

For catalyst activity and stability tests, materials were placed in an M-R-10A micro-reactor (Kunlun Yongtai Company, China) under temperature-programmed reaction conditions. The catalyst bed was heated in nitrogen gas flow to 600 °C with a 10 °C/min ramping rate and then heated up to 850 °C at 5 °C/min. The gas products were sampled and analysed via online gas chromatography (Shimadzu, GC-2014) every 30 min at 850 °C to test the stability of the catalyst samples. Catalysts were tested under the flow of a calibrated mixed gas that comprised CH_4, CO_2 and N_2 (internal standard). The amount of catalyst was kept at 0.5 g.

With the online GC measurements of outlet gases from the reactor (N_2, CH_4, CO_2, H_2 and CO), the conversions of CH_4 and CO_2, the yields of products (H_2 and CO), and the carbon balance of the reactions can be determined with the following equations:

$$\text{Conversion} = C_i = 1 - \frac{X_i^{\text{outlet}} X_{N_2}^{\text{inlet}}}{X_i^{\text{inlet}} X_{N_2}^{\text{outlet}}} \ (i = CH_4 \, \text{or} \, CO_2)$$

$$\text{Volume of gas in product} = \frac{X_i^{\text{inlet}} \, \text{Volume of} \, N_2}{X_{N_2}^{\text{outlet}}} \ (i = H_2 \, \text{or} \, CO)$$

$$\text{Yield of } H_2 = Y_{H_2} = \frac{\text{volume of } H_2 \text{ in products}}{2 * \text{volume of methane in reactants}}$$

$$\text{Yield of CO} = Y_{CO} = \frac{\text{volume of CO in products}}{\text{volume of } C \text{containing reactants}}$$

$$\text{Product Ratio} = H_2/CO = \frac{\text{volume of } H_2 \text{ in products}}{\text{volume of } CO \text{in products}}$$

$$\text{Carbon Balance} = B_{\text{carbon}} = \frac{(X_{CH_4}^{\text{outlet}} + X_{CO_2}^{\text{outlet}} + X_{CO}^{\text{outlet}})X_{N_2}^{\text{inlet}}}{(X_{CH_4}^{\text{inlet}} + X_{CO_2}^{\text{inlet}})X_{N_2}^{\text{outlet}}}$$

Characterization of catalyst pre- and post- reaction

All the samples, including ZrO_2 support, the various precursors and carbides, were characterized by high-resolution X-ray diffraction (XRD) using a PANalytical X'Pert PRO diffractometer with CuKα radiation (45 kV, 40 mA). The samples were flat loaded in the custom-built sample holders and scanned from 10° to 70° 2θ with a step size of 0.0084° and a scanning speed at 0.017778° s^{-1}.

Raman spectra were collected (resolution of 2 cm^{-1}) using a PerkinElmer Raman Stage 400F with a 785-nm laser. The samples were pressed onto a microscope slide and the scanning time was set to 20 s (repeated 10 times) within the spectral range from 100 to 2,500 cm^{-1}.

Brunauer, Emmett and Teller (BET) analysis via nitrogen adsorption was carried using a Micrometrics Chemi-Sorb 2720 to measure the surface area and pore volume of

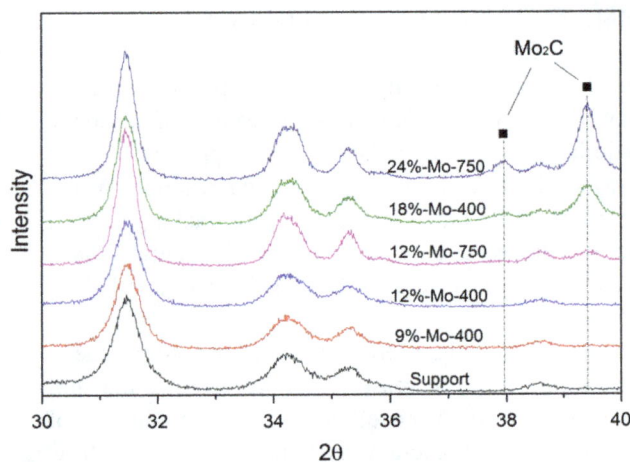

Fig. 1 X-ray diffraction patterns of ZrO_2 supported Mo_2C with different loading level (9–24 wt%) and on ZrO_2 supports with different pre-heating temperature (750 °C, 400 °C, as indicated)

samples both before and after the DRM reactions at the temperature of liquid nitrogen (−195.8 °C).

The carbon deposits of catalysts pre- and post- reaction were measured via thermo-gravimetric analysis (TGA). The instrument employed was a PerkinElmer, TGA-7, using flowing air at 10 ml/min from 50 to 1,000 °C with a ramp rate of 10 °C/min.

Results and discussion

DRM over supported Mo_2C catalysts

Figure 1 shows the XRD patterns of ZrO_2-supported Mo_2C with different loading levels at different pre-heating temperatures. It is apparent that Mo_2C is dispersed differently over the surface of ZrO_2 supports which are pre-heated at 400 and 750 °C, respectively. The peak of Mo_2C at $2\theta = 39.49°$ can be clearly observed on 12 %-Mo-750 carbides while the peaks at the same position can hardly be observed on 12 %-Mo-400, indicating that Mo_2C has a considerably enhanced dispersion on the 400 °C-pre-heated support (this has been shown to have a higher surface area according to Table 2).

In Fig. 2, Laser Raman spectra of Mo_2C 6–24 % loaded on 400 °C-pre-heated ZrO_2 were presented. It was difficult to observe the ZrO_2 support when the loading level was above 12 %, which indicates that the ZrO_2 support had been fully covered by Mo_2C, demonstrating that the availability of the ZrO_2 surface for potential catalytic activation decreased as the loading level increased.

During the process of DRM reaction, CH_4 absorption and subsequent dissociation is usually recognized to be the first step, as well as the rate determining step (1) [13, 20,

21, 23], and this is followed by the oxidation of carbon species (adsorbed carbon and CH_x) on a catalyst surface by CO_2, forming CO (2):

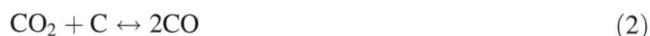

$$CH_4 \leftrightarrow CH_x + (4 - x)H \leftrightarrow C + 2H_2 \qquad (1)$$

$$CO_2 + C \leftrightarrow 2CO \qquad (2)$$

Besides the DRM, a side reaction called "reverse water gas shift" (RWGS, (3)) can lead to a higher conversion of CO_2 and the consumption of generated H_2.

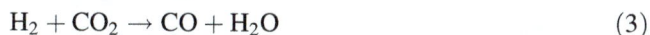

$$H_2 + CO_2 \rightarrow CO + H_2O \qquad (3)$$

The activity test results presented in Table 1 show very poor performance of the bulk Mo_2C to activate and convert both CH_4 and CO_2 at atmospheric pressure, and the conversions are even lower than the ZrO_2 support which is "carburized" under the same conditions to prepare carbide catalysts. This can be attributed to the low surface area of the bulk Mo_2C, which leads to a low availability of Mo_2C

Fig. 2 Laser Raman spectra of Mo_2C 6 %-24 % loaded on 400 °C-pre-heated ZrO_2 (notice: the appearance of MoO_3 in Laser Raman spectra is because the outer layer Mo_2C was oxidized during the passivation step after carburization, and the very thin layer of MoO_3 cannot be observed in the XRD patterns in Fig. 1)

active sites on the surface of material (which is different from the high-surface-area metal carbide materials prepared by Claridge et al. [9].). The activity in converting both CH_4 and CO_2 over pure ZrO_2 support is attributed to its amphoteric property [8]. Both Lewis acid sites and base sites can enhance the adsorption of CH_4 and CO_2 on ZrO_2 surface [4] where the reactants can be activated and converted. According to the product ratio, significantly more CO is generated than H_2, and this is due to the RWGS (reaction (3)).

Importantly, higher activities and stabilities were observed over supported Mo_2C compared to the bulk Mo_2C. Both reactant's conversions and product yields were enhanced due to the Mo_2C dispersion effect, and there is the possibility to attribute this improvement to a synergy between Mo_2C and ZrO_2 support.

Comparison of the catalysts with the same pre-treated supports (400 or 750 °C) revealed that a higher Mo_2C loading level led to a lower conversion of CH_4 and CO_2. This may be due to high levels of Mo_2C dispersed on the surface decreasing the availability of the ZrO_2 support acid sites themselves (shown in Fig. 2, the higher loading level of Mo_2C leads to a lower intensity of ZrO_2 peaks) which as started previously is active for the DMR over its Lewis acid and base sites. Despite the lower CH_4 and CO_2 conversions, the samples with higher loading levels (18 %-Mo-400 and 24 %-Mo-750) also had a similar or slightly higher product yields compared to low loading level. It could be concluded that the higher availability of ZrO_2 surface led not only to a higher conversion but also a higher degree of side reactions (e.g. RWGS) which affected the yields of products.

The catalytic performances over 12 %-Mo-400 and 12 %-Mo-750 (Table 1) were shown to be close to each other at 4 h even though the conversions of CH_4 and CO_2 over 12 %-Mo-400 were higher than the other ones. This phenomenon can be explained by a further sintering effect on the materials when reacting temperature reached to 850 °C, and both catalysts reached the similar Mo_2C

Table 1 Conversion of CH_4 and CO_2, yield of H_2 and CO, product ratio and carbon balance over different samples (bulk Mo_2C, pure ZrO_2 support and ZrO_2 supported Mo_2C catalysts) during DRM reactions (conditions: 850 °C, atmospheric pressure, 4-h duration, CH_4/CO_2 1, GHSV 4.8×10^3 ml h^{-1} g_{cat}^{-1})

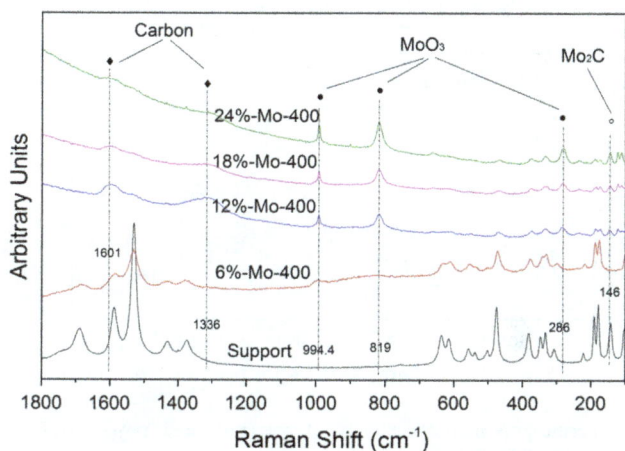

Sample	C_{CH_4} (%)		C_{CO_2} (%)		Y_{H_2} (%)		Y_{CO} (%)		H_2/CO		B_{carbon} (%)	
	0 h	4 h	0 h	4 h	0 h	4 h	0 h	4 h	0 h	4 h	0 h	4 h
Bulk-Mo_2C	12.99	14.78	36.89	20.61	7.42	5.65	27.86	19.16	0.26	0.29	106.4	103.8
ZrO_2-support	19.44	23.14	31.35	33.31	8.95	9.38	24.26	25.19	0.35	0.35	98.8	98.2
9 %-Mo-400	41.77	37.56	57.05	52.09	29.17	24.80	43.59	39.39	0.65	0.61	96.6	97.2
12 %-Mo-400	41.46	36.94	56.08	51.50	27.22	22.58	41.03	38.51	0.63	0.57	97.8	98.6
18 %-Mo-400	40.80	36.45	55.09	48.01	31.14	24.47	43.76	38.80	0.66	0.59	97.3	98.0
12 %-Mo-750	40.31	37.42	55.26	51.26	27.82	23.34	40.80	38.28	0.63	0.58	94.9	94.5
24 %-Mo-750	30.93	28.19	46.54	42.85	27.24	23.44	41.99	39.22	0.60	0.55	99.0	99.5

Table 2 Surface area and pore volume of samples (ZrO_2 supported Mo_2C catalysts) pre- and post- reactions

Sample name	Surface area (m^2/g)		Pore volume (cm^3/g)	
	Pre	Post	Pre	Post
9 %-Mo-400	55.77	35.45	0.2612	0.2318
12 %-Mo-750	29.87	24.35	0.2158	0.1624

dispersion level (even though they looked different before the reaction, as shown in Fig. 1).

The carbon balance during reaction over each sample was also displayed in Table 1. It was observed that the carbon balance over Bulk-Mo_2C was above 100 %, revealing that Mo_2C was consumed by CO_2 to form MoO_2 and CO (4). The numbers of carbon balance over ZrO2 support and ZrO2 supported catalysts were slightly below 100 %, which indicates that amorphous carbon was generated during the reaction.

$$Mo_2C + 5CO_2 \rightarrow 2MoO_2 + 6CO \tag{4}$$

The results of the surface area and pore volume of samples both prior to ("pre-") and post-reaction ("post-") are presented in Table 2. A noticeable difference between the pre-reaction surface areas of the two samples was observed, indicating a more highly sintered ZrO_2 support after being heated at 750 °C.

Moreover, the difference between the post-reaction surface areas of the two samples is much smaller, which is caused by a further sintering effect when the catalyst was heated up to 850 °C during the reaction, and this sintering effect was possibly one of the main routes of deactivation during the dry reforming of methane.

Comparison of the XRD patterns of the supported carbides pre- and post-reaction is presented in Fig. 3. They showed that the MoO_2 peaks at $2\theta = 26.11°$ and $37.12°$ were observed in the post-reaction catalysts. This phenomenon indicated the oxidation of Mo_2C to MoO_2 with CO_2 (4), which is known to be another main deactivation mechanism of DRM [9].

A new peak corresponding to the Mo_2C peak at $2\theta = 39.49°$ was observed. This indicated that Mo_2C crystallized when the surface area of ZrO_2 support decreased during the 4-h activity test (shown in Table 2). The crystallite size of ZrO_2 was also increased as calculated with XRD patterns shown in Fig. 3. (Pre-reaction 340–370 nm; post-reaction 570–630 nm), which is another evidence of minimized sintering effect existing on ZrO_2 support.

In order to elucidate whether the catalysts experienced carbon deposition over the catalysts during the dry reforming of methane, TGA (thermo-gravimetric analysis) was undertaken. Results for the pre- and post- reaction 18 %-Mo-400 catalysts are displayed in Fig. 4 with

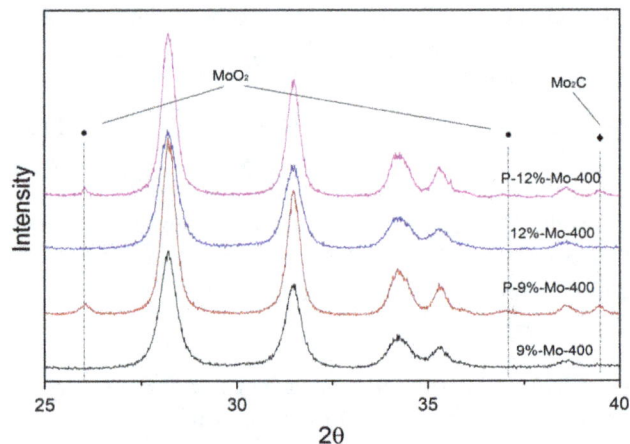

Fig. 3 X-ray diffraction patterns of the ZrO_2 supported Mo_2C catalysts before and after DRM (*P* post-reaction)

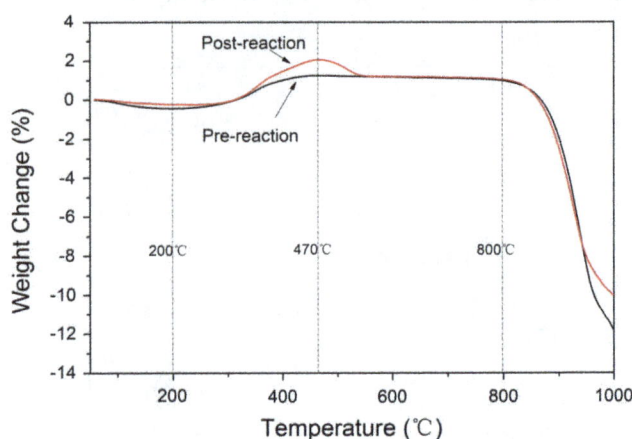

Fig. 4 Thermo-gravimetric analysis of the pre- and post- reaction 18 %-Mo-400 catalysts

Table 3 Weight changes in thermo-gravimetric analysis of the pre- and post- reaction 18 %-Mo-400 catalysts at different temperature ranges

Sample	200–470 °C (%)	470–800 °C (%)	800–1,000 °C (%)
Pre-reaction	1.6784	−0.26314	−13.0005
Post-reaction	2.28482	−1.01721	−11.1111

the calculation of weight change at each temperature range exhibited in Table 3. The weight increased between 200 and 470 °C, which indicated the oxidation of Mo_2C to MoO_3. Weight loss was observed starting at 470 and 800 °C and related to the combustion of amorphous carbon [14] and the sublimation of MoO_3 [15], respectively. The graphitic carbon which is expected to be combusted between 600 and 800 °C [5] can only be observed at an

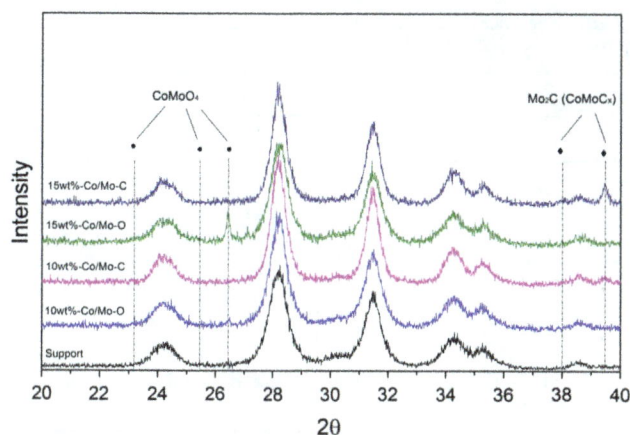

Fig. 5 X-ray diffraction patterns of ZrO$_2$ supported 10/15 % Co–Mo bimetallic oxides and carbides (*C* carbide, *O* oxide)

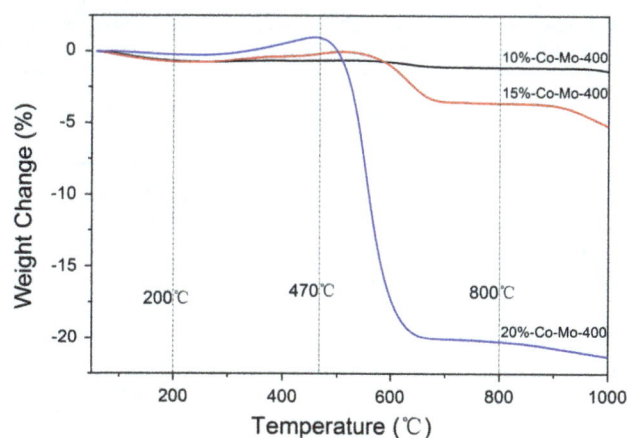

Fig. 6 Thermo-gravimetric analysis of the pre-reaction Co–Mo bimetallic catalysts

arguably low level. Upon comparison of these two TGA curves, more amorphous carbon was observed on the post-reaction catalyst, which means the generation of carbon via the decomposition of CH$_4$ (reaction (1)) occurred on the surface of catalyst during the dry reforming of methane.

DRM on supported Co–Mo carbide catalysts

As an extension of this work based on the results of mono-metallic carbides, bimetallic carbide catalyst systems were also investigated to enhance the activity and stability of catalysts in the dry reforming of methane. The results can be summarized as follows.

In the XRD patterns after incorporating cobalt with the molybdenum catalyst in Fig. 5, the peaks corresponding to the CoMoO$_4$ structure at $2\theta = 23.33°$, $25.50°$ and $26.50°$ were observed. In addition, the Co$_3$O$_4$ phase was not observed. This indicated that only CoMoO$_4$ had been formed during the calcination process. After carburisation, phases of Mo$_2$C at the angle of $2\theta = 38.101°$ and $39.492°$ were observed clearly and there appeared to be no evidence of Co metal particles or Co carbide in the XRD patterns. According to Xiao et al. [22], this phenomenon is attributed to the direct incorporation of cobalt into the lattice of Mo$_2$C carbide, and due to the comparable atomic radius of Co ($r \approx 126$ pm) compared with that of Mo ($r \approx 154$ pm), the framework of Mo$_2$C did not change significantly even though the lattice was partially substituted by Co atoms.

The TGA results of the pre-reaction bimetallic catalysts are exhibited in Fig. 6 with the calculation of weight change at each temperature range exhibited in Table 4. From the weight loss observed starting around 470 °C (which is related to the combustion of amorphous carbon), it indicates that the increased loading level of the bimetallic carbide enhanced the formation of carbon during the carburization step. The weight loss after 800 °C was much

Table 4 Weight changes in thermo-gravimetric analysis of pre-reaction Co–Mo bimetallic catalysts at different temperature ranges

Sample	200–470 °C (%)	470–800 °C (%)	800–1,000 °C (%)
10 %-Co–Mo-400	0.0248	−0.495	−0.24795
15 %-Co–Mo-400	0.5874	−3.46087	−1.63518
20 %-Co–Mo-400	1.12482	−21.1706	−1.02829

less than that exhibited by the monometallic catalysts (as shown in Fig. 4).

The catalyst activity test results over 10, 15 and 20 % bimetallic carbides are shown in Table 5. With the incorporation of Co into Mo$_2$C, there was a significant increase in both CH$_4$ and CO$_2$ conversion: The conversion of CH$_4$ was nearly complete and the conversion of CO$_2$ also reached above 95 % when the loading level of the bimetallic carbide reached 15 % or higher. This very large activity improvement was due to the introduction of Co which has been proven to be very active metal in the DRM reaction. In conjunction with the above, the stabilities of both CH$_4$ and CO$_2$ conversions were significantly improved with the addition of Co components. In addition, the stability of both H$_2$ and CO yields were enhanced as the loading level increased.

The H$_2$/CO product ratio over all the three bimetallic carbide catalysts (as shown in Table 5) were above 1, which indicated the CH$_4$ decomposition (reaction 1) was favoured over the bimetallic catalysts. As mentioned in monometallic carbide catalyst discussion, the availability of ZrO$_2$ contributes to the conversions of both CH$_4$ and CO$_2$, and on the other hand, it also improves the side reactions which can lead to the observed difference

Table 5 Conversions of CH_4 and CO_2, yields of H_2 and CO, product ratio and carbon balance over bimetallic carbide samples during DRM reactions (conditions 850 °C, atmospheric pressure, 4 h duration, CH_4/CO_2 1, GHSV 4.8×10^3 ml h^{-1} g_{cat}^{-1})

Sample	C_{CH_4} (%)		C_{CO_2} (%)		Y_{H_2} (%)		Y_{CO} (%)		H_2/CO		B_{carbon} (%)	
	0 h	4 h	0 h	4 h	0 h	4 h	0 h	4 h	0 h	4 h	0 h	4 h
10 %-Co–Mo-400	98.28	97.13	87.04	85.29	94.1	87.9	88.0	82.8	1.13	1.13	98.2	94.4
15 %-Co–Mo-400	97.46	97.46	97.66	95.50	97.9	95.4	89.8	86.2	1.09	1.10	94.2	91.8
20 %-Co–Mo-400	97.75	97.50	98.02	97.07	97.1	95.2	89.8	86.9	1.09	1.09	92.5	92.7

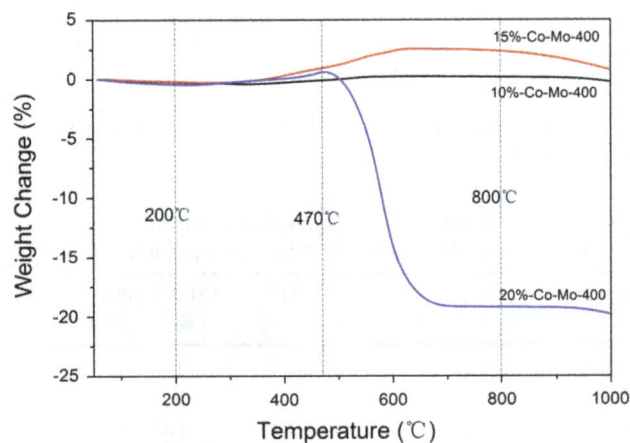

Fig. 7 Thermo-gravimetric analysis of the post-reaction Co–Mo bimetallic catalysts

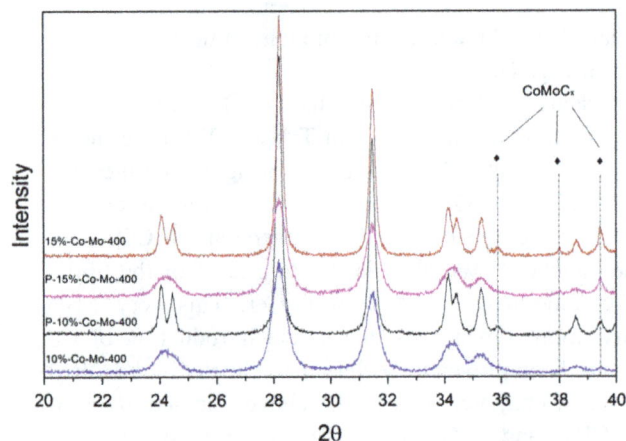

Fig. 8 X-ray diffraction patterns of ZrO_2 supported 10/15 % Co–Mo bimetallic carbides before and after the dry reforming of methane (P post-reaction)

Table 6 Weight changes in thermo-gravimetric analysis of pre-reaction Co–Mo bimetallic catalysts at different temperature ranges

Sample	200–470 °C (%)	470–800 °C (%)	800–1,000 °C (%)
10 %-Co–Mo-400	0.15786	0.25653	−0.40511
15 %-Co–Mo-400	1.15481	1.37135	−1.60592
20 %-Co–Mo-400	1.05181	−19.8175	−0.60463

The TGA results of the post-reaction bimetallic catalysts are exhibited in Fig. 7 with the calculation of weight change at each temperature range exhibited in Table 6. No weight losses are observed on 10 %-Co–Mo-400 and 15 %-Co–Mo-400, and arguably, there is a reduction of weight loss from 470 °C on 20 %-Co–Mo-400 (the gain of weight from 200 °C over 20 %-Co–Mo-400 is masked by the great weight loss from 470 °C, which means the carbon hasn't been totally consumed during DRM). This phenomenon on bimetallic catalysts indicated the consumption of carbon by CO_2 ($CO_2 + C \leftrightarrow 2CO$, reaction (3)) during the dry reforming of methane, favouring to the CO production. However, the carbon balances of reaction over bimetallic catalysts (Table 5) were below 100 %, which indicates that carbon-containing product apart from CO and coke was generated during the reaction, and the unexpected product is hypothesized as light hydrocarbons.

The very high stabilities of Co–Mo bimetallic carbide catalysts were also exhibited in their XRD patterns after being tested in the dry reforming of methane. In Fig. 8, only the peaks corresponding to $CoMoC_x$ were observed on the ZrO_2 support and there appeared to be no evidence of MoO_2 or $CoMoO_x$ generation. Compared to the formation of MoO_2 over ZrO_2 supported Mo_2C catalysts (shown in Fig. 3), these catalysts exhibited significantly enhanced oxidation resistance as the bi-metallic carbide phase, this could be explained by the presence of cobalt in the carbide, which may directly enhance the oxidative resistance of the carbide, or provide a re-carburisation route for the oxidized molybdenum (Table 6).

between the conversions and the product yields. As exhibited in Table 5, the conversions of CH_4 and the yields of H_2 were arguably at the same level. This phenomenon indicated that the incorporation of Co into the Mo_2C catalysts can significantly depress the Reverse Water Gas Shift (reaction (3)).

Conclusions

These investigations have shown that bulk Mo_2C did not exhibit a significantly high activity in the dry reforming of methane (DRM); this was attributed to the very low surface area when using the preparation method as described. On the other hand, pure ZrO_2 support was treated with the same method to give ZrO_2 supported Mo_2C catalyst, and this exhibited a relatively higher activity compared to the bulk Mo_2C. It appeared that Lewis acid and base sites on the support ZrO_2 had contributed to this enhanced activity.

The catalytic activity decreased slightly with time on stream for the supported Mo_2C over a period of 4 h. It was thought that this deactivation was caused by two effects: one was the oxidation effect on Mo_2C by CO_2, and the other one was the loss of surface area caused by sintering effect of the ZrO_2 support.

ZrO_2 supported cobalt and molybdenum bi-metallic carbide catalysts were also prepared and examined for catalytic activity. Significant enhancements were observed for both the activity and the stability of the catalysts in the DRM. It was observed also that the incorporation of cobalt atoms into the Mo_2C lattice significantly improved the materials' resistance to aerial oxidation. Compared to the mono-metallic carbide, the bi-metallic catalysts clearly have significantly higher CH_4 conversion. Moreover, an increase of Co loading in the mixed carbide also reduced the influence of Reverse Water Gas Shift, which improved the stability of H_2 yields.

Acknowledgments The authors would like to thank King Abdulaziz City of Science and Technology (KACST) for funding. Thanks are due to Guangzhou Boxenergytech Ltd for BET analysis.

References

1. Al-Fatesh ASA, Fakeeha AH (2011) Reduction of green house gases by dry reforming: effect of support. Res J Chem Environ 15(2):259–268
2. Ashcroft AT, Cheetham AK, Green MLH, Vernon PDF (1991) Partial oxidation of methane to synthesis gas using carbon dioxide. Nature (London) 352(6332):225–226
3. Basini L, Sanfilippo D (1995) Molecular aspects in syn-gas production: the CO2-reforming reaction case. J Catal 157(1):162–178
4. Bhattacharyya K, Danon A, Vijayan BK, Gray KA, Stair PC, Weitz E (2013) Role of the surface lewis acid and base sites in the adsorption of CO2 on titania nanotubes and platinized titania nanotubes: an in situ FT-IR study. J Phys Chem C 117(24):12661–12678
5. Bom D, Andrews R, Jacques D, Anthony J, Chen B, Meier MS, Selegue JP (2002) Thermogravimetric analysis of the oxidation of multiwalled carbon nanotubes: evidence for the role of defect sites in carbon nanotube chemistry. Nano Lett 2(6):615–619
6. Bradford MCJ, Vannice MA, Ruckenstein E (1999) CO2 reforming of CH4. Catal Rev Sci Eng 41(1):1–42
7. Brungs AJ, York APE, Claridge JB, Marquez-Alvarez C, Green MLH (2000) Dry reforming of methane to synthesis gas over supported molybdenum carbide catalysts. Catal Lett 70(34):117–122
8. Budiman AW, Song S-H, Chang T-S, Shin C-H, Choi M-J (2012) Dry reforming of methane over cobalt catalysts: a literature review of catalyst development. Catal Surv Asia 16(4):183–197
9. Claridge JB, York APE, Brungs AJ, Marquez-Alvarez C, Sloan J, Tsang SC, Green MLH (1998) New catalysts for the conversion of methane to synthesis gas: molybdenum and tungsten carbide. J Catal 180(1):85–100
10. Courty P, Durand D, Freund E, Sugier A (1982) C1-C6 alcohols from synthesis gas on copper-cobalt catalysts. J Mol Catal 17(2–3):241–254
11. Darujati ARS, Thomson WJ (2005) Stability of supported and promoted-molybdenum carbide catalysts in dry-methane reforming. Appl Catal A 296(2):139–147
12. Erdoehelyi A, Cserenyi J, Papp E, Solymosi F (1994) Catalytic reaction of methane with carbon dioxide over supported palladium. Appl Catal A 108(2):205–219
13. Erdohelyi A, Cserenyi J, Solymosi F (1993) Activation of methane and its reaction with carbon dioxide over supported rhodium catalysts. J Catal 141(1):287–299
14. Fakeeha AH, Khan WU, Al-Fatesh AS, Abasaeed AE (2013) Stabilities of zeolite-supported Ni catalysts for dry reforming of methane. Cuihua Xuebao 34(4):764–768
15. Halawy SA, Mohamed MA, Bond GC (1993) Characterization of unsupported molybdenum oxide-cobalt oxide catalysts. J Chem Technol Biotechnol 58(3):237–245
16. Luisetto I, Tuti S, Di Bartolomeo E (2012) Co and Ni supported on CeO2 as selective bimetallic catalyst for dry reforming of methane. Int J Hydrogen Energy 37(21):15992–15999
17. Naito S, Tsuji M, Miyao T (2002) Mechanistic difference of the CO2 reforming of CH4 over unsupported and zirconia- supported molybdenum carbide catalysts. Catal Today 77(3):161–165
18. Shao H, Kugler EL, Ma W, Dadyburjor DB (2005) Effect of temperature on structure and performance of in-house cobalt-tungsten carbide catalyst for dry reforming of methane. Ind Eng Chem Res 44(14):4914–4921
19. Steinhauer B, Kasireddy MR, Radnik J, Martin A (2009) Development of Ni–Pd bimetallic catalysts for the utilization of carbon dioxide and methane by dry reforming. Appl Catal A 366(2):333–341
20. Takayasu O, Hongo N, Matsuura I (1993) Spillover effect for the reducing reaction of CO2 with CH4 over SiO2-supported transition-metal catalysts physically mixed with MgO. Stud Surf Sci Catal 77:305–308
21. Tsipouriari VA, Zhang Z, Verykios XE (1998) Catalytic partial oxidation of methane to synthesis gas over Ni-based catalysts I. Catalyst performance characteristics. J Catal 179(1):283–291
22. Xiao T-C, York APE, Al-Megren H, Williams CV, Wang H-T, Green MLH (2001) Preparation and characterization of bimetallic cobalt and molybdenum carbides. J Catal 202(1):100–109
23. Yan Z-F, Ding R-G, Song L-H, Qian L (1998) Mechanistic study of carbon dioxide reforming with methane over supported nickel catalysts. Energy Fuels 12(6):1114–1120
24. York APE, Claridge JB, Brungs AJ, Tsang SC, Green MLH (1997) Molybdenum and tungsten carbides as catalysts for the conversion of methane ot synthesis gas using stoichiometric feedstocks. Chem Commun (Cambridge) 1:39–40

Permissions

All chapters in this book were first published in APR, by Springer; hereby published with permission under the Creative Commons Attribution License or equivalent. Every chapter published in this book has been scrutinized by our experts. Their significance has been extensively debated. The topics covered herein carry significant findings which will fuel the growth of the discipline. They may even be implemented as practical applications or may be referred to as a beginning point for another development.

The contributors of this book come from diverse backgrounds, making this book a truly international effort. This book will bring forth new frontiers with its revolutionizing research information and detailed analysis of the nascent developments around the world.

We would like to thank all the contributing authors for lending their expertise to make the book truly unique. They have played a crucial role in the development of this book. Without their invaluable contributions this book wouldn't have been possible. They have made vital efforts to compile up to date information on the varied aspects of this subject to make this book a valuable addition to the collection of many professionals and students.

This book was conceptualized with the vision of imparting up-to-date information and advanced data in this field. To ensure the same, a matchless editorial board was set up. Every individual on the board went through rigorous rounds of assessment to prove their worth. After which they invested a large part of their time researching and compiling the most relevant data for our readers.

The editorial board has been involved in producing this book since its inception. They have spent rigorous hours researching and exploring the diverse topics which have resulted in the successful publishing of this book. They have passed on their knowledge of decades through this book. To expedite this challenging task, the publisher supported the team at every step. A small team of assistant editors was also appointed to further simplify the editing procedure and attain best results for the readers.

Apart from the editorial board, the designing team has also invested a significant amount of their time in understanding the subject and creating the most relevant covers. They scrutinized every image to scout for the most suitable representation of the subject and create an appropriate cover for the book.

The publishing team has been an ardent support to the editorial, designing and production team. Their endless efforts to recruit the best for this project, has resulted in the accomplishment of this book. They are a veteran in the field of academics and their pool of knowledge is as vast as their experience in printing. Their expertise and guidance has proved useful at every step. Their uncompromising quality standards have made this book an exceptional effort. Their encouragement from time to time has been an inspiration for everyone.

The publisher and the editorial board hope that this book will prove to be a valuable piece of knowledge for researchers, students, practitioners and scholars across the globe.

List of Contributors

Hongwei Yang
Department of Chemistry, Wolfson Catalysis Centre, University of Oxford, Oxford OX1 3QR, UK

Shik Chi Edman Tsang
Department of Chemistry, Wolfson Catalysis Centre, University of Oxford, Oxford OX1 3QR, UK

Rasha S. Kamal
Department of Petroleum Applications, Egyptian Petroleum Research Institute, Nasr City, Cairo, Egypt

Nehal S. Ahmed
Department of Petroleum Applications, Egyptian Petroleum Research Institute, Nasr City, Cairo, Egypt

Amal M. Nasser
Department of Petroleum Applications, Egyptian Petroleum Research Institute, Nasr City, Cairo, Egypt

Devendra Pakhare
Department of Chemical Engineering, Louisiana State University, Baton Rouge, LA 70803, USA

Hongyi Wu
Department of Chemistry, Southern University, Baton Rouge, LA 70816, USA

Savinay Narendra
Department of Mechanical Engineering, Indian Institute of Technology, Kharagpur 721302, India

Victor Abdelsayed
National Energy Technology Laboratory, U.S. Department of Energy, Morgantown, WV 26507, USA

Daniel Haynes
National Energy Technology Laboratory, U.S. Department of Energy, Morgantown, WV 26507, USA

Dushyant Shekhawat
National Energy Technology Laboratory, U.S. Department of Energy, Morgantown, WV 26507, USA

David Berry
National Energy Technology Laboratory, U.S. Department of Energy, Morgantown, WV 26507, USA

James Spivey
Department of Chemical Engineering, Louisiana State University, Baton Rouge, LA 70803, USA

Tara Shirvani
Smith School of Enterprise and the Environment, University of Oxford, Oxford, UK

Javed Ali
Department of Chemistry, Centre for Catalysis Research, WestCHEM, University of Glasgow, G12 8QQ Glasgow, Scotland, UK

S. David Jackson
Department of Chemistry, Centre for Catalysis Research, WestCHEM, University of Glasgow, G12 8QQ Glasgow, Scotland, UK

M. H. Wilson
Center for Applied Energy Research, University of Kentucky, Lexington, KY 40511, USA

J. Groppo
Center for Applied Energy Research, University of Kentucky, Lexington, KY 40511, USA

A. Placido
Center for Applied Energy Research, University of Kentucky, Lexington, KY 40511, USA

S. Graham
Center for Applied Energy Research, University of Kentucky, Lexington, KY 40511, USA

S. A. Morton III
Department of Engineering, James Madison University, Harrisonburg, VA 22807, USA

E. Santillan-Jimenez
Center for Applied Energy Research, University of Kentucky, Lexington, KY 40511, USA

A. Shea
Department of Biosystems and Agricultural Engineering, University of Kentucky, Lexington, KY 40506, USA

M. Crocker
Center for Applied Energy Research, University of Kentucky, Lexington, KY 40511, USA

C. Crofcheck
Department of Biosystems and Agricultural Engineering, University of Kentucky, Lexington, KY 40506, USA

R. Andrews
Department of Chemical and Materials Engineering, University of Kentucky, Lexington, KY 40506, USA

Adil A. Mohammed
Chemical Engineering Department, Faculty of Engineering, Cairo University, Giza 12613, Egypt

Seif-Eddeen K. Fateen
Chemical Engineering Department, Faculty of Engineering, Cairo University, Giza 12613, Egypt

Tamer S. Ahmed
Chemical Engineering Department, Faculty of Engineering, Cairo University, Giza 12613, Egypt

Tarek M. Moustafa
Chemical Engineering Department, Faculty of Engineering, Cairo University, Giza 12613, Egypt

Peter Frank
Department of Chemistry, University of New Hampshire, Durham, USA

Alka Prasher
Department of Chemistry, University of New Hampshire, Durham, USA

Bryan Tuten
Materials Science Program, University of New Hampshire, Durham, USA

Danming Chao
Alan G. MacDiarmid Institute, College of Chemistry, Jilin University, Changchun, People's Republic of China

Erik Berda
Department of Chemistry, University of New Hampshire, Durham, USA
Materials Science Program, University of New Hampshire, Durham, USA

Adam F. Lee
European Bioenergy Research Institute, Aston University, Aston Triangle, Birmingham B4 7ET, UK

A. K. Wahab
SABIC-Corporate Research and Innovation (CRI) at KAUST, Thuwal, Saudi Arabia

S. Bashir
SABIC-Corporate Research and Innovation (CRI) at KAUST, Thuwal, Saudi Arabia

Y. Al-Salik
SABIC-Corporate Research and Innovation (CRI) at KAUST, Thuwal, Saudi Arabia

H. Idriss
SABIC-Corporate Research and Innovation (CRI) at KAUST, Thuwal, Saudi Arabia

Said Salah Eldin Elnashaie
Department of Chemical and Environmental Engineering, Faculty of Engineering, Universiti Putra Malaysia, 43400 Serdang, Selangor, Malaysia
Chemical and Biological Engineering Department, University of British Columbia (UBC), Vancouver, Canada

Firoozeh Danafar
Department of Chemical Engineering, College of Engineering, Shahid Bahonar University of Kerman, 7618891167 Kerman, Iran

Fakhru'l-Razi Ahmadun
Department of Chemical and Environmental Engineering, Faculty of Engineering, Universiti Putra Malaysia, 43400 Serdang, Selangor, Malaysia

Zi-Feng Yan
State Key Laboratory of Heavy Oil Processing, CNPC Key Laboratory of Catalysis, China University of Petroleum, Qingdao, China

Hamid A. Al-Megren
State Key Laboratory of Heavy Oil Processing, CNPC Key Laboratory of Catalysis, China University of Petroleum, Qingdao, China

Ehsan Javadi Shokroo
PART-SHIMI Company, Martyr Fahmideh Talent Foundation, Campus of Knowledge Based Companies, 87, Mirzaye Shirazi, 71 888 41111 Shiraz, Fars, Iran

Mohammad Shahcheraghi
PART-SHIMI Company, Martyr Fahmideh Talent Foundation, Campus of Knowledge Based Companies, 87, Mirzaye Shirazi, 71 888 41111 Shiraz, Fars, Iran

Mehdi Farniaei
PART-SHIMI Company, Martyr Fahmideh Talent Foundation, Campus of Knowledge Based Companies, 87, Mirzaye Shirazi, 71 888 41111 Shiraz, Fars, Iran

Jianxiang Wu
College of Chemistry and Chemical Engineering, Shanghai University of Engineering Science, Shanghai 201620, China

Yilong Gao
College of Chemistry and Chemical Engineering, Shanghai University of Engineering Science, Shanghai 201620, China

Wei Zhang
College of Chemistry and Chemical Engineering, Shanghai University of Engineering Science, Shanghai 201620, China

Yueyue Tan
College of Chemistry and Chemical Engineering, Shanghai University of Engineering Science, Shanghai 201620, China

Aomin Tang
College of Chemistry and Chemical Engineering, Shanghai University of Engineering Science, Shanghai 201620, China

Yong Men
College of Chemistry and Chemical Engineering, Shanghai University of Engineering Science, Shanghai 201620, China

Bohejin Tang
College of Chemistry and Chemical Engineering, Shanghai University of Engineering Science, Shanghai 201620, China

Syed Ahmed Ali
Center of Research Excellence in Petroleum Refining and Petrochemicals, King Fahd University of Petroleum and Minerals, Dhahran 31261, Saudi Arabia

Chris Llewellyn Smith
Oxford University, Oxford, UK

Madhav Ghanta
Center for Environmentally Beneficial Catalysis, University of Kansas, Lawrence, KS 66045-7609, USA
Department of Chemical and Petroleum Engineering, University of Kansas, Lawrence, KS 66045-7609, USA

Darryl Fahey
Center for Environmentally Beneficial Catalysis, University of Kansas, Lawrence, KS 66045-7609, USA

Bala Subramaniam
Center for Environmentally Beneficial Catalysis, University of Kansas, Lawrence, KS 66045-7609, USA
Department of Chemical and Petroleum Engineering, University of Kansas, Lawrence, KS 66045-7609, USA

Yu Huang
Guangzhou Boxenergy Technology Ltd, Guangzhou Hi-Tech Development Zone, Guangzhou, People's Republic of China

Haoyi Chen
Guangzhou Boxenergy Technology Ltd, Guangzhou Hi-Tech Development Zone, Guangzhou, People's Republic of China

Jixin Su
Environmental Engineering College, Shandong University, No. 27 Shanda Nanlu, Jinan, People's Republic of China

Tiancun Xiao
Guangzhou Boxenergy Technology Ltd, Guangzhou Hi-Tech Development Zone, Guangzhou, People's Republic of China

Henry Martinez
Department of Chemistry, University of Minnesota, 207 Pleasant St. SE, Minneapolis, MN 55455-0431, USA

Jihua Zhang
Department of Chemistry, University of Minnesota, 207 Pleasant St. SE, Minneapolis, MN 55455-0431, USA

Shingo Kobayashi
Department of Chemistry, University of Minnesota, 207 Pleasant St. SE, Minneapolis, MN 55455-0431, USA

Yuewen Xu
Department of Chemistry, University of Minnesota, 207 Pleasant St. SE, Minneapolis, MN 55455-0431, USA

Louis M. Pitet
Department of Chemistry, University of Minnesota, 207 Pleasant St. SE, Minneapolis, MN 55455-0431, USA

Megan E. Matta
Department of Chemistry, University of Minnesota, 207 Pleasant St. SE, Minneapolis, MN 55455-0431, USA

Marc A. Hillmyer
Department of Chemistry, University of Minnesota, 207 Pleasant St. SE, Minneapolis, MN 55455-0431, USA

A. K. Wahab
SABIC: T&I Riyadh and CRI-KAUST, Riyadh, Saudi Arabia

T. Odedairo
SABIC: T&I Riyadh and CRI-KAUST, Riyadh, Saudi Arabia

J. Labis
KAIN, Riyadh, Saudi Arabia

M. Hedhili
KAUST, Thuwal, Saudi Arabia

A. Delavar
KAUST, Thuwal, Saudi Arabia

H. Idriss
SABIC: T&I Riyadh and CRI-KAUST, Riyadh, Saudi Arabia

Oliver R. Inderwildi
Smith School, University of Oxford, South Parks Road, Oxford OX1 3QY, UK

Fabian Siegrist
Deloitte LLP, 30 Rockefeller Plaza, New York City, NY 10111, USA

Robert Duane Dickson
Deloitte LLP, 30 Rockefeller Plaza, New York City, NY 10111, USA

Andrew J. Hagan
World Economic Forum, 91–93 route de la Capite, 1223 Cologny, Geneva, Switzerland

X. Du
Inorganic Chemistry Laboratory, Department of Chemistry, KACST-Oxford Petrochemical Research Centre (KOPRC), University of Oxford, Oxford OX1 3QR, UK

L. J. France
Inorganic Chemistry Laboratory, Department of Chemistry, KACST-Oxford Petrochemical Research Centre (KOPRC), University of Oxford, Oxford OX1 3QR, UK

V. L. Kuznetsov
Inorganic Chemistry Laboratory, Department of Chemistry, KACST-Oxford Petrochemical Research Centre (KOPRC), University of Oxford, Oxford OX1 3QR, UK

T. Xiao
Inorganic Chemistry Laboratory, Department of Chemistry, KACST-Oxford Petrochemical Research Centre (KOPRC), University of Oxford, Oxford OX1 3QR, UK

P. P. Edwards
Inorganic Chemistry Laboratory, Department of Chemistry, KACST-Oxford Petrochemical Research Centre (KOPRC), University of Oxford, Oxford OX1 3QR, UK

Hamid AlMegren
Petrochemicals Research Institute (PRI), King Abdulaziz City of Science and Technology (KACST), P.O. Box 6086, Riyadh 11442, Saudi Arabia

Abdulaziz Bagabas
Petrochemicals Research Institute (PRI), King Abdulaziz City of Science and Technology (KACST), P.O. Box 6086, Riyadh 11442, Saudi Arabia